AQUACULTURE

ELLIS HORWOOD SERIES IN AQUACULTURE AND FISHERIES SUPPORT
Series Editor: DR L.M. LAIRD, University of Aberdeen

AQUACULTURE
Biology and Ecology of Cultured Species

Edited by
Gilbert Barnabé

ELLIS HORWOOD
NEW YORK LONDON TORONTO SYDNEY TOKYO SINGAPORE

First published 1994 by
Ellis Horwood Limited
Campus 400, Maylands Avenue
Hemel Hempstead
Hertfordshire, HP2 7EZ
A division of
Simon & Schuster International Group

Translated from the original French title:
Bases Biologiques & Écologique de l'Aquaculture
First published by Lavoisier Technique & Documentation,
© the copyright holders, Paris in 1991

Printed and bound in Great Britain by
Hartnolls, Bodmin

Library of Congress Cataloging-in-Publication Data

Available from the publisher

British Library Cataloguing in Publication Data

A catalogue record for this book is available from the British Library

ISBN   0-13-482324-9 (hbk)   0-13-482316-8 (pbk)

1 2 3 4 5   98 97 96 95 94

Table of contents

PART III   CRUSTACEAN FARMING: THE BIOLOGICAL BASIS
A. Van-Wormhoudt and C. Bellon-Humbert

PART IV   BIOLOGICAL BASIS OF FISH CULTURE
G. Barnabé

PART V   AQUACULTURE DISEASES
F. Baudin Laurencin and M. Vigneulle

Introduction

1. DEFINITION

The term aquaculture is defined as 'the art of increasing and rearing aquatic animals and plants'. It covers activities whose main object is the production of fresh-water, brackish and marine species by man under controlled or semi-controlled conditions.

The production of living matter from the aquatic medium is fundamental to all aquaculture activities; aquaculture is based on the manipulation of the natural or artificial aquatic environment for the production of species which are useful to man. It therefore involves all aspects of production of living matter in water.

There is no hard and fast distinction between the culture of plants and the rearing of animals as on land where plant crops and animals can be considered separately, but there are different constraints imposed by the characteristics of the species being cultured. The ongrowing of fish at high density has all the characteristics of animal rearing while that of mussels or oysters, depending more on natural seasonal rhythms, is more similar to terrestrial plant crops.

In the widest sense, aquaculture includes activities associated with the preparation of animals and plants for sale and their marketing. There are four main areas of aquaculture production: algae (mainly in South-East Asia), molluscs, crustaceans and fish.

2. THE BIRTH OF MODERN AQUACULTURE

The earliest kinds of aquaculture developed both from observations of the behaviour of wild animals such as the migrations of fish, and the settlement of molluscs on a support, and of some types of fishing based on the capture of fish as they moved from lagoons to the sea. After capture these fish were placed in protected rearing areas, and molluscs were stocked in special shallow pools (see Chauvet (1989) and Héral (1990) for a more detailed account).

Marine fish farming began at the start of the 20th century in Europe but lacked technology and the results of scientific research which are now available (plastic materials, electricity, hormones) and, moreover, was taking place at a time when it was inconceivable that marine resources should ever be over-exploited (Fabre-Domergue and Biétrix (1905), reviewed by Shelbourne (1964)). In other countries, especially

South-East Asia, aquaculture has more ancient roots (a work on fish culture has been attributed to Fan-Li in China around 1400 BC).

Since the 1970s the overall limits of primary ocean production (plant production based on photosynthesis) have been known and those of fishing almost achieved (fishing world wide removed 90 million tonnes of the 100 million tonnes which, it has been calculated, could be supplied by the world's oceans). Thus aquaculture has been accorded a new importance as a method of producing food: the ocean receives over 70% of the energy reaching the earth from the sun (this energy is the basis of life on earth), but produces less than 10% of the animal protein for human nutrition and 1.3% of overall food supplies (Duvigneaud 1980).

However, the understanding of the management of aquatic ecosystems is much harder for man than that of terrestrial ecosystems:

— The extent of the aquatic domain, the dispersal and mobility of animals living there and the fact that people operating there are frequently working 'blind', working from the surface in three dimensions, compared to the accessibility of land crops and animals; agriculture has developed at the soil–air interface.
— Interactions between aquatic species and the medium in which they are immersed (importance of the level of oxygen, pH, etc.) should be well understood by the farmer. The use of even simple instruments is necessary (thermometer, oxygen meter, etc.). Contrary to popular belief, the requirements of a good fish farmer are not necessarily the same as those of a good shepherd; they are more complex and require skills in areas where humans have only limited skills or sensory capacities.

In this context, scientific and technical progress offers the means of developing a new form of aquaculture based on a combination of biology, oceanography and many other disciplines. This new aquaculture has been born at the crossroads of various scientific and technical disciplines and has now become a subject in its own right with its own meetings, publications and popularizers. The work of the pioneers has led to a rapid increase in the number of farmers. The new aquaculture should not be seen as opposing fishing or traditional aquaculture (mollusc farming, pond fish culture) as it is a complementary activity alongside these.

3. OBJECTIVES

The objectives of aquaculture are varied and depend on the economic context in which they are found: these have been listed elsewhere (Barnabé 1990).

In industrialized countries the main objective is the production of high-value products for human consumption which are not obtained in sufficient quantity from fishing. Thus in Japan, the farmed production of yellowtails and sea bream, which only started twenty years ago, now exceeds the fishing catch. A similar picture exists in Europe for salmon and sea bass.

More recently, research has focused on products with certain dietary qualities (low level of fats, high levels of vitamins and trace elements) adding to the demand already mentioned in the richer countries. In Japan, as in Western Europe, some products of marine culture are extremely expensive (algae, prawns, bass, pearls). Treatment of obesity with cultured chlorella or alginic acid (from cultured algae), identification of new antibiotics in many marine species (hypotensis) in seaweeds and anti-inflammatory compounds in a New Zealand mussel are a few examples of the results of the extensive research on the use of marine products, many of which are already being cultivated or are known to be suitable for cultivation.

The benefit of a diet based on sea food on cardiovascular diseases, rheumatism, diabetes and even the brain (Castell 1988) has been clearly demonstrated and is changing the composition of our diet. For example, with oils we know that a high level of certain unsaturated fatty acids in marine animals is the basis of some of these properties. Table 1 compares the composition of various products from agriculture and aquaculture.

Table 1. Composition of different farmed products (per 100 g)

Product	Carp	Eels	Shrimp	Beef	Egg	Milk
Organic matter (g)						
Proteins	22	20	16	21	13	3
Lipids	9	18	1	6	11	3
Carbohydrate	3	3	15	3	0	45
Mineral (mg)						
Calcium	72	150	50	4	65	100
Sodium	—	—	—	90	90	36
Phosphorus	180	180	200	190	230	90
Iron	20	10	50	36	26	10
Vitamins (International units or mg)						
A (IU)	20	3000	100	40	800	120
Carotene (IU)	0	0	0	30	10	120
D (IU)	0	150	0	0	10	0
B1 (mg)	40	1	4	6	10	4
B2 (mg)	8	10	5	13	30	15
Niacine (mg)	200	300	250	400	10	20
C (mg)	0	150	200	0	0	200

Adapted from Liao, 19 World Aquaculture Society meeting, Honolulu, Hawaï, January 1988.

In developing countries (particularly in the tropical belt) the objective is the production of animal protein which cannot be supplied by traditional culture in sufficient quantity to support the ever-growing population and compensate for the desertification of the land. For example, in India (Jhingran 1990), where work in this area has been in progress for many years, results have been positive. But these countries also seek to produce high-value species such as penaeid shrimps or the fresh-water prawn *macrobrachium* for export.

Thus, aquaculture has many facets; the production of fish for commercial or leisure fishing or of juveniles for restocking natural beds (scallops), introduction of new species (Manilla clams) and production of ornamental fish are examples showing the diversity of new techniques in a whole range of types of aquatic systems.

However, it may be in the protection of the aquatic environment, highly polluted by man's activities, that aquaculture will play a major role. Techniques of water purification and recycling of water are based on the same processes as those used in aquaculture.

4. SITES

In many countries the marine environment is under some form of collective jurisdiction. The idea of ownership is therefore a little different from that known on land although this is not always the case; Japanese fish farmers are, for example, proprietors of their rearing sites (Doumenge 1990); this is also true in Italy for extensive ongrowing operators of fish in private lagoons and ponds (valliculture), an ancient practice.

The occupation of space includes volume, not just area as in the case of land; this biases comparisons as it is necessary to express yields in terms of volume rather than area.

A further difference from terrestrial rearing which, in a given geographical unit, will have many common characteristics (climate, soil type, etc.) is that each aquaculture site may differ significantly from its neighbours (presence or absence of shelter, current, depth, level of dissolved gases and salts, turbidity, etc.). Each site must be considered as a separate entity. This implies that each new enterprise will require a survey of the above factors. The impact on the environment of similar rearing operations in the neighbourhood may be very different, which is rarely the case on land.

In spite of this variation, the natural aquatic environment constitutes either the basis of the rearing operation or the origin of the water used in culture, as is the case for earth ponds. Constraints (which may be ecological, biological or legal) linked to sites are often impossible to circumvent.

5. THE SPECIES

The ecological constraints associated with the relationship between the species being cultured and its environment may also be limiting. The environmental conditions offered by a natural site should correspond as closely as possible to the biological requirements of the species which is to be cultured, and both of these are difficult to change (although techniques and technology are constantly improving).

The relationship between a species and its environment determines its 'ecological niche' and is very diverse in the aquatic medium, integrating many environmental factors (temperature, salinity, pH, oxygen, etc.) while the birds and mammals on which agriculture is based have a constant internal temperature, live on one type of substrate (soil) and breathe the same air; environmental constraints are therefore less variable. Also, even today there are already over one hundred species of aquatic animal being cultivated and this number is continuing to grow. However, in contrast to the land, very few of these animals can be described as truly domesticated, i.e. the whole life cycle is managed with exclusive dependence on broodstock which have themselves been reared in captivity. Trout and carp are probably the only examples.

A situation noticed by Iversen as long ago as 1974 still characterizes the aquaculture of the 1990s: 'the production of juveniles of marine species at low price has not yet been achieved'. This is a problem which weighs heavily on the economics of rearing animals where production is understood and also in the farming of new species, especially crustaceans and fish.

In spite of these problems, the overall economic outlook for aquaculture is positive; a study carried out in Taiwan and reported by Arrignon (1982) established that the cost of aquaculture production per kilogram live weight is less than that of a pig which is itself considered to be an excellent converter of feed to flesh. On the basis of an arbitrary cost of 1 for a pig, the costs for aquaculture are:

Fresh-water aquaculture	0.43
Brackish water aquaculture	0.53
Marine aquaculture	0.24

6. MAN'S INTERVENTION

Man's activities are faced with the difficulty that water is not his natural environment (although if the water is led to the tanks, rearing is more similar to that on the land). To venture into this environment man uses various techniques which vary according to the species being farmed:

— The first of these must be access to the subjects being cultured in a three-dimensional environment; fishing is purely a capture activity but the disposal of fish in a small, privately owned pond is decided by the owner. The containing of the subjects being reared is therefore the greatest concern for the fish farmer. This is relatively simple for various of the algal and mollusc species which become firmly attached to the substrate. Fish and crustaceans are, however, considerably more mobile and it is therefore necessary to keep them in cages anchored in their natural environment or in tanks or ponds, either sunk into the land or built on it.

— The modes of action of man on the development of the aquatic environment or on the species being farmed, even in tanks and in fresh water is far from being controlled satisfactorily. Available techniques (variation in the rate of water turnover, use of aerators, changes in the quantity of nutrients (feed) and fertilizers, control of the density and biomass of the stock) are far from having been standardized and are, as yet, not reproducible. The situation is therefore very different from that of terrestrial culture or rearing. This lack of understanding of the operating conditions of the production ecosystem is characteristic of aquaculture and is having a major effect on its development (Laubier and Barnabé 1990).

7. THE GROWTH OF PRODUCTION

In spite of the above limitations, the threat of pollution, uncertain profitability and the various disasters which have dogged aquaculture, the industry continues to develop. In 15 years, the world production from aquaculture (6 million tonnes in 1976) has more than doubled, reaching 11 million tonnes in 1985 and over 16 million tonnes in 1992. While supplies from fishing have increased at 0.3% per annum, those from aquaculture have increased at 5.5% each year (Nash 1987).

If these figures are projected forwards to the year 2000 and the economics of the producing countries are taken into account, production should reach 22 million tonnes, double the 1984 production. At present, aquaculture produces 13% of the production from fisheries; by AD 2000 this will be 25%. Culture of finfish produces only 7% of the figure for the worldwide capture fishery (82 million tonnes in 1986) and crustacean farming, 4% of the fishery output. However, 75% of mollusc supplies come from culture.

Expressing this production in tonnage alone fails to show the true economic value of aquaculture production, especially for finfish and crustaceans. Man is able to select the species which are to be farmed but is not always able to choose which ones to capture. Thus 50 tonnes of farmed bass sold at £10 per kilogram fetches as much as 2500 tonnes of sardines sold by a fisherman for 20p per kilogram (or even returned to the sea where the product is in surplus). Between 10 and 20 million tonnes of fish captured (one-fifth of the world's catches) are transformed into fishmeal because the species caught have little or no market value for direct consumption by man. Three to four kilograms of fish are needed for each kilogram of fishmeal produced; this is sold at the low price of £300–400 per tonne and used for animal feed. Because the fish cultured are the most valuable species it is possible that by the year 2000 the value of cultured aquatic species will exceed the value of the products from the commercial capture fishery.

In terms of employment, taking France as an example, marine fisheries produce 600 000 tonnes of fish with 20 000 people directly employed while aquaculture, with a production of 200 000 tonnes (33% of aquatic production) has 23 000 employees. French imports of aquatic products have a value of over £1000 million. Exports are much lower; the deficit is over £800 million.

It is clear that fishing cannot supply the demand for luxury products as the natural stocks are overexploited and the hunt for these species has a detrimental effect on the species which are caught at the same time and rejected. Fishing for shrimps involves a return to the sea of four-fifths of the captured animals, mostly various species of fish.

In Europe, aquaculture produces 1 300 000 tonnes (making this region the second highest producing area in the world) and 150 000 people are employed directly (aquaculture production by country and by species is given by Barnabé (1990)).

The growth of production depends on the progress of research on the aquatic environment and the species being cultured. The objective of this book is to present a summary of the information on which the production of various groups of aquatic animals is based.

REFERENCES

Arrignon J. C. V., 1982. Regards sur l'aquaculture mondiale. *Aquaculture*, **27**: 165–186.

Barnabé G., 1990. Quelques données sur les productions de l'aquaculture. In: *Aquaculture*, G. Barnabé (ed.) Lavoisier—Tec. & Doc., Paris: 1257–1262.

Castell J. D., 1988. Fish as brain food. *Word Aquaculture*, Sept.: 21–22.

Chauvet C., 1990. L'aménagement des milieux lagunaires méditerranéens. In: *Aquaculture*, G. Barnabé (ed.). Lavoisier—Tec. & Doc., Paris: 857–887.

Doumenge F., 1990. L'aquaculture au Japon. In: *Aquaculture*, G. Barnabé (ed.). Lavoisier—Tec. & Doc., Paris: 949–1068.

Duvigneaud P., 1980. *La synthèse écologique*. Doin, Paris: 350 pp.

Fabre-Domergue et Biétrix E., 1905. Introduction à l'étude de la pisciculture marine. In: *Travail du Laboratoire de Zoologie Marine de Concarneau*. Vuibert and Nony, Paris: 205–243.

Héral M., 1990. L'ostréiculture française traditionnelle. In: *Aquaculture*, G. Barnabé (ed.). Lavoisier—Tec. & Doc., Parish: 347–397.

Iversen E. S., 1972. Farming the edge of the sea. *Fishing News Books*, Farnham, UK: 436 pp.

Jhingran V. G., 1989. Aquaculture en Inde. In: *Aquaculture*, G. Barnabé (ed.). Lavoisier—Tec. & Doc., Paris: 1117–1152.

Laubier L., Barnabé G., 1990. Conclusions. In: *Aquaculture*, G. Barnabé (ed.). Lavoisier—Tec. & Doc., Paris: 1117–1152.

Nash C., 1987. Aquaculture attracts increasing share of development aid. *Fish Farming International*, 14 (6): 22–24.

Shelbourne J. E., 1964. The artificial propagation of marine fish. In: *Advances in marine biology*, F. S. Russel (ed.), Vol. 2: 1–83.

Part I
The production of aquatic organisms

G. Barnabé

1

The aquatic environment

1. INTRODUCTION

All water, whether fresh, brackish or marine, is populated by living creatures, the most well known and frequently eaten of which are fish, molluscs and crustaceans. However, these creatures do not form the basis of the production of living matter in water; just as on land, only plants (autotrophs) are able to synthesize the organic compounds (proteins, lipids, carbohydrates) which make up all living matter.

Before examining the biological processes which bring about this production, it is necessary to examine physical aspects of the environment. In most examples marine and fresh waters will be considered together.

Oceans and seas cover 71% of the surface of the planet but, apart from a few offshore developments, aquaculture is confined to coastal zones. Fig. 1 shows the major ocean zones. Away from the shore the sea bed generally slopes gently downwards over the continental shelf; this zone extends outwards for tens of kilometres. At a depth of 150–200 m the sea bed begins to slope away steeply to deep water. Ecologically, the continental slope separates the neritic zone waters of the continental shelf from oceanic waters. In some places the continental shelf is very narrow; this has a major influence on aquaculture (e.g. the Mediterranean, French Côte d'Azur).

Continental waters are subdivided into habitat types: lotic (running waters) and lentic (natural and artificial lakes and ponds). The traditional home of aquaculture is the wet lands bordering sea coasts and estuaries, transitional zones between the mouths of rivers and the sea. The table below provides some of the major facts relating to the world's water resources.

The continental shelf has a greater surface area than that of the moon.

Aquatic ecosystems are characterized by one dominating feature: permanent motion in three dimensions, movement not only induced by the forces of waves and currents but also by other vertical and horizontal forces.

The characteristics which affect aquaculture installations are shown in Fig. 2.

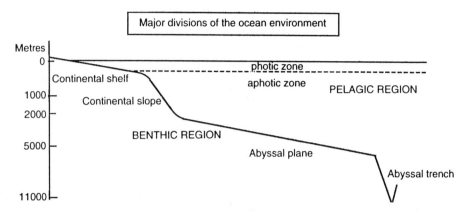

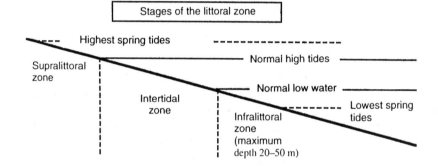

Fig. 1.

	Area (millions of km^2)	Volume (millions of km^3)
Salt water		
Atlantic Ocean	106	354
Pacific Ocean	180	720
Indian Ocean	74	291
Total	360	1365
Fresh water		
Lakes	0.70	0.10
Ice caps	18	30

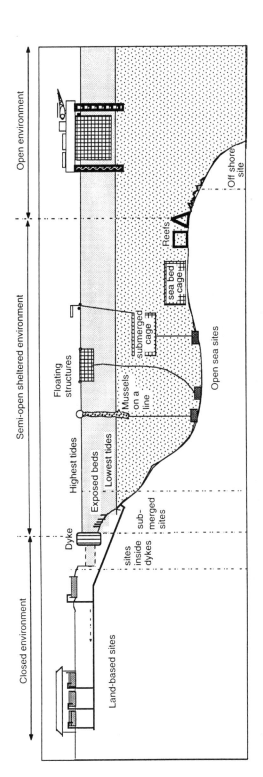

Fig. 2. Sites used for sea water culture.

2. WATER MOVEMENTS (HYDRODYNAMICS)

2.1 Tides

This periodic oscillation of the level of the sea is caused by the attraction of the moon and sun on water bodies; there are two periods of high water and two periods of low water every day. Two tidal cycles are completed in around 24 hours and 50 minutes, therefore between one day and the next there is a delay of 50 minutes for each low and high tide. Tidal height varies from place to place and tidal dynamics are extremely complex. Tides cause considerable displacement of water; this obviously has an effect on aquaculture installations. In many localities (mollusc culture in the intertidal zone, for example) this turnover of water is responsible for bringing food and oxygen to the cultured animals and allows temporary access by farmers to the rearing sites.

The Baltic and Mediterranean seas are separated from the oceans and have reduced tides (apart from the Adriatic). This changes the way coastal aquaculture operates.

2.2 Currents

The coastal fringe can be described as a layer of water of limited depth with constantly changing level (tides), slow horizontal movements (currents) and undulations (waves, swell). Structures used in aquaculture in the open sea must be able to withstand these two types of movement, which are sometimes associated.

Arising from the differences in the density of the water and from tides and winds, currents have a massive role in the world's oceans (climate, fisheries, etc.). They play an essential role in aquaculture in the turnover of water in rearing units, bringing in plank-tonic nutrients, to such an extent that equipment designed to create currents through aquaculture units is now being specially made.

Currents exert a rectilinear force on the stationary structures used in farming the open sea, which can be expressed by Morrison's formula:

$$F_t = 1/2D.T.S.V^2$$

F_t = Drag force
D = Density of water, around 1.025 kg m^{-3}
T = Coefficient of drag (variable, between 0.5 (sphere) and 1.5 (flat surface))
S = Area, m^2, equivalent to the projection of the area subject to the current
V = Current velocity, m s^{-1}

Observation has shown that current tends to cause flexible structures in open water to bend towards the sea bed; sea weeds, nets, net moorings and their buoys are thus bent by strong currents. This effect can be understood from the above formula; if structures are aligned parallel to the current the effective surface area, S, is reduced, as is the coefficient of drag, T (which is an expression of the hydrodynamic profile of the structure). A position of equilibrium is therefore reached, as the current speed usually decreases rapidly with increased depth. Wind creates waves and also currents (drift). The surface velocity is 1–2% of the wind speed, and this must be taken into account for aquaculture installations.

One type of current is of particular significance for production in sea water: this is the phenomenon where enriched waters are brought to the surface (upwelling or resurgence) and has been studied in Galicia, a major mussel production area (Hanson *et al.* 1986). Under the influence of north-east winds, which push the surface waters from the rias of north-west Spain into the open sea, deep water from the central Atlantic is drawn upwards. The water rises up the continental shelf towards the rias (deep waters are characterized by their salinity, temperature and level of nutrients). This periodic enrichment is responsible for the rich nutrient load of the waters and, in consequence, the high productivity of mussels. This is a direct demonstration of the interactions which can be established, even at long distance in this environment under the influence of climate, where there are no physical barriers.

2.3 Ocean swell and waves

The short-term oscillations in sea level (the time between two succeeding waves) cause rotational displacement of water molecules, the speed of which is already as high as 2 m s^{-1} for 2 m high waves with a period of around 4 seconds. Such a swell brings about rotational displacement speeds corresponding to currents of 4 knots (7.4 km h^{-1}), reversing each half cycle. Waves with a height of 2 m are not unusual in the open sea and these are responsible for the most rapid water movements. Bearing in mind the inertia present in rearing structures (mussel ropes, cages, etc.), the constraints are clear. When currents and waves are associated, their speeds add together when they are in the same direction. The forces on the rearing structure are increased in proportion.

The amplitude of waves or the ocean swell decreases with depth in an even more marked fashion than current, especially in coastal waters (Fig. 3). This decrease in amplitude is accompanied by flattening, causing particles to move to and fro on the sea bed (at around 30–40 m depth such movements are very slight).

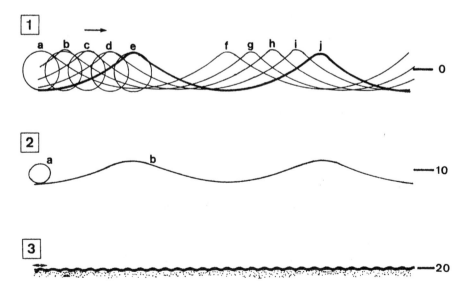

Fig. 3. Simplified representation of the ocean swell on the sea bed, 20 m down.

2.4 The role of hydrodynamic phenomena

It has long been known that the exchanges which take place at the water surface (air–water interface) have effects on many climatic or hydrological phenomena, but the nature of these exchanges is poorly understood. Thorpe & Hall (1987) provide the beginnings of the understanding of these phenomena: under the action of winds over 3 m s^{-1}, the waves breaking at the surface carry bubbles of air down to a depth of 10 m and also cause warm water to be brought upwards. These waves transmit kinetic energy downwards. There are other processes such as the formation of convection cells. This group of phenomena are shown diagrammatically in Fig. 4 (adapted from Weller 1987). In practice the surface of the oceans is a constantly mixing layer but the same basic laws characterize the surface of all water bodies, varying with intensity linked to size, depth, etc.

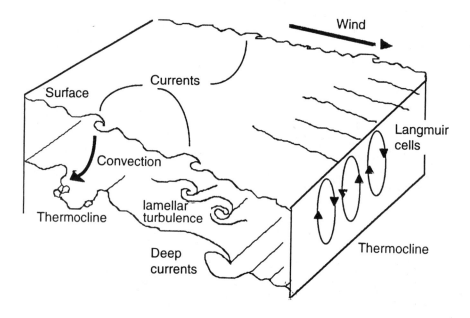

Fig. 4. Superficial stratification of waters or 'mixing zone'.

3. PHYSICO-CHEMICAL PROPERTIES OF SEA WATER

3.1 Temperature

The study of the distribution of living things, both on land and in water, is a scientific discipline: zoogeography. In the aquatic environment, temperature is the major controlling factor.

   Temperature is also of major importance for aquaculture: mean temperatures at a site will determine the species which can be cultured. There are warm-water species (penaeid shrimps, fish such as tilapia and some of the carp and mussel species), temperate species (sea bass, sea bream, catfish, common carp, oysters, mussels, etc.) and cold-water species (salmon, cod, halibut).

This is only a crude classification as temperate species (e.g. bass) can breed between 10 and 14 °C but have an optimum growth temperature between 23 and 25 °C and can survive at all temperatures between 1 and 30 °C. It is possible to rear most species over a wide range of temperatures but, away from the optimum, growth is so slow that the time to reach market size is too great for rearing to be profitable.

Latitude and altitude play a role in water temperature. In the sea, currents dominate, bringing warm and cold water far from their place of origin and allowing the rearing of exotic species. For example, the Gulf of Mexico receives warm water from the south but the coast of California receives cold water from the north.

It is also important to consider depth because deep ocean waters are rich in nutrients and are the subject of an unusual but promising form of aquaculture. There are cold waters, below 1000 m down, with a temperature of around 4 °C which might in the future be used in the rearing of salmonids (preferred temperatures below 18–20 °C) in the tropics. Other aspects are the use of thermal energy from the seas (in America) or the use of cold, enriched waters brought to the surface.

The temperatures of coastal and shallow inland waters (lagoons, rearing ponds, reservoirs) may cause problems for aquaculture; their temperature is close to that of the atmosphere since there is considerable mixing of water due to the action of wind and currents, but these may also have the effect of bringing water of different temperature up to the surface (see section 2 above). Wild animals such as fish can move away from unfavourable temperatures; captive animals are unable to do this. Rapid changes in temperature may therefore occur in spite of the high specific heat of water (see 'Specific heat', this chapter). Because of their lower thermal inertia these shallow zones tend to be more similar in temperature to the air and to be warmer in summer and colder in winter than deeper sea or ocean water.

This variation in temperature with depth is usually not a gradual change but, most frequently, there is a sudden discontinuity which is known as the thermocline. This forms as a result of the sun heating the upper layers of water and because the action of wind and waves does not extend deeper than 40 m (see section 2.4 above). This heating of surface water is accompanied by a decrease in specific density (see 'Specific density', this chapter). Warm water, which is lighter, does not mix with the colder, *more dense* water below. This results in stratification, the thermocline presents a barrier which may be biological (e.g. tuna will not pass through a thermocline). An illustration of the development of this phenomenon is presented in Fig. 5. In Norway, such stratification has been induced artificially (introducing fresh water which lies above salt water in a lake without mixing allows the capture of solar energy by the salt water which is insulated from the cold air). Temperatures of up to 30 °C have been achieved and flat oysters (*Ostrea edulis*) have been farmed (Korringa 1976).

The full effects of temperature on living organisms are not examined here but all farmed aquatic species are poikilotherms and their metabolism is therefore dependent on ambient temperature. Optimum temperatures are often different for species which must be cultured together (e.g. plankton and planktonophagic fish).

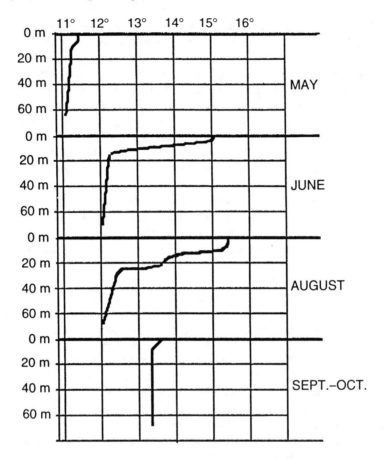

Fig. 5. Development of the thermocline in the waters of the Channel (after Péres & Deveze 1963).

3.2 Salinity, dissolved salts

Primary constituents

Marine water is saline; salinity is expressed in parts per thousand ($S‰$), representing the mass, in grammes, of solids contained in sea water when iodide and bromide ions are replaced by their chloride equivalent, carbonates converted to oxides and all organic matter oxidized.

Chlorinity (Cl ‰) represents the mass of chlorine equivalent to the total quantity of halogens in a kilogram of sea water. Salinity and chlorinity are linked: $S‰ = 1.805 \, Cl‰ + 0.030‰$. The mean level of salts is 35 g l^{-1} in the ocean (33–37‰). For a chlorinity of 20‰ at atmospheric pressure the number of cubic centimetres of oxygen which can be dissolved (saturation) in 1 litre of sea water varies with temperature:

5°C, 7.07;   10°C, 6.35;   15°C, 5.79;   20°C, 5.1;   25°C, 4.86;   30°C, 4.46.

The solubility also varies with salinity but in a less marked fashion; thus for a chlorinity of

0‰, 6.57 cm^3;   5‰, 6.26;   10‰, 5.95;   15‰, 5.63;   20‰, 5.31.

Fig. 6 (Alzieu 1989) shows the relationship between oxygen saturation, salinity and temperature.

The percentage of salts in ocean sea water is as follows:

Cl$^-$	(Chloride	55.2% (18.98 g l^{-1})
Na$^+$	(Sodium)	30.4% (10.56 g l^{-1})
SO$_4^{--}$	(Sulphate)	7.7%
Mg^{++}	(Magnesium)	3.7%
Ca^{++}	(Calcium)	1.16%
K$^+$	(Potassium)	1.1%
Br$^-$	(Bromine)	0.1%
Sr^{++}	(Strontium)	0.04%
H$_3$BO$_3$	(Boric acid)	0.07%
H$_2$CO$_3^-$, CO$_3^{--}$	(Carbonic acid and carbonates)	0.035%

In addition to the above constituents, sea water contains minute traces of many other elements but their level never exceeds 0.025% of the principal constituents.

Secondary constituents

Secondary constituents (0.02–0.03%) may be of disproportionately great significance because they include the mineral elements (nitrogen (N) and phosphorus (P)) which are essential for the production of living matter in water. The nitrogen compounds (nitrites, nitrates, ammonium salts) and phosphorus compounds (various phosphates) are used with carbon dioxide (see 'Dissolved gases', below) by phytoplankton in the synthesis of living matter under the action of sunlight (photosynthesis). Their level is variable; biological processes may tie up all the nutrients.

The oligoelements (substances present at very low levels) such as iron (1–50 mg.m^3), copper (4–10 mg.m^3), manganese (1–10 mg.m^3), zinc (5–14 mg.m^3), selenium (4 mg.m^3) are essential, as are some of the vitamins and other organic substances (amino acids, carbohydrates, etc.) which are present in sea water but whose role is still not understood. Silicon (10–1000 mg.m^3) is a constituent of the exoskeleton of diatoms.

Artificial sea water is used in aquaria but its composition only corresponds approximately to that of natural sea water. The formula is given by Spotte (1973).

Salinity has a clear effect on the distribution of animals which usually tolerate only a narrow range of values (the exception is the brine shrimp, *Artemia salina*, with a salinity tolerance range of 0–200‰). Few species, apart from eels and salmonids, are found in fresh and salt waters; salinity and variations in salinity are therefore important features of any aquaculture site.

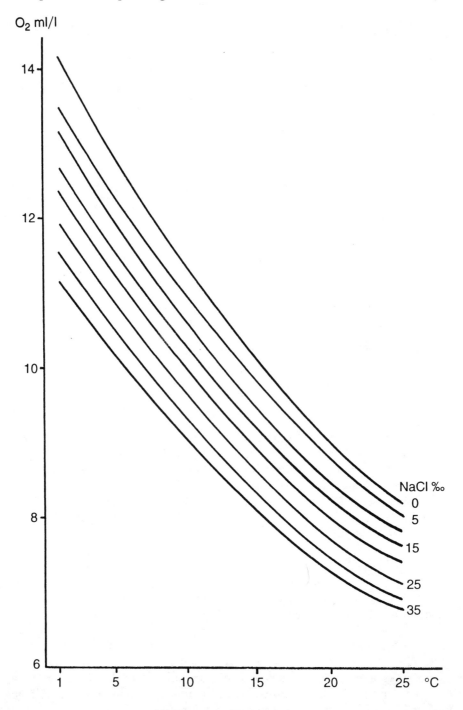

Fig. 6. Oxygen saturation level in relation to salinity and temperature.

Composition of the synthetic sea water 'Instant Ocean' giving a salinity of
33.70‰. All values are in parts per million (ppm).

Constituent	ppm	Constituent	ppm	Constituent	(ppm)
Cl	18 400	H_3BO_3	25	I[*]	0.07
Na	10 200	Sr	6	EDTA[*]	0.06
SO_4	2 518	SiO_3	3	Al[*]	0.04
Mg	1 238	PO_4	1.3	Zn[*]	0.02
Ca	390	F	1.0	V[*]	0.02
K	370	MoO_4	0.6	Co[*]	0.01
HCO_3	140	S_2O_3	0.3	Fe[*]	0.01
Br	65	Li	0.2	Cu[*]	0.003
				Rb[*]	0.1

[*] Present as traces in solution.

The aquatic environment can be characterized by salinity:

Fresh water   Salinity ≤ 0.5‰

Oligohaline   Salinity   0.5–3‰ (brackish water)

Mesohaline   Salinity   3–16.5‰ (brackish water)

Polyhaline   Salinity   16.5–30‰

Marine   Salinity > 30‰

Hypersaline   Salinity > 40‰

The internal environment of aquatic organisms has a different composition of dissolved substances from that of the surrounding environment. To maintain a constant internal environment, living organisms have mechanisms for osmoregulation: methods differ between groups.

3.3 Other physical factors
Specific density
The presence of dissolved salts results in sea water having a higher density (1.04 g l^{-1}) than fresh water (1.000 g l^{-1}); this has an effect on the flotation of aquatic organisms and also on water stratification. In an estuary fresh water flows over salt water which can push several kilometres up river as a saline wedge or plug. In the sea, in summer, the surface water is warmer (and less dense) than deeper water; the two are separated by the thermocline, a narrow band of water (<1 m). This is most frequently found at depths between 20 and 50 m although it can be as deep as 200 m, and it is important for aquaculture in the open sea to know where it is as there may be a 5 or 10°C difference between the temperature of the water above and below. The thermocline thus acts as a physical barrier. In winter the mixing caused by storms leads to homogenization, at least to a depth of 40–50 m.

On a reduced scale, the same phenomena occur in lakes: the thermocline separates the upper layer (epilimnion) which is illuminated, aerated, and 'living' from the hypolimnion, the deep, dark deoxygenated zone. In the hypolimnion there may sometimes be a complete absence of oxygen (anoxia) in the lower layers which are thus unsuitable for aquaculture (for example, in cages) but also for the wild fish in these water bodies. The natural seasonal mixing by winds may not be enough and it may be necessary to install mechanical equipment to destratify lakes (Steichen *et al.* 1979) and aerate the deep water. This also happens under sea-water salmon cages in some Norwegian fjords.

Specific heat

This is the quantity of energy necessary to increase the temperature of a mass of a substance by 1°C (the calorie is the energy necessary to increase the temperature of 1 g of water from 14.5 to 15.5°C). The specific heat is thus 1 cal g^{-1} °C^{-1} or 4.186 J g^{-1} °C^{-1}. This is an exceptionally high value and varies little for waters of different salinities and temperatures. Sudden variations in temperature in the aquatic environment are thus cushioned. The thermal stability (and also the physical and chemical stability) of the marine environment, as well as the biological stability, is very different from the land; it is interesting to note that poikilotherms dominate the sea.

The importance of this for the fish farmer is that a great deal of energy is required to change the temperature of the aquatic environment.

Hydrostatic pressure

A further physical factor, hydrostatic pressure, is, as yet, of little interest to the fish farmer although buoyancy problems have been noted (undoubtedly linked to the swim bladder) with trout held in cages submerged a few tens of metres down. It is essential that the trout have access to the surface (or to an air pocket). A 10 m increase in depth means an increase in pressure of one atmosphere.

Atmospheric precipitation

Rain may affect aquaculture by reducing the salinity of the water and thereby eliminating marine species in shallow sites which are normally saline. There may be a proliferation of pathogenic bacteria from fresh water which cannot survive in full-strength sea water (this leads to a prohibition on the sale of shellfish as in 1989 in the Etang de Thau because of Salmonella). Acid rain and runoff carry nitrogen, sulphur and aluminium (Solbé 1988); no body of water can be considered to be pristine. Rain can even trigger plankton blooms as floods carry alluvial deposits and fertilizers; these can also influence fisheries and aquaculture by improving larval survival and the growth rate of molluscs.

3.4 Dissolved gases

Oxygen (O_2)

Oxygen is essential for all living things (aerobes). It is used in the oxidation of food, liberating the energy necessary for all vital activities (swimming, hunting, reproduction, growth, etc.). This oxygen is removed from the surrounding environment during respiration, which is expressed by the equation:

$$CH_2O + O_2 = H_2O + CO_2$$

The aquatic environment contains relatively little oxygen (less than $10 \text{ cm}^3 \text{ l}^{-1}$, contrasting with $200 \text{ cm}^3 \text{ l}^{-1}$ for air). It is therefore essential for all forms of aquaculture to maintain an adequate level of oxygen in the rearing water.

This concentration of oxygen is close to saturation (and sometimes supersaturation) in the natural environment, and varies from $8 \text{ cm}^3 \text{ l}^{-1}$ in cold water to $4.5 \text{ cm}^3 \text{ l}^{-1}$ in tropical waters, being less in marine waters and shallow fresh waters because of the photosynthetic activities of plants. In eutrophic zones there may be excessive demands for oxygen (anoxia), pushing levels down; this is dystrophia. In high-density rearing units (intensive farms such as for salmonids) failure of water exchange leads to depletion and even exhaustion of oxygen in the environment and the death of the animals being farmed.

The decomposition of organic matter by the numerous microorganisms living in the water column or on substrates (see Chapter 2) also consumes oxygen. This requirement for oxygen and the respiration constitutes the biological oxygen demand (BOD) and may be important when the decomposing animals and plants are in closed systems (for example in a pond). Using herbicides in water bodies is not always practical. The chemical oxygen demand (COD) is a measure of the chemical reactions which consume oxygen; these are low in the waters used in aquaculture.

In general it is estimated that an oxygen level of 5 mg l^{-1} is sufficient for most species. Certain African catfish (*Clarias*) can survive without oxygen dissolved in the water; tilapias can survive a level of 1 mg l^{-1}. Low oxygen levels are a particular problem in summer (high temperatures, abundance of plants, decreasing day length, high food consumption by the farmed animals, high rates of respiration and excretion in natural environments which are shallow or stagnant (stratified waters)). The consumption of oxygen at night pushes down the level before dawn and the species which are most susceptible die; this itself increases the BOD (because of the decomposition by microorganisms) at a moment when there is no photosynthesis; this is dystrophia.

Nitrogen
Natural waters are saturated by this gas and there is a supersaturation of 2% near to the surface. If levels exceed 3–4% supersaturation, a non-infectious disease may ensue; this is gas bubble disease and results, among other things, in exophthalmia. This supersaturation can be caused by the gas dissolving when air is drawn through badly maintained pumps.

Hydrogen sulphide (H_2S)
This is the result of anaerobic decomposition and leads to anoxia in the environment and water flowing through culture units. It is found frequently in the organic-rich sediments under fish cages or beneath suspended mollusc culture systems; if such sediments are stirred up and become suspended in the water column they are toxic to the farmed species.

3.5 pH and the carbon dioxide system
pH is an expression of the hydrogen ion concentration of water ($pH = -\log_{10}[H^+]$). It is

a measure of the acidity or alkalinity of water which is expressed on a scale between 0 and 14. Between 0 and 7 the water is acid, it is neutral at 7 and basic above this value. The pH of sea water varies between 7.9 and 8.3 depending on the zone.

pH has a very important influence on the chemical environments, for example the equilibrium between NH_4^+ and NH_3 in water is shifted towards the formation of NH_3, which is extremely toxic for fish, as pH rises. The same water which is non-toxic at pH 6.7 may kill animals at pH 8.

The pH equilibrium depends on other interactions, chiefly the carbon dioxide carbonate system and pH governs the carbonate content of waters.

The carbon dioxide system

Carbon dioxide gas (molecules of undissociated CO_2) is scarce in the sea (a few parts of a millilitre per litre) because it is part of a reversible chemical equilibrium with carbonate and bicarbonate ions and carbonic acid.

$$2H_2O + 2CO_2 \rightleftharpoons 2H_2CO_3 \rightleftharpoons H^+ + 2HCO_3^- \rightleftharpoons 2H^+ + 2CO_3^-$$

carbon dioxide   carbonic acid   bicarbonate   carbonate

The total level of carbon dioxide varies between 40 and 50 mg l^{-1}. Molecules of carbonic acid make up only a few percent of the level of undissociated carbon dioxide molecules. Therefore, most carbon dioxide is bound up in either carbonate or bicarbonate ions.

Introducing carbonates (in the form of calcareous chips as used, for example, in the biological filtration of water) pushes the reaction to the right and thus increases pH. In practice, pH in the sea depends on the above equilibrium but other compounds may have a buffering effect on the pH of sea water.

The world's oceans contain enormous quantities of CO_2 (350×10^{12} tonnes of carbon).

In lakes the situation is vary variable and depends on the geology of the watershed. Volcanic regions have acid waters (sometimes pH 6) because of the low level of carbonates which pushes the equilibrium to the left. In calcareous regions the high availability of carbonates leads to high pH levels (8.5) and to hard waters; hardness is directly related to the carbonic acid. Aquaculturists must pay close attention to the hardness of water.

Various indices are used to measure hardness:

— Total hardness. This is the sum of the calcium and magnesium ions. It is an expression of the temporary hardness (which disappears when the water is boiled because of the deposition of calcium) and permanent hardness (which remains after boiling; unprecipitated calcium and magnesium).
— Hardness expressed as either the level of calcium or of magnesium respectively.
— Alkalinity. This gives information on the level of carbonates, bicarbonates and hydroxides but it is necessary to distinguish:

— simple alkalinity, which measures the total level of alkaline hydroxides but only half the level of carbonates
— total alkalinity, which measures the total level of hydroxides, carbonates, alkaline bicarbonates and alkaline earths.

Details of methods for the analysis of fresh water can be found in a work by Terver (1989), and Aminot & Chaussepied (1983) give methods for sea water. Some constituents must be analysed by different methods in the two environments (e.g. ammonia compounds).

Kits for the analysis of parameters which are important in aquaculture are being developed and will remove the need for complex, costly analysis.

4. PENETRATION OF SOLAR RADIATION INTO WATER

Solar radiation is the foundation of all life on land and water as it enables the production of living matter through photosynthesis (the only exception is the community of chemosynthetic animals associated with submarine hydrothermal vents, which are of minor importance). Apart from this, solar radiation is the major influence on the thermal regime of the seas and consequently of climate.

The solar energy received on the surface of the earth is of the order of 5×10^{20} kcal (1.73×10^{17} W). Oceans receive 71% of the radiant energy; in intermediate latitudes this represents 9–10 billion kcal ha^{-1} yr^{-1} (Duvigneaud 1980) or 1×10^6 kcal m^{-2} yr^{-1}. Only a part of this can be used because of absorption by the atmosphere, reflection at the surface and differential absorption of radiation that can be used for photosynthesis, which means that photosynthesis can only take place in the top 15–50 m of water, depending on turbidity.

Fig. 7 shows the fate of a ray of light reaching the surface of a body of water: if the sun has an angle at the horizon below 48.5° it is reflected from the surface without penetrating the water. At a greater angle it penetrates the surface but the path is deflected; this angle of refraction (r) is calculated from the angle of the sunlight with the vertical;

$$\sin r = \tfrac{3}{4} \sin i$$

Day length under water is thus much shorter than on land.

The surface of water is seldom flat thus waves tend to reflect more of the incident light; and while it has been proved that the changing light levels lead to improved photosynthetic efficiency (Walsh & Legendre 1983), this is of little overall importance.

The intensity of such radiation varies with latitude, season, time of day, cloud cover, etc., as on land but the penetration varies with other parameters.

— Transparency of the water is very variable and relates to the level of suspended particles (plankton). The relative transparency is measured with a Secchi disc (a white disc, 30 cm in diameter). The depth at which the disc disappears is measured. A spectrophotometer can be used to measure the absorption of light passing through a sample of water. Chinese fish farmers plunge their forearms into their ponds to assess the density of algal plankton which colours the water green. Fig. 8 (after Almazan & Boyd 1978) shows the relationship between the depth where the Secchi disc disappears and the level of planktonic chlorophyll in the water.

In the Vigo ria, Fraga (1979) showed a decrease of 26% in productivity (from the concentration in plankton) linked to a decrease in transparency because of the increase in suspended particles caused by human activity. These particles either reflect or

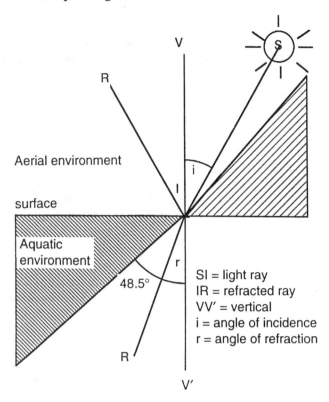

Fig. 7. Refraction of light at the water surface.

absorb light. Water in the open sea is much more transparent and is classed as blue water.

— The coloration of water, either by run-off from the land or by planktonic organisms, reduces the penetration of solar radiation. In coastal regions (neritic) yellow-green substances apparently linked with humic acids colour the waters slightly, but this can be distinguished from planktonic blooms which give a marked red or brown tinge and are referred to as 'red tides'. The 'marine humus' is made up of organic molecules (dissolved organic matter) (Chapter 2). Figure 9 (after Péres 1976) shows the absorption of light as a function of depth in marine coastal waters and in the open sea. In ponds used for rearing, the absorption of light by phytoplankton is much more rapid as the transparency of the water is less than one metre.

The differential absorption of the different coloured light (and different wavelengths) varies; blue is the dominant colour because it penetrates deepest (Fig. 10).

The euphotic zone is defined in relation to the penetration of light (the layer of water above depths receiving less than 1% of the incident light at the surface). Photosynthesis can take place within this layer which extends from the surface to a depth which varies between 20 and 100 m. The oligotrophic zone (400–600 m) is only penetrated by blue-violet rays, the deepest waters (aphotic) remain dark.

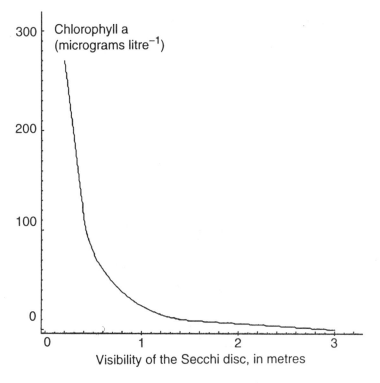

Fig. 8. Relationship between the visibility of a Secchi disc and the level of chlorophyll a.

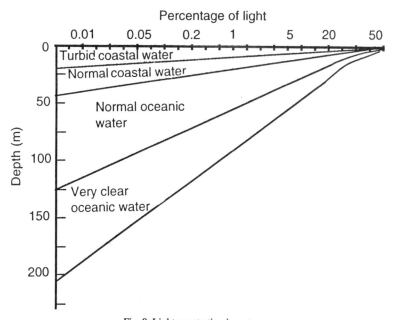

Fig. 9. Light penetration in water.

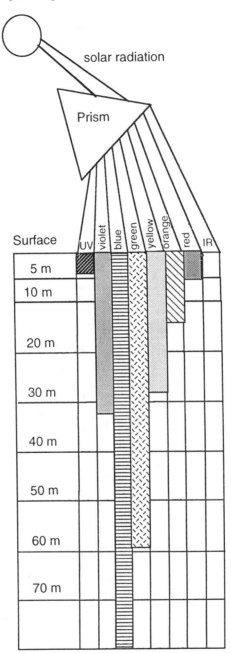

Fig. 10. Absorption of light of different wavelengths.

REFERENCES

Almazan G., Boyd C., 1978. An evaluation of Secchi disk visibility for estimating plankton density in fish ponds. *Hydrobiologia*, **61** (3): 205–208.

Alzieu C., 1990. L'eau milieu de culture. In: *Aquaculture*, G. Barnabé (ed.). Lavoisier Ed., Paris: 17–45.

Aminot A., Chaussepied M., 1983. Manuel des analyses chimiques en milieu marin, CNEXO ed., Brest: 395 pp.

Duvigneaud P., 1980. *La synthèse écologique*. Doin Ed., Paris: 380 pp.

Fraga F., 1979. La profundidad de vision del disco de Secchi y su relacion con las concentrationes de fitoplancton y arcilla. *Invest. Pesq.*, **43** (2): 519–528.

Hanson R. B., Alvarez-Ossorio M. T., Cal R., Campos M. J., Roman M., Santiago G., Varela M., Yoder J. A., 1986. Plankton response following a spring upwelling event in the ria de Arosa. *Spain. Mar. Ecol. Prog. Ser.*, **32**: 101–113.

Korringa P., 1976. Farming the European flat oyster (*Ostrea edulis*) in a Norvegian poll. In: *Farming the flat oyster of the genus*, Ostrea, P. Korringa (ed.). Elsevier Publ.: 187–204.

Péres J. M., Deveze L., 1963. *Océanographie biologique et biologie marine*. Tome II: La vie pélagique. P.U.F., Paris: 514 pp.

Péres J. M., 1976. *Précis d'océanographie biologique*. P.U.F., Paris: 247 pp.

Solbé J., 1988. Water quality. In: *Salmon and trout farming*, L. Laird & T. Needham (eds). E. Horwood Publ., Chichester (England): 68–86.

Spotte S., 1973. *Marine aquarium keeping, the Science Animals and Art*. J. Wiley & Sons, New York: 171 pp.

Steichen J. M., Garton J. E., Rice E. C., 1979. The effect of lake stratification on water quality. *Journal AWWA*, **71** (4): 219–225.

Terver D., 1989. *Manuel d'aquariologie*; 1. L'aquarium, eau douce eau de mer (4° Ed.). Réalisation Editoriales Pédagogiques, Paris: 303 pp.

Thorpe S. A., Hall A. J., 1987. Bubble clouds and temperature anomalies in the upper ocean. *Nature*, **328**: 48–51.

Walsh P., Legendre P., 1983. Photosynthesis of natural phytoplankton under high frequency light fluctuations simulating those induced by sea surface waves. *Limnol. Oceanogr.*, **28** (4): 688–697.

Weller R. A., 1987. Mixing in the upper ocean. *Nature*, **328**: 13–14.

2

The production of living matter in the aquatic environment

1. MAJOR CATEGORIES OF AQUATIC ORGANISMS

The organisms inhabiting this environment fall into two categories:

— Those which live on the bottom (resting, attached or burrowing); these make up the benthos and are called benthic creatures. They are in contact with the sea bed and fall into two categories; attached (oysters, mussels, attached seaweeds) and mobile (scallops).
— Those which live in the water column; these are pelagic creatures. They move through the water column with varying abilities. The nekton are species capable of major changes of position (e.g. fish); this distinguishes them from plankton.

According to Hamner (1988) plankton are a vast assemblage of organisms, mainly of small size, from bacteria to larval fish which are found in water, most of them are capable of active swimming or are transported passively by changing their flotation.

The plankton is subdivided into size groups:

— Megaplankton. This may reach several metres in length (or diameter)—medusae, sea squirts.
— Macroplankton, >5 mm. Krill, larval fish, sea squirts.
— Mesoplankton, 200–5000 μm. Large diatoms, diatom chains, rotifers, copepods, larval molluscs and worms, fish eggs, etc.
— Microplankton, 20–200 μm. Unicellular algae, ciliates, copepod nauplii and copepodites, larval molluscs.
— Nanoplankton, 2–20 μm. Including ultraplankton (2–10 μm), unicellular algae, flagellates and small ciliates.
— Picoplankton, 0.2–2 μm. Algae and bacteria making up the bacterio-plankton. Picoplankton are retained by filters with a 0.2 μm pore size.

Ultraplankton comprises living organisms of less than 10 μm diameter.
Other terms refer to different biological characteristics:

— Haloplankton. This is the permanent plankton.
— Meroplankton. Temporary plankton (larval crustaceans, molluscs and fish for example).
— Tripton. Particulate inert material (non-living).

This vocabulary may seem out of place in aquaculture but the subdivisions will be found in the population of a pond as much as in the open sea. The biological processes taking place are fundamentally the same.

2. CHARACTERISTICS OF THE ENVIRONMENT AND AQUATIC LIFE

One of the characteristics of aquatic ecosystems is that, unlike on land, water can never be limiting and other limiting factors are temporary. There are no physical barriers and no shelter in the three-dimensional underwater world. There are no reference points equivalent to forests in the sea which can be studied in isolation (Hamner 1988). Variations in temperature are never above 30°C while on land they may exceed 70°C. Because of this there are major differences between terrestrial and aquatic ecosystems.

All aquatic creatures have certain common characteristics which distinguish them from those on land:

— Mechanically, aquatic animals live in an environment with a density close to their own, which means that supporting structures (the skeleton) can be reduced and also sometimes that the locomotory apparatus disappears (attached molluscs).
— Metabolism is based on proteins and not on carbohydrates.
— The size of particles determines the structure of the food chains.
— The absence of thermal regulation (poikilothermic animals) considerably reduces metabolic costs and allows more effective conversion of ingested food to flesh.
— The majority of aquatic species have a very high reproductive potential; a mussel or an oyster releases many millions of eggs each year, a prawn or a fish many hundreds of thousands and the number of spores released by seaweeds is even greater.
— Modes of reproduction are often different from those of species cultured or reared on land: that of seaweeds is very different from that of cultivated plants. The reproduction of molluscs, crustaceans and fish is dissimilar from the birds and mammals which make up the bulk of terrestrial culture. Eggs of aquatic species hatch out to produce larvae which undergo a very different metamorphosis and mode of life from that of the adults. Only rarely is any parental care given to the juvenile stages which are dispersed passively within the vast aquatic environment. The rearing of these planktonic stages is poorly managed as the fundamentals are not yet well understood; it is nevertheless essential for the fish farmer as it provides the individuals which start off the rearing process.
— Filter-feeding animals such as bivalve molluscs (oysters and mussels) have no terrestrial equivalent; they continually sift sea water to extract phytoplankton or particles which occur naturally and provide food.
— The final point is that the relative abundance of living creatures in the aquatic environment declines rapidly as the size of individuals increases.

3. PHOTOSYNTHESIS AND THE TROPHIC WEB

Under the action of solar radiation the inorganic materials in the environment (carbon dioxide gas, nitrogen and phosphorus compounds) are transformed into organic matter (in living cells) by plants; this process is termed photosynthesis. It takes place in the water mass and is carried out by autotrophs which may be planktonic or benthic. This type of production is termed primary production as it is the starting point for the production of living matter both in water and on land.

The photosynthetic process is a vast complex of physical, chemical and biological interactions which affect the relations between the primary producers and their environment; in summary at a physico-chemical level light energy (photons) induces the transport of electrons by chemical compounds (chlorophyll pigments [chlorophyll a, b and c], carotene and derivatives [peridinine, fucoxanthine], phycocyanine, phycoerythtine). This energy is then transferred by specific transport agents (ATP) and used in the formation of organic molecules needed by living organisms (carbohydrates then proteins and lipids). The process of photosynthesis takes place at the level of the plant pigments and certain bacteria in the euphotic zone, i.e. shallow waters and the upper layers of deeper waters. In simple terms photosynthesis effects the conversion of inorganic carbon (see the carbon dioxide system, Chapter 1) into organic carbon.

Photosynthesis in the aquatic medium is dependent on the degree of penetration of light into the environment (Fraga 1979) and the availability of certain essential minerals (see Chapter 1). The principal essential elements are dissolved in the water at varying concentrations which allows direct exchange across cell walls. We shall see (section 5 below) that other organic substances present in infinitesimal quantities in the water also play an essential role. The abundance of chlorophyll (linked with the density and the size of photosynthetic organisms) regulates the intensity of primary production but the speed of these processes and the dependence on temperature and many other factors play their part (floatability of phytoplankton, movement of water masses, etc.).

This living plant material provides food for animals known as 'herbivores', which are themselves incapable of synthesizing the basic materials of life (carbohydrates, proteins and fats). They are the second link in the food chain and are termed secondary producers. These herbivores are, in their turn, the prey of carnivores, the third link in the food chain. Synthesis of living matter from organic matter by phytoplankton is a characteristic of autotrophs, and is also a characteristic of plants and certain bacteria. All other species are called heterotrophs and depend on autotrophs for their nutrients.

The transfer of living matter (and consequently energy) in food chains is carried out with an efficiency of around 10% at each change of level. In the classic scheme, 1000 g of phytoplankton are converted to 100 g of sardine which in their turn make 10 g of mackerel and then 1 g of tuna.

The transfer of living matter through successive links of a simple food chain is not the only pathway in water. Recent basic research has uncovered other processes for both primary production and interactions in the microscopic environment which had previously passed unnoticed; because of this it is now normal to refer to aquatic trophic webs where relations are numerous and complex between the consumer and the consumed (Fig. 1).

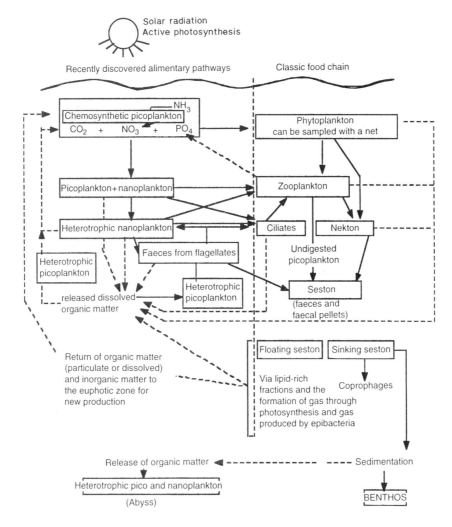

Fig. 1. Schematic representation of the complexity of the aquatic nutrition network. After a figure of Davis (1982) reported by Stockner (1988).

Some types of aquaculture production consist of nothing more than diverting the transfer of living matter in the aquatic environment towards human consumption. The trophic web thus forms the basis of this type of production.

4. PRIMARY PELAGIC PRODUCTION

The processes regulating primary production are identical in principle in both fresh and salt water (Goldman 1979a, b; Goldman & Mann 1980). They are associated with autotrophic organisms which can be separated on the basis of size.

Phytoplankton is the dominant community populating fresh and salt waters, spreading through the pelagic zone, the most extensive habitat type on the planet; but algae, both benthic and floating (sargassum, filamentous algae, attached diatoms) and certain types of attached bacteria are also primary aquatic producers. Macrophytic algae (seaweeds) make only a small contribution to oceanic production in comparison to plankton (0.2 billion tonnes compared with 35–70 billion tonnes from the phytoplankton). However, they are significant in aquaculture as they are directly consumed by man and they have a role in other types of aquaculture.

In the ocean productivity (in grams of carbon per square metre per year (g C m^{-2} yr^{-1})) is of the order of 25–50 g C m^{-2} yr^{-1} in the open sea. This natural production is low because the level of essential nutrients in surface waters is too low (0.001–0.08 mg N l^{-1}, 0.002–0.003 mg P l^{-1}) for good primary production (which requires 0.05 mg N l^{-1} and 0.005 mg P l^{-1}). Deep waters are different and do not show these low levels of nutrients (0.2–0.5 mg N l^{-1}, 0.05–0.09 mg P l^{-1}).

The most productive neritic zone will produce between 100 and 400 g C m^{-2} yr^{-1}, and upwelling zones between 300 and 500 g C m^{-2} yr^{-1}.

Living things probably make up no more than 10% of the total organic material present in the ocean.

The description of the plankton is outside the remit of this book, and more information on this subject can be found in the works of Péres (cited in the preceding chapter) and in the *Atlas of marine phytoplankton* (Volume 1, Sournia, 1986; Volume 2, Ricard 1987; Volume 3 in preparation).

4.1 Autotrophic phytoplankton and bacterioplankton

The phytoplankton are unicellular algae typified in the ocean by *Synechococcus*, a small oval cell measuring between 1 and 5 μm. In littoral and lagoon waters species measuring 2–7 μm predominate (*Chlorella nana*, *Nannochloris* sp, etc.); these are cultured in aquaculture. Coccolithophores are autotrophic flagellates, 4–5 μm, with a calcareous skeleton. They are characteristic of warm waters of poor nutrient status. According to Péres and Deveze (1963) they make up 30–98% of unicellular planktonic organisms in the Mediterranean and the genus *Coccolithus* predominates.

The bacterial plankton is represented by photosynthetic cyanobacteria measuring 0.2–0.4 μm. Such organisms may reach densities of several tens of million per litre in the water. In spite of this abundance their importance was not appreciated for a long time.

These species (except for the coccoliths which are identified by their calcareous skeleton) are identified by fluorescent or electron microscopy. Towards the end of the 1970s they were identified in the sea and in fresh water, opening up a debate on the true primary production of the aquatic environment. In practice the bulk of primary pelagic production is through these very small cells. They show optimum growth and photosynthesis at very low light levels, either towards the bottom of the photic zone or where only blue and violet rays penetrate (blue–violet rays of 455 nm). In mid-Atlantic the maximum chlorophyll level is 90 m below the surface of the euphotic zone where light intensity is only 1% of that at the surface (Glover *et al.* 1986). Such plankton are even found in the dark and are found at depths of 1000 m. They use nitrate, ammonium and urea and are responsible for 50–80% of the total nitrogen uptake by the whole of the plankton.

Primary production by the phytoplankton goes from 0.01 to 31 mg C m^{-3} h^{-1}. The biomass, expressed in chlorophyll content, represents 1–90% of the total phytoplankton chlorophyll. The contribution of the picoplankton to primary production is greater in the deep ocean in oligotrophic waters (50–80%) than in coastal waters which may be eutrophic (2–25%). In the Barents Sea and other northern waters up to 90% of the chlorophyll biomass in the water is contained in the phytoplankton (0.05–1 mg chlorophyll m^{-3}). In the Mediterranean the phytoplankton dominate in winter and spring. In fresh water, picoplankton contribute 16–70% of carbon production and 6–43% of chlorophyll biomass. This contribution increases with depth, reflecting the efficiency with which the pigments are able to utilize blue light.

The biomass of plankton is greater by an order of magnitude in temperate than in tropical (Sargasso Sea) waters. Tropical seas can be considered to be deserts; however, differences in production are less marked as the rate of turnover of biomass is three times per day (Sheldon 1984) not 0.2 as had been previously thought.

This was a major discovery as it showed that, in the absence of predators and of mortality, 1 g of plankton produces 3 g of living matter each day. This information is in agreement with results for zooplankton whose biomass is not especially low (see below). The growth of tropical oceanic plankton is therefore very rapid; the generation is estimated as 6.5 h, taking into account a 12 h night (i.e. a generation time of 3 h during the day). With a generation time of 6.5 h and equilibrium between production and predation, 12 h of growth leads to a production of 23.4 mg C l^{-1} day^{-1} or 1.2 g C m^{-2} day^{-1}. Thus, although the comparison has been made between tropical waters and deserts, the only resemblance is in the biomass. In their other characteristics (rapid growth and turnover rates) they can be compared favourably with tropical forests (Sheldon 1984).

While it is well established that pico- and ultraplankton (<10 μm) are an essential component of planktonic communities, their abundance is limited by predation by flagellates, ciliates and rotifers (in fresh water) but not by copepods which take larger prey. They are also consumed by other plankton and benthic microphages (see section 9 below). Such results shed a new light on the factors determining production in aquatic ecosystems and are clearly of relevance to aquaculture.

4.2 Autotrophic nanoplankton and microplankton

Diatoms are often the main constituent of the larger phytoplankton, especially in temperate and cool oceanic and coastal waters. They are able to concentrate near the surface as many are positively buoyant. In spite of this they mainly use light at the blue-green end of the spectrum (red, which only penetrates the uppermost layers of the water, may have an inhibitory effect). Their maximum rate of photosynthesis takes place at depths of 20–30 m. The size of individuals is between 20 and 200 μm but they frequently form chains several millimetres long. The genera *Chaetoceros*, *Skeletonema*, *Phaeodactylum*, *Thalassiosira*, *Nitzschia* dominate planktonic populations in coastal or lagoon waters. They proliferate in eutrophic waters (coastal ponds, lagoons, lakes) and the first three species mentioned above are often cultivated in aquaculture. The blue coloration of oysters raised in special beds is caused by the presence of some blue pigmented *Navicula* in the rearing beds.

Peridinians (dinoflagellates or Dynophyceae) are generally less abundant in temperate waters, but are the dominant groups in this size class in oligotrophic tropical waters and in temperate waters in the summer. Species without an outer case (naked dynoflagellates) have been reared to feed captive larval anchovies (*Gymnodinium*). Armoured peridinians have an outer case made up of cellulose plates. They serve as food for many species of plankton-eating fish (anchovies (Hunter 1976), sardines). These are the organisms responsible for red tides. Even at a low density (200 l^{-1}) some species produce toxins which are accumulated by molluscs and are toxic to those eating shellfish (see review by Alzieu 1990). This poses problems for the culture of marine molluscs throughout the world.

Pelagic production is characterized by fluctuations linked to hydrological conditions: this in turn is dependent on climatic conditions. Thus, in the Mediterranean during winter and early spring the genus *Chaetoceros* and ultraplankton make up the biggest proportion of plant production. In late spring there is a proliferation of *Rhizoselenia* and *Nitzschia* in surface waters. In summer, dinoflagellates predominate but diatoms are found at the base of the photic zone (30 m). There is a second plankton bloom in the autumn; this, too, is largely composed of diatoms (Margalef 1969).

There are many other plankton belonging to different groups which contribute to aquatic production, but to a lesser extent. It has been noted that the abundance of the different types is linked to the nutritional richness of the waters. Diatoms are found in nutrient-rich waters, peridinians are confined to oligotrophic waters, and ultra- and nanoplankton to ultra-oligotrophic waters.

From a theoretical point of view, Martinez *et al.* (1983) showed that nitrogen could be fixed in pelagic waters by diatom masses with symbiotic intracellular bacteria. This is a new source of nitrogen for oligotrophic waters. Thorpe & Hall (1987), studying the availability of atmospheric gases in the superficial layers of the ocean, showed that carbon and mineral nitrogen are present throughout and are therefore unlikely to limit production.

4.3 Sedimentation, resuspension and productivity

While only a small proportion of the phytoplankton and the particulate wastes produced have positive buoyancy (because of gas bubbles or a high level of lipids), most is likely to become sediment. There is a constant elimination of phytoplankton through sedimentation; losses are compensated by growth and the organisms maintain their position in the upper layers by regulation of osmotic pressure, movements of appendages, flagellae and synthesis of lipids. However, sedimentation takes place at a rate of several metres per day when the density is around 1.10 and that of sea water 1.025.

Decomposition under bacterial action of dead phytoplankton and various zooplankton wastes (faeces or from moulting) to their basic mineral constituents is slow and waters rich in such nutrients are therefore usually at great depths. Production is limited to the upper, photic zone where photosynthesis can take place; this is above the mineralization zone. There are thus two zones (Fig. 2), an upper zone which exports to the lower zone, according to the classical scheme put forward by Margalef.

The persistence of pelagic life is due to the mixing processes which take place in the upper layers (Chapter 1) and the input of dust from the atmosphere. These are far from

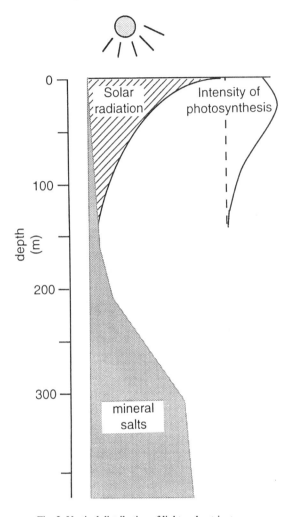

Fig. 2. Vertical distribution of light and nutrients.

negligible but the abundance of plankton is only made possible through a continuous or periodical return of mineral salts from deep waters to the euphotic zone. This happens in places of 'upwelling'. The speed at which salts ascend can go from a few metres to 30 metres per day. The enrichment acts as a fertilizer to the euphotic zone; this is the main parameter determining productivity.

Upwelling is caused either by large-scale ocean currents or by winds which regularly come from one direction and therefore can induce localized currents. These are the best areas in the world for fishing (33% of production comes from 1% of the oceans) and production of phytoplankton can reach 10 tonnes dry weight ha^{-1} yr^{-1}. Upwelling zones, such as the one off the coast of Peru, produce around 15 million tonnes of anchovies per year in an area of 60 000 km^2; this is 2.5 t ha^{-1} yr^{-1} wet weight. We have shown (Chapter 1) that upwelling affects the growth of mussels in culture in the rias of Galicia (Hanson *et al.* 1986). Such rich waters are most frequently characterized by an abundance of diatoms. Waters low in nutrients (oligotrophic) are characterized by an abundance of flagellates.

Minas *et al.* (1986) calculated primary production in various upwelling zones from chemical and hydrographic data.

— in Peru, production is 0.6 g C m^{-2} day^{-1} (219 g C m^{-2} yr^{-1})
— off south-west Africa: 1.1 g C m^{-2} day^{-1} (401 g C m^{-2} yr^{-1})
— off north-west Africa: 2.3 g C m^{-2} day^{-1} (839 g C m^{-2} yr^{-1})

The table below shows the mean productivity of various zones.

Zone	Production (primary)	Trophic levels	Production of fish
Oceanic	50	5	0.005
Coastal	10	3	0.34
Upwelling	200	1.5	36

Production may sometimes be limited by the absence of a single substance. Martin *et al.* (1990) suggested that in the Antarctic the absence of iron limited primary production in an area rich in other nutrients. Two zones comparable for their levels of nitrogen and phosphorus—one a coastal zone rich in iron (7.4 nmol kg^{-1}), probably of terrestrial origin—had a primary production of 3 g C m^{-2} day^{-1}; the other—in the open sea, deficient in iron (0.16 nmol kg^{-1}) and in manganese—had a production of 0.1 g C m^{-2} day^{-1} as the phytoplankton is unable to use more than 10% of the principal nutrients present in the water.

These examples from oceanography have their equivalent in aquaculture as the processes are basically similar but on a different scale.

5. THE MICROBIAL 'LOOP'

5.1 Inert particulate material
It has been shown that, alongside primary production and the food chain described above, there exists particulate material (dead phytoplankton, protozoan tests, moulted copepod exoskeletons, various dead organisms, excreta, mucus globules, etc.) which are colonized by microorganisms (protozoa, bacteria) dependent on these, at least partially, for their food. These 'flocs' may grow, either through absorption of organic substances dissolved in the water or through the activity of the constituent organisms. This inert matter is characterized by its level of carbon, which forms the foundation of the basic molecules of living things (carbohydrates, lipids, proteins). This is organic carbon in contrast to the inorganic carbon of the carbon dioxide gas–carbonate system.

Most authors agree that this particulate organic matter (Seston, Fig. 1) is an important means for energy and the flux of materials in the epipelagic zone as it supports microbial activity (bacteria, protozoa, detritivores). 10–20% of bacteria are attached to particles, the rest are free in the water, constituting the heterotrophic bacterioplankton (0.3–1.2 μm).

We shall see that this material may also be consumed directly by zooplankton (section 6.1 below).

Progress in scientific instrumentation has allowed Koike *et al.* (1990) to show that a part of the material generally considered as dissolved is, in fact, made up of very small inert organic particles (between 0.38 and 1 μm) which they were able to distinguish from bacteria and whose abundance is of the order of 10 million ml^{-1}. They are non-rigid and can be passed through filters with a 0.25 μm pore size. They are especially abundant at the surface and their discovery has led to a new approach to the recycling of organic matter in the sea.

5.2 Dissolved organic matter and chemical mediators

Particulate organic matter represents no more than 9% of the organic carbon in the ocean. The majority (89%) is in the dissolved state (the remaining 2% is bound up in living aquatic organisms (Fiala-Médioni & Pavillon 1988).

This dissolved organic matter is made up of the products of excretion (faeces, secretions, etc.) or the products of the decomposition of living organisms after death. Near to the coast material transported from the land by rivers or the atmosphere provides a small supplement. Organic matter can be found as simple molecules (amino acids, for example) and also in the form of complex molecules (marine humus), the composition of which has been known for around 10 years through modern techniques of analysis. The concentration of dissolved organic carbon in surface waters goes from 1 to several milligrams per litre and the world's oceans constitute a huge reserve of organic carbon (Chapter 1).

Azam *et al.* (1983) estimated that 5–50% of the carbon fixed during primary production (transformed into phytoplankton) are found as organic matter dissolved in water. These organic molecules dispersed through the environment play several roles in spite of their low concentration.

—Bacteria are able to extract them from the water at very low concentrations and use them to increase their numbers (see section 5.3 below); this is also true for certain species of planktonic algae as well as other small species such as algae and ciliates.

—It has been shown that substances such as biotin, thiamin and Vitamin B12 are, at very low doses, the initiators of primary production.

—Other substances whose nature is not completely understood (polysaccharides, lipids or proteins for example) are secreted by certain types of phytoplankton and act as inhibitors for other types of phytoplankton. Diatoms thus send out substances which control the development of dinoflagellates and vice versa, but certain dinoflagellates inhibit the development of other species of dinoflagellates, etc.

Chemical communication, based on the detection of infinitesimally low concentrations (chemoreception), has been demonstrated for many species and plays a major role in behaviour: feeding (appetence of prey), reproduction (attraction of sexes at the time of fertilization), etc. While it is known that these mediators ensure a control of the abundance of planktonic species, many other interactions remain unknown or are difficult to demonstrate.

Water therefore contains substances which are capable of affecting living organisms even when found at very low levels. In some instances this information may have an application in aquaculture (for example, there are substances capable of inducing the fixation of planktonic mollusc larvae onto a substrate and others which stimulate feeding). These are significant for aquaculture and illustrate the importance of continued research. It might, for example, be possible to control the development of algal cultures in the natural environment. This is impossible at present. The ideal would be to limit the development of the dinoflagellate blooms which secrete toxins which are in turn accumulated by edible molluscs.

5.3 Bacteria

Organic matter is first utilized by bacteria. Their small size and large surface area allows the absorption of nutrients present at very low concentrations. There is therefore a commensalism in the production of dissolved organic matter by phytoplankton and its use by bacteria. The production of dissolved organic matter is influenced by the availability of mineral nutrients. Numbers of bacteria are very variable but relatively low in marine waters, as was shown by Azam et al. (1983):

Environment	Number ($\times 10^8$ l^{-1})	Biomass (μg C l^{-1})
Estuaries	50	
Coastal waters	10–50	5–200
Open sea	0.5–10	1–5
Deep waters	0.1	

The majority (70–80%) of these bacteria (0.3 to 1–2 μm) are free and mobile in the water column. The rest are associated with particulate matter; they show a kinesis towards plant cells but maintain a distance of around 10 μm around living cells (antibiotic emission?) while they become attached to moribund algae.

Newel (1984) reported experimental observations on free living bacteria (in open water) found around wastes from macrophytes or dead phytoplankton cells (these used carbon spreading out from this detritus). Pulverized dry plant material (30 mg l^{-1}) attracts a proliferation of bacteria approaching 10^7 cells ml^{-1} in one week (Fig. 3). Such bacterial populations may provide food for many species. The density of bacteria remains at around 10^7 ml^{-1}.

5.4 Protozoa

These are the first and the principal consumers of bacteria in both the pelagic (Sherr et al. 1986 and Figs 1 and 3) and other aquatic environments such as water treatment works (Curds 1975).

In the example reported by Newel (1984) the population of bacteria declined rapidly as it was consumed by the bacteria-eating flagellates (Monas, Oikomonas, Bodo, etc.) and ciliates. Ciliates consume bacteria at a rate of 500–600 h^{-1} for Tetrahymena for example. The ciliate Uronema marinum (with a volume of 1000 μm^3) consumes six times its volume every day but smaller flagellates (12 μm^3) are able to ingest 10–30 times their

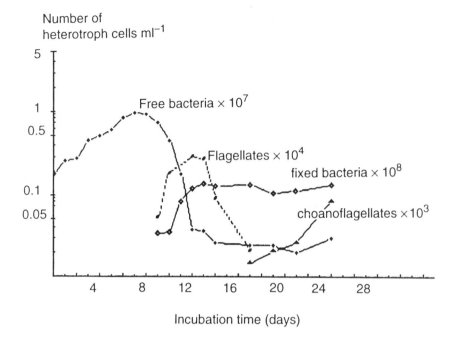

Fig. 3. Number of microheterotrophs colonizing (at 10°C) leaves and pulverized aquatic vegetation (14.5 mg carbon per litre of water) (after Newel 1982).

weight each day. Thus these protozoan 'grazers' of bacteria have a major impact on the biomass of bacteria in the natural environment. According to Sheldon *et al.* (1986), zooflagellates (flagellate heterotrophs) between 3 and 12 μm consume particles between 0.5 and 2 μm. They reach densities of $3-10^3$ ml^{-1}. Ciliates between 12 and 15 μm consume particles from 1 to 3 μm. Oligotrich ciliates (tintinnides), larval metazoans of 20–600 μm, can capture particles from 3 to 15–20 μm. The density of bacteria is limited (to around 10×10^6 ml^{-1}) by heterotrophic flagellates of 3–10 μm. Information given by Fenchel (1982) and Newel (1984) gives an estimate of the consumption of bacteria by microflagellates in the pelagic environment: on average, these flagellates consume 17 times their weight in bacteria, of which 27% is transformed into flagellates. This yield is above the 10% which characterizes the mean transfer rate between levels in the classic trophic web.

Populations of zooflagellates and ciliates have been estimated in the natural environment. Ciliates constitute a flux (measured by capture on submerged traps set at different depths) of 0.32 million to 2 million cells m^{-2} day^{-1} and a biomass representing 5–158 times that of the bacteria. The flux of zooplankton ranges from 25 to 955 million m^{-2} day^{-1}. This biomass is characteristic of the euphotic zone and decreases with depth.

The predators of bacteria are able to filter 12–67% of the mass of water each day (consuming bacteria but not the bigger phytoplankton).

It is interesting to note that the different links in this 'microbial loop' use, as food, particles whose size differs from their own by an order of magnitude. Size is the first factor determining the structure of the alimentary chains. The transfer of biomass in the pelagic trophic webs, towards the largest organisms, is partly by direct transfer through predation implying a difference in size between the eater and the eaten. This is taken into account in the global models used in fundamental oceanography (Azam *et al.* 1983; Borgman 1983; Sheldon *et al.* 1986).

It must, however be remembered that aquatic species always play a role not only as consumers but also as the food of other species; this is the sense in which one refers to a 'loop'. The excretion of faeces and emission of sexual products or juveniles often constitutes a direct opportunity for other forms of life (Fig. 1). Thus reproduction slows the transfer of biomass to the upper end of the alimentary chain, while gametes (small size) increase the biomass at the base of the food chain (Borgman 1983). This may be significant, for example, in oysters. Héral (1989) estimated that 78% of the total energy absorbed is used for the production of gametes.

Aquaculture production depends directly on the processes of production and biological transfer in the aquatic environment.

6. ZOOPLANKTON (Secondary pelagic production)

6.1 Microzooplankton

Flagellates and other protozoa, as well as phytoplankton, are in their turn food for small microzooplankton (around 20–200 μm). These are mainly larval and adult copepods, ciliates and larvae from a wide range of species which are particularly abundant in the euphotic zone. The transfer of matter between flagellates and copepods takes place with an efficiency of 30–40% according to Sheldon *et al.* (1984). In the Sargasso Sea for example the mean density of zooplankton varies between zones from 1900 to 5400 individuals per cubic metre (Böttger 1982). Copepods make up 90% of this total (molluscs and Appendiculates make up the rest). Nauplii represent 51–68% and copepodites 28–42% of the copepod portion. Microplankton (<200 μm) make up 52–68% of the total production.

Because of the huge supply of faecal bacteria (from copepods and other marine species) bacterial aggregates formed by faecal bacteria have an important role as direct nutrients for zooplankton, as well as in decomposition processes in the ocean. Zooplankton filter feeders such as copepods are able to graze on particles made up of bacterial aggregates. Ogawa (1977) fed various species of copepods with faecal bacteria, from various marine species, in suspension. The importance of bacteria as a food item has been demonstrated for krill by Tanoué and Hara (1986). The bacteria–choanoflagellate–krill–vertebrate chain exists alongside the traditional diatom–krill–vertebrate chain. Other types of detritus are utilized (dead plankton cells and other corpses), especially where the supply of food is limited, but assimilation is less efficient (9% for detritus, 24% for bacteria and 54% for phytoplankton).

Copepods filter water by creating a current with their appendages which may beat at a rate of up to 600 movements per minute. The current passes over a filter trap (Fig. 4), guarded by the maxillae; particles are directed towards the mouth. An individual *Calanus* filters between 1.5 and 3 litres of water per day, and it has been estimated that the filtration of water by zooplankton in the ocean is between 3 and 6 million $km^3 \, day^{-1}$

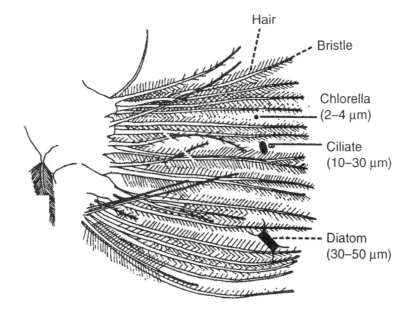

Fig. 4. Copepod maxilla (*Calanus helgolandicus*) and prey organisms consumed (after Conover 1976).

which adds up to a separation of 220–400 thousand million tonnes of suspended matter per year (Savenko 1988). This demonstrates the continuous process of resuspension of matter.

The significance of plankton for fisheries has been studied in Louisiana (Mulkana 1970). Up to 83% of the fish captured in the fishery were filter feeders, dependent on plankton.

6.2 Other zooplankton

This comprises all individuals above 200 μm in size (mega-, macro- and mesoplankton). Adult forms of most copepods belong to either the micro- or mesoplankton. *Calanus* is between 3 and 5 mm in length and is amongst the biggest of the copepods, but many small copepods and all larval forms, rotifers and many other species slip through the traditional large-meshed plankton nets. Bernhard *et al.* (1973) showed that a 70 μm mesh size was necessary to capture nauplii. Mulkana (1970) showed that the biomass of microzooplankton represents (in terms of dry weight) 113 times the weight of plankton captured in traditional nets with a 330 μm mesh size. The majority of the studies carried out in the past used identical nets and have been the basis for the calculation of production. Microzooplankton and large zooplankton are part of the same community which was poorly understood by workers in the past because of the problem of the selectivity of the large-meshed plankton nets.

Microplankton and larger zooplankton can be collected for feeding the larvae of cultured species (Barnabé 1980, 1986 and Part IV, Chapter 4, this volume). Prey of increasing size are used as the mouth size of the farmed predators increases as they grow.

Several species of plankton over 20 mm in length are captured in industrial fisheries (Omori 1980): these are the Scyphomedusae in China, Japan and Korea, krill (Euphausiids and Mysids) in the Antarctic, small decapods in various places and copepods in Canada, Japan and Norway. These species are used as bait or in the feeding of larval fish (Norway). This provides an example of a direct link between fisheries and aquaculture.

7. SPATIAL HETEROGENEITY AND 'SWARMS'

As has been shown in Chapter 1, physical factors control the abundance and distribution of marine planktonic species; the thermocline or tidal currents are thus determinants of plankton production (Loder & Pratt 1985). Light, nutrients and mixing determine the species composition of the phytoplankton communities, but two biological factors, competition and predation, regulate the types of structure of planktonic communities (e.g. the abundance of one species and scarcity of another). There are numerous interactions between the various factors, which are not always well understood. In consequence, the intensity of primary production is not uniform and depends on hydrodynamics, the level of dissolved substances, etc. (Lasker 1984).

Even phytoplankton cells (microalgae) in passive suspension in the water are not distributed in a homogeneous fashion throughout the water body, although most are unable to move; they form patches whose size varies depending on the dominant species (1–12 m according to McAlice (1970). He demonstrated that different densities were shown in samples taken more than 10 cm apart). Other authors sampled phytoplankton at stations 10 or so metres apart and found 215 cells ml^{-1} of 20 species and then, successively, 2 cells of a single species and, lastly, nothing at all. 'Elongation' of these swarms by the wind has sometimes been noticed as has concentration at the interfaces between bodies of water of different density and temperature. Because of the stratification of water due to thermoclines or picnoclines, the distribution of planktonic communities has been closely studied at these levels; there is both a concentration of phytoplankton and an aggregation of zooplankton. The biomass of prey and predators makes the thermocline a place of predation and competition (Harris 1987).

Similar results have been reported by Matsushita *et al.* (1988) who concluded that the distribution of 'swarms' of larval fish and their zooplanktonic prey are influenced by oceanographic phenomena (upwelling and the boundaries between bodies of water known as a 'front'). These two discontinuities are characterized my major turbulence in the water. Zones where there are swirling, turbulent currents are favoured sites for the settlement and fixation of larval molluscs (Raimbault, pers. comm.). These oceanographic phenomena therefore influence the major fluctuations in variations of stock numbers of pelagic species. In the 1970s there was a huge drop in anchovy production off Peru because of the persistence of an abnormal warm current known as 'El Nino'; this is part of a more widespread phenomenon which has been studied (Sherman and Alexander 1986).

Where passive concentrations of phytoplankton form through hydrodynamic phenomena there is the greatest chance for the survival of the less mobile juvenile copepod stages; this group makes up between 80 and 90% of the marine zooplankton. Adults which are capable of much greater movement and which undergo vertical migrations have a greater chance of survival, whatever the spatial distribution of the phytoplankton. This provides an explanation for the mortality which affects juveniles. If the dynamic movement of water is responsible for the heterogeneity of distribution of living creatures, such discontinuities should be demonstrated in an aquarium and may also occur in aquaculture ponds and even more so in cages.

Islands and headlands induce such turbulence (Wolanski & Hamner 1988) and it is a good idea, before installing aquaculture operations which are dependent on plankton production, to map out the location of such turbulent areas and to choose sites in relation to them. Because of the link of production to certain hydrological conditions for both aquaculture and fisheries, in some instances artificial initiation of water movement may be worth investigating.

In spite of the dominant influence of hydrography on the spatial heterogeneity in plankton, some planktonic concentrations have a different origin. Tanaka *et al.* (1987a, b) observed concentrations of *Acartia* in the form of a mat (30 cm thick) or an ellipsoid (long axis, 50 cm), each containing a single species, located on the sea bed in a Japanese bay. The different shapes of the swarms show that they are formed through a biological process: that of behaviour. Concentrations of 326–511 individuals per litre have been found, ten times higher than elsewhere. Larval Japanese bream hunt in these swarms which are near to the coast at water depths of less than 15 m.

An extraordinary phenomenon of plankton concentration has been reported by Alldredge *et al.* (1984) who made direct observations (in a one-man submarine) of concentrations of copepodites of the species *Calanus pacificus* at depths of 450 m, 100 m above the bed of the Santa Barbara Basin, California. The concentrations had a depth of 17–23 m. The mean density of copepods per cubic metre in this layer is 14×10^6 ($\pm 11 \times 10^6$) but falls to 500 m^{-3} above and below it. These concentrations are not associated with any discontinuity of physical or chemical factors. The authors concluded on the basis of biochemical analysis that the organisms were in a state of diapause during periods where there was a low availability of nutrients in surface waters. This can therefore be classified as a behavioural phenomenon.

There are also major differences between species; it has been shown in the laboratory that *Acartia tonsa* and *Centropages typicus* need a constant prey density and are very sensitive to concentrations of prey (dinoflagellates). On the other hand, *Pseudocalanus minutus* and *Calanus finmarchicus* are able to survive long periods of starvation. In the interior of the swarms caused by hydrodynamic forces some species may therefore have an advantage over others but it is likely that the swarms disperse quickly enough to avoid one species becoming dominant (this is suggested by the results of experiments carried out in large volume containers or ponds).

The uniformity of pelagic life is, therefore, only illusory but the plankton remain a most difficult, but very interesting, group to study.

8. PLANKTONOPHAGIC NEKTON

It is important to consider these spatially heterogeneous plankton which constitute the food of larval molluscs, crustaceans and fish, and also some adults of these groups, including farmed species. Fish belong to the plankton during their larval stage (reduced mobility), then to the nekton when they reach a length of several centimetres and swim actively, showing that there can be a transition between the two groups (see Part IV, Chapter 4).

While linked to water masses plankton are capable of horizontal displacements of low amplitudes and some (copepods) nycthemeral variations of a few hundred metres. The relationship between this distribution on a small scale and the probability of predation of larval fish has been studied in the sea by Jenkins (1988) for two species of plaice: larval plaice meet greater variations in prey density when moving a few tens of metres in a vertical direction than they do when moving roughly 10 metres horizontally. In the vertical plane, the density changes rapidly. This author considered the presence of multi-species swarms (responding positively to the same environmental factors) and also nega-tive response resulting in the separation of the different species. Although the abundance of microplankton on the study site was well below the density used in the rearing of larval fish, the larval plaice showed the same growth rate as those raised in the laboratory at prey densities which are ten times as high. Not only is the hypothesis concerning the feeding of larval fish in swarms unconfirmed but rearing procedures overestimate the abundance of prey necessary for the successful rearing of larvae. In any case there is a great divergence between results obtained in the natural environment and those in culture (see Part IV, Chapter 4).

Contradictory information is given by Owen (1981) from the study of plankton aggregations in the environment of larval anchovies in southern California. He concluded that mobile and non-mobile species are both affected by small-scale aggregating effects but there is a greater amplitude of distribution in the vertical plane than in the horizontal (the swarms have a lens-shaped appearance). This phenomenon shows its greatest ampli-tude of variation at the level of the picnocline.

Many fish eat plankton throughout their lives; such species are mostly small and pelagic (sardines, anchovies, herring, etc.) but there are notable exceptions (basking sharks!).

The food of herring has been the subject of many studies and Last (1989) brought these studies together. Although individuals of this species can reach lengths of 30 cm their food is mainly copepods then euphausiids (krill), sand eels, eggs of various species of fish (15% by number but 1.8% by weight) and sprat larvae. Food varies as a function of the size of the fish; this illustrates the way that reproduction, through the production of gametes and larvae, contributes to the recycling of organic matter in pelagic food chains. Predation on larvae by planktonophagic species is high and is the principal cause of mortality in the wild (Hunter 1984).

9. BENTHIC PRODUCTION

9.1 Primary production

Around 200 million tonnes of higher aquatic plants and attached algae are produced each year; they therefore play a marginal role in global oceanic production. However, these algae are cultured for human consumption in South-East Asia (Japanese people each consume 50 g per day) or for the agro-food industries where they are used in the manufacturing process.

Production is confined to places where light can reach the sea bed and thus takes place in lagoon or estuarine zones, as well as along shallow coasts and in fresh waters (ponds, reservoirs, lakes, rivers, etc.). Some of these sites are particularly well suited to aquaculture.

The depth of the layer of water in which photosynthesis can take place is less than that of the photosynthetic zone in the deep ocean and it might therefore be thought that there was less photosynthesis. However, in the areas where macrophytes (whether as stands of *Laminaria* or as marine flowering plants such as the Posidonids or Zostera become attached to the sea bed) overall production (phytoplankton and macrophytes) can be as high as the highest levels due to phytoplankton alone in upwelling zones (Newel 1982). The overall production is in the region of 1.5% of the incident light energy in both places. This author shows that the most productive aquatic ecosystems of the pelagic zone (upwelling zones and coastal lagoon regions) reach a known maximum level of primary production of $(0.3–0.5 \times 10^5$ kJ m^2 yr^{-1} according to latitude).

The analysis also shows that the primary production of the plants exceeds their consumption by herbivores. Detritivores associated with the macrophyte layer utilize the plant detritus, giving rise to secondary production which has been estimated to approach 10% of the primary production of the plants. These figures show that plant waste is used directly (without decomposing or mineralizing). Living phytoplankton particles and particles of detritus are thus a major food resource for the consumers in lagoons, estuaries and coastal waters.

Information given by Newel (1984) allows the comparison of a chain using phytoplankton wastes and a chain using seaweed wastes (Fig. 5).

One particular type of primary producer is worthy of special attention because of its role in aquaculture; this consists of symbiotic dinoflagellates (zooxanthellae) which participate in the feeding of the tridacnids (bivalve molluscs (giant clams)) by fixing carbon through photosynthesis; this can then be absorbed by the mollusc. This has been studied by Fitt *et al.* (1986), among others: the survival and growth of juveniles is better in the presence of symbionts although their presence is not essential. However, the adults extract a considerable amount of the carbon necessary for growth and metabolism from their zooxanthellae. In tropical waters where primary production comes from small-sized organisms which are difficult for man to utilize, the symbiotic production of dinoflagellates–tridacnid constitutes a short trophic chain with primary producers directly accessible (reproduction of these molluscs can be controlled in experimental conditions). These edible bivalves are in great demand and they form the basis of a promising aquaculture industry in the Pacific and the Caribbean.

9.2 Benthic filter feeders

Animals on the sea bed filter and ingest particles suspended in the water above the sea bed. Of all the fixed species, the filter-feeding molluscs undoubtedly make up the highest production in aquaculture. In general, filter feeders retain suspended particles between 5 and 25 μm but, if the density of particulate matter is low, smaller particles (1 μm) and bigger ones (up to 40 μm) are ingested (Mook 1981). In Part II of this book, Lubet

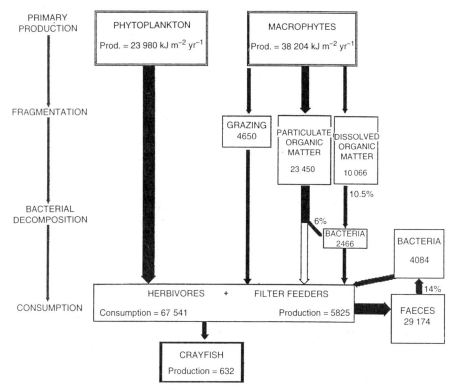

Fig. 5 Representation of the energy flux in open water and in a wrack bed.

describes the biological background to the rearing of the species which are used as food. Note that the same particles may be recycled through several species before falling to the sea bed.

Thus, the consumers of pelagic protozoa (themselves predators on bacteria) can be species that have been cultivated; for mussels, transformation of energy from primary production has been calculated as $62190\ kJ\ m^{-2}\ yr^{-1}$ per mussel. With a yield of 9.3% this gives a mussel production of $5825\ kJ\ m^{-2}\ yr^{-1}$. These mussels may in their turn serve as food for crayfish (transfer rate 10.8%), giving $632\ kJ\ m^{-2}\ yr^{-1}$. The best illustration is that of Newell (1984) (Fig. 5) which shows where losses of energy occur at different parts of the trophic web.

In spite of the complexity of these trophic webs it can be calculated that the 'ecological yield' is 10% for the largest carnivores which constitute the top link in the chain. This is comparable with the yield in terrestrial ecosystems.

10. VARIATIONS IN PRODUCTION

10.1 Periodic variations

Primary production requires light: the cycle is therefore nycthemeral. The study of daily variations in production shows that maximum production is observed between 11 and

14 h (maximum solar radiation). Major variations in the level of oxygen, pH and many other factors affect shallow waters; aquaculture can make its own contribution to these variations and is subject to them. As a consequence of this, ponds where rearing takes place often have their own equipment to counteract such variations (e.g. aerators). Taking into account the high specific heat of water (Chapter 1), this nycthemeral periodicity in the temperature oscillation is only a few degrees (usually <3°C between night and day). Seasonal variations affect the temperature and the length of day; it is important to emphasize that temperature regulates growth rate (see preceding chapter) and also that periodic variations affect other features such as sexual maturation, migration and diapause. In the North Sea primary production goes from 10 mg C m^{-2} day^{-1} in December to 700 mg in August and 70 mg in November. In the Tunis Lake this production goes from 500 mg C m^{-2} day^{-1} in spring to 10 mg in summer and 2000 mg in autumn. In warm places seasonal variations are less distinct.

10.2 Irregular variations
Most forms of aquaculture, which are dependent on the production of the natural ecosystem (e.g. mollusc culture), show variations with a periodicity of between a few years and a few tens of years. A chronological series has been developed in France for the production of oysters, which demonstrates this point clearly (Héral 1990).

These variations also affect fisheries (Lasker 1984). Oceanographic studies have shown in some instances that there is a relationship between hydrographic phenomena and fish production, but in other examples the fluctuations remain unexplained; we shall return to these in Part IV.

11. POLLUTION

Pollution introduces organic carbon, mineral salts and new molecules into the aquatic ecosystem. The role played in the production process, even at small doses, has been observed but pollutants introduced into the aquatic ecosystem by man (pesticides, detergents, hydrocarbons, heavy metals) also have an effect on chemical transmitters whose action may be highly modified and on the metabolism of living things. A herbicide such as pyridazinone is, for example, capable of stimulating the growth of *Chlorella*, even at very low concentrations, increasing diameter by 30% and stimulating the synthesis of fatty acids (Herczeg *et al.* 1980); and while some other substances synthesized by man are toxic, the effects of the majority are unknown.

Even though pollution in the sea is relatively low it should be remembered that even at infinitesimally small doses some pollutants interfere with metabolic pathways, changing the equilibrium of phytoplankton communities which may, for example, lead to proliferations of toxic dinoflagellates. This may limit the development of aquaculture.

Waters in mollusc-culture beds are regularly analysed to determine the density of bacteria which are pathogenic to man, but the effect of pollution on bacterial pathways may be reversible (through purification). However, such pollution can have catastrophic consequences for aquaculture, as has been shown by the prohibition of the sale of shellfish from the Etang de Thau in 1989–1990 because of the presence of Salmonella.

Aquaculture is incompatible with polluted waters and this activity can thus be viewed as a way of protecting coastal waters.

12. CONCLUSION

The first stage of the production of living matter in the aquatic environment takes place with very high numbers of small organisms distributed in three dimensions. This is different from the terrestrial ecosystem; a further difference is that there is no necessity for massive support structures such as tree trunks and stems.

Some of these producer organisms have only been known for the last 20 years through the development of sophisticated techniques. Because of their small size these organisms can easily exchange nutrients and other substances with the aquatic environment in which they are immersed. The exchange can be very rapid, bringing about a daily production which can be several times their biomass. In the oceans there is an organic soup made up of small particles (<1 μm) whose density near the surface is around 10 million per millilitre (Koike *et al.* 1990).

This information on the dynamics of plankton productivity dramatically changes traditional oceanography, but aquaculture, which rests on the production of living matter in the aquatic environment, is one area where the application of disciplines is important, although at present there seems little contact between aquaculture and the study of plankton and oceanography. The nutrition of molluscs, of fish and crustaceans and of extensive aquaculture, and the rearing of larvae of most species, requires one or other constituent part of the plankton. There is a need for organizations concerned with the management of the sea and its resources to bring information together to improve our understanding.

As a result of man's ability to synthesize molecules which are not found in nature, and the dispersal of these molecules together with small quantities of other potent chemicals (nitrates, trace elements, etc.) all the oceans are influenced by man's activities. The contamination of the seas and beaches with oils, hydrocarbons and other wastes, which has already happened, is the most visible indication of pollution. However, there is even more serious pollution which is less clearly visible; chemical pollution from agriculture and industry is spread through most coastal zones. According to the German environmental protection group 'Bund', the Baltic Sea will (soon?) be dead up to 9 m deep as a result of pollution from Poland and the countries of the former Soviet Union (800 000 tonnes of waste fertilizer every year from agriculture).

While the sensitivity of aquatic organisms to infinitesimally small concentrations of various substances can modify the subtle equilibria which regulate the biological processes of production, these substances can also be concentrated along food chains and may bring about changes in the natural composition of aquatic organisms, making them toxic when consumed by humans (e.g. *Dynophysis* with dangerous toxins; Minamata disease due to the accumulation of mercury) or causing the total mortality of animals in rearing systems (e.g. microalgae in Norwegian inlets where salmon are reared).

Aquaculture is in the unusual position of being able to mitigate the effects of pollution by recycling the wastes which are present in the aquatic environment. Some forms of aquaculture can help turn wastes into production which is useful to man. The production of microalgae and fish culture in ponds are fundamentally purification systems. Their use for this purpose has already become a reality although much progress in the understanding of the methodology remains to be made.

REFERENCES

Alldredge A. L., Robinson B. H., Fleminger A., Torres J. J., King J. M., Hamner W. M., 1984. Direct sampling and in situ observation of a persistent copepod aggregation in the mesopelagic zone of the Santa Barbara Basin. *Mar. Biol.*, **80**: 75–81.

Alzieu C., 1990. L'eau milieu de culture. In: *Aquaculture*, G. Barnabé (ed.). Technique of Documentation (Lavoisier) Ed., Paris: 17–45.

Aubert M., 1988. Eutrophie et dystrophie en milieu marin: origine et évolution. *Rev. Int. Océanogr. Med.*, **LXXXXI**: 3–16.

Aubert M., Revillon P., Aubert J., 1985. Sels organiques, eutrophie et dystrophie en milieu marin. *Rev. Int. Océanogr. Méd.*, **LXXVII–LXXVIII**: 3–45.

Azam F., Fenchel T., Field J. G., Gray J. S., Meyer-Reil L. A., Thingstad F., 1983. The ecological role of water column microbes in the sea. *Mar. Ecol., Prog. Ser.*, **10**: 257–283.

Barnabé G., 1980. Système de collecte du zooplancton à l'aide de dispositifs autonomes et stationnaires. In: *La pisciculture en étang*, R. Billard (ed.). INRA Publ., Paris: 215–220.

Barnabé G., 1990. La collecte du zooplancton. In: *Aquaculture*, G. Barnabé (ed.). Technique of Documentation (Lavoisier) Publ., Paris: 259–270.

Bernhard M., Nassogne A., Zattera A., Moller F., 1973. Influence of pore size of plankton nets and towing speed on the sampling performance of two high-speed samplers (Delfino I and II) and its consequences for the assessment of plankton populations. *Marine Biology*, **20**: 109–136.

Borgman U., 1983. Effect of somatic growth and reproduction on biomass transfer up pelagic food webs as calculated from particle-size conversion efficiency. *Can. J. Fish. Aquat. Sc.*, **40** (11): 2010–2018.

Böttger R., 1982. Studies on the small vertebrate plankton of the Sargasso Sea. *Helgolander Meeresunters*, **35**: 369–383.

Curds C. R., 1975. Protozoa. In: *Ecological aspect of used-water treatment*, Curds and Hawkes (eds). Academic Press Publ., London, Vol. 1: 203–268.

Dagg M., 1977. Some effects of patchy food environments on copepods. *Limnol. Océanogr.*, **22** (1): 99–107.

Fraga F., 1979. Descenso de la productividad en la ria de Vigo a causa de la atenuacion de la luz por la arcilla en suspension. *Inv. Pesq.*, **43** (2): 529–532.

Fiala-Médioni A., Pavillon J. F., 1988. La matière organique dissoute; *Synthèse. Océanis*, **14** (2): 295–303.

Fitt W. K., Fisher C. R., Trench R. K., 1986. Contribution of the symbiotic dinoflagellate *Symbiodinium microadriaticum* to the nutrition growth and survival of larval and juvenile Tridacnid clams. *Aquaculture*, **55**: 5–22.

Glover H. E., Keller M. D., Guillard R. R. L., 1986. Light quality and oceanic ultraphytoplanktons. *Nature*, **319** (6049): 142–143.

Goldman J. C., 1979a. Outdoor algal mass culture—I. Applications. *Water Research*, **13**: 1–19.

Goldman J. C., 1979b. Outdoor algal mass culture—II. Photosynthetic yield limitations. *Water Research*, **13**: 119–136.

Goldman J. C., Mann R., 1980. Temperature-influenced variations in speciation and chemical composition of marine phytoplankton in outdoor mass cultures. *J. exp. mar. Biol. Ecol.*, **46**: 29–39.

Hamner W. M., 1988. Behavior of plankton and patch formation in pelagic ecosystems. *Bull. Mar. Sci.*, **43** (3): 752–757.

Hanson R. B., Alvarez-Ossorio M. T., Cal R., Campos M. J., Roman M., Santiago G., Varela M., Yoder J. A., 1986. Plankton response following a spring upwelling event in the Ria de Arosa, Spain. *Mar. Ecol., Prog. Ser.*, **32**: 101–113.

Harris R. P., 1987. Spatial and temporal organization in marine plankton communities. In: *Organization of communities*, Gee J. H. R. and Giller P. S. (eds). Blackwell Sci. Publ., Edinburgh: 327–346.

Héral M., 1990. L'ostréiculture française traditionnelle. In: *Aquaculture*, G. Barnabé (ed.). Technique of Documentation (Lavoisier) Publ., Paris: 347–397.

Herczeg T., Lehoczki E., Rojik I., Vass I. S., Farkad T., Szalay L., 1980. Stimulatory effects of pyridazinone herbicides on Chorella. *Plant Science Letters*: 285–294.

Hunter J. R., 1976. Culture and growth of northern anchovy *Engraulis mordax*. *Fish. Bull.*, **74** (1): 81–88.

Hunter J. R., 1984. Inference regarding predation on the early life stage of Cod. In: *The propagation of cod*, L. Gadus Morhua, E. Dah, D. S., Danielsen, Moksnes, P. Solemdal (eds). *Flodevingen rapportser.*, **1**: 533–562.

Jenkins G. P., 1988. Micro- and fine-scale distribution of microplankton in the feeding environment of larval flounder. *Mar. Ecol., Prog. Ser.*, **43**: 233–244.

Koike I., Shigemitsu H., Kazuki T., Kazuhiro K., 1990. Role of sub-micrometre particles in the ocean. *Nature*, **345**: 242–244.

Lasker R., 1984. The role of stable ocean in larval fish survival and subsequent recruitment. In: *Marine fish larvae*, R. Lasker (ed.). Washington Sea Grant Progam Publ., Univ. of Washington Press, Washington: 80–87.

Last J. M., 1989. The food of herring Clupea harengus in the North Sea (1983–1986). *J. Fish Biol.*, **34**: 489–501.

Loder J. W., Pratt T., 1985. Physical controls on phytoplankton production at tidal fronts. In: *Proceedings of the Nineteenth European Marine Biology Symposium*, P. E. Gibbs (ed.). Cambridge Univ. Press, Cambridge: 3–21.

Margalef R., 1969. Composicion especifica del fitoplancton de la costa catalano-levantina (Méditerraneo occidental) en 1962–1967. *Inv. Pesq.*, **33** (1): 345–380.

Martin J. H., Gordon R. M., Fitzwater S. E., 1990. Iron in Antarctic waters. *Nature*, **345**: 156–158.

Martinez L., Silver M. W., King J. M., Allridge A. L., 1983. Nitrogen fixation by floating Diatoms mats: A source of new nitrogen to oligotrophic ocean water. *Science*, **221** (4606): (tiré à part non paginé).

Matsushita K., Shimizu M., Nose Y., 1988. Food density and rate of feeding larvae of anchovy and sardine in patchy distribution. *Nippon Suissan Gkkaishi*, **54** (3): 401–411.

McAlice B. J., 1970. Observations on the small-scale distribution of estuarine phytoplankton. *Mar. Biology*, **7**: 100–111.

Minas H. J., Minas M., Packard T. T., 1986. Productivity in upwelling area deduced from hydrographic and chemical field. *Limnol. Océanogr.*, **31** (6): 1182–1206.

Mook D. H., 1981. Removal of suspended particles by fouling communities. *Mar. Ecol., Prog. Ser.*, **5**: 279–281.

Mulkana M. S., 1970. Significance of nanoplankton to commercially important finfish and shellfish in Barataria Bay, Louisiana. *Proceed. Louisiana Acad. Science*, **XXXIII**: 38–43.

Newel R. C., 1982. The energetics of detritus utilisation in coastal lagoons and nearshore waters. *Océanol. Acta*, No. Sp.: 347–355.

Newel R. C., 1984. The biological role of detritus in the marine environment. In: *Flow of energy and materials in marine ecosystems*. Fasham (ed.), Plenum Publishing Corporation: 317–343.

Ogawa K., 1977. The role of bacterial floc as food for zooplankton in the sea. *Bull. Jap. Soc. Scient. Fish.*, **43** (3): 395–407.

Omori M., 1980. Etude du zooplancton appliqué aux pêches dans le monde. *Enseign. Océanogr. Biol. Univ. Paris* **VI**. Doc. Ronéo: 3 pp.

Owen R. W., 1981. Microscale plankton patchiness in the larval anchovy environment. *Rapp. P.-V, Réun. Cons. int. Explor. Mer*, **178**: 364–368.

Péres J. M., Deveze L., 1963. *Océanographie biologique et biologie marine*. P.U.F. Ed., Paris: 514 pp.

Ricard M., 1987. *Atlas du phytoplancton marin: Vol. 2*. Editions du CNRS, Paris: 297 pp.

Savenko V. S., 1988. The role of biosedimentation. In: *The formation of bottom deposits. Vodn. Resur.*, **4**: 12–129 (in Russian).

Sheldon R. W., 1984. Phytoplankton growth rates in the tropical ocean. *Limnol. Océanogr.*, **29** (6): 1342–1346.

Sheldon R. W., Nival P., Rassoulzadegan F., 1986. An experimental investigation of a flagellate-ciliate-copepod food chain with some observations relevant to the linear biomass hypothesis. *Limnol. Océanogr.*, **31** (1): 184–188.

Sherman K., Alexander L. M. (ed.), 1986. Variability and management of large marine ecosystems. *AAS Selected Symposium*. Westview Press Publ., Boulder, Colorado: 319 pp.

Sherr E., Sherr B., Paffenhöfer G. A., 1986. Phagotrophic protozoa as food for metazoans: a missing trophic link in marine pelagic foodwebs system. *Marine Microbial Food Webs*, **1** (2): 61–80.

Stockner J. G., 1988. Phototrophic picoplankton: An overview from marine and freshwater pecosystems. *Limnol. Océanogr.*, **33** (part 2): 765–775.

Sournia A., 1986. *Atlas du phytoplancton marin: Vol. 1*. Editions du CNRS, Paris: 219 pp.

Tanaka M., Ueda H., Azeta M., 1987a. Near-bottom copepod aggregations around the nursery ground of the juvenile Red Sea bream in Shijiki Bay. *Nippon Suisan Gakkaishi*, **53** (9): 1537–1544.

Tanaka M., Ueda H., Azeta M., Sudo H., 1987b. Significance of near-bottom copepod aggregations as food resources for the juvenile Red Sea bream in Shijiki Bay. *Nippon Suisan Gakkaishi*, **53** (9): 1545–1552.

Tanoué E., Hara S., 1986. Ecological implications of fecal pellets produced by Antarctic krill *Euphausia superba* in the Antarctic ocean. *Mar. Biol.*, **91** (3): 359–369.

Thorpe S. A., Hall A. J., 1987. Bubble clouds and temperature anomalies in the upper ocean. *Nature*, **328** (2): 48–51.

Wolanski E., Hamner W. M., 1988. Topographically controlled fronts in the ocean and their biological influence. *Science*, **241**: 177–181.

3

Plankton culture

In the natural aquatic environment, whether fresh water or marine, the first link in the food chain is always planktonic organisms. As has been shown in Chapter 2, phytoplankton (microalgae) represent the start of the planktonic food chain; they are the primary producers. Planktonic filter feeders, including larval molluscs, rotifers, copepods and artemia, and benthic filter feeders (molluscs), are the first level of consumers (herbivores); larval fish or crustaceans which consume rotifers and artemia are carnivores, representing the second level of consumption. The intensive fish hatchery (Part IV, Chapter 4) attempts to model the trophic web, raising each link separately. In extensive culture of fry or adult fish in ponds there is no such separation.

Man has an interest in the planktonic level both in relation to the control of the entire aquatic ecosystem (e.g. pond fish culture, Part IV, Chapter 5) and in the control of each of the trophic links. Many of the techniques in the culture of phytoplankton have been in use for many years. However, the need to produce zooplankton for feeding fish and crustaceans led to a resurgence in research on phytoplankton culture in the 1970s.

1. THE CULTURE OF MICROALGAE

This form of aquaculture has two purposes: one is the production of microalgae which are used directly to feed farm animals and humans; the other is to feed larval forms of cultured species which consume phytoplankton directly (molluscs, the first larval stages of crustaceans) in hatcheries, when this is not the case the phytoplankton provide the diet required by zooplankton which are then fed to other species of larval fish.

The forms of primary production in the natural environment were described briefly in Chapter 2. Although the principles are the same, research aimed at optimizing production has led to the development of techniques which are often far removed from the processes in the natural environment. While this manipulation of rearing conditions is the route most frequently followed, there is another way which consists of finding, isolating and then testing stocks of microalgae in the natural environment whose potential has not yet been determined. The culture of these algae is important for another reason: as a means of studying the primary processes in the aquatic environment.

1.1 Factors determining the processes of production

The dynamics of the development of microalgae are similar to those of bacteria but the difference from the latter is the dependence on light energy. This factor is completely independent of the composition of the culture medium. As with bacteria, after an exponential growth phase in young populations there is a plateau stage where a stable population is maintained, then finally a decline. In the natural environment this development occurs spontaneously and results in phytoplankton blooms.

Goldman (1979) made a detailed study of the limitations in the conversion of photosynthetic energy during in-depth research into the conditions of the mass production of phytoplankton in open waters: the conversion of solar energy to algal biomass is controlled by the amount of light available, the 'photosynthetic machinery' of algae, nutrient salts, temperature and the fundamental characteristics of the culture system.

In practice, the maximum dry weight production in ideal conditions is 30–40 g m^{-2} day^{-1} ($14\,600$ g m^{-2} yr^{-1}) and the efficiency of conversion of solar energy to living matter reaches a maximum of 5%. Such results may be achieved in small volumes but never in large ones. The best figures have been obtained for short periods at various sites throughout the world. Lower figures are achieved over longer periods or in northerly latitudes. In Belgium De Pauw (1981) achieved a production of 10 g m^{-2} day^{-1} in summer but only 1 g in winter.

1.1.1 Light intensity

Solar radiation reaches the outer layers of the atmosphere at a constant rate of 1.94 cal cm^{-2} min^{-1} but before it reaches the land (or water) much has been absorbed by the atmosphere; radiant energy is around 780 cal cm^{-2} day^{-1}. The available light energy varies with latitude and also with season (Fig. 1). The intensity is also lower when there is cloud cover and there are further losses at the surface of the water (depending on the angle of incident light and the wind); these losses have been estimated at 10% (Goldman 1979). Only the visible wavelengths from the spectrum (400–700 nm) are used for photosynthesis; this is no more than 45% of the total energy (Goldman 1979). Ultraviolet in the short wavelengths and infrared beyond the long wavelength end of the visible spectrum make up the rest. Different algae absorb different fractions of the spectrum. Natural light or normal domestic lighting are used for cultures.

1.1.2 Levels of nutrient salts

This factor limits growth rate (Fig. 2). In culture, an attempt is made to keep the concentration of salts above the optimum level in order to make maximum use of the energy (which cannot be managed) and which is often the limiting factor for production. Levels of nutrients used in culture are much higher than those in the natural environment, even in upwelling zones (see examples below). This level is calculated as a function of the speed of absorption of microalgae which depends on the growth rate of the population (itself regulated by a range of environmental factors (Fig. 2)).

1.1.3 Temperature

Results obtained by several authors show that there is a slow increase in the growth rate

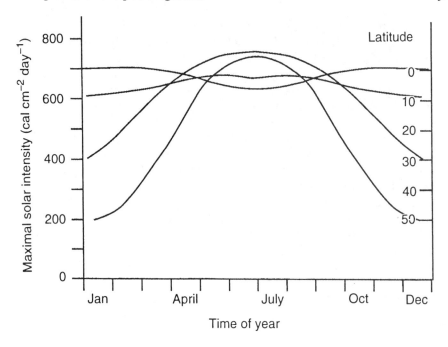

Fig. 1. Maximal solar intensity as a function of latitude and month.

of the culture with temperature, an optimum and then a linear decrease before collapse (Fig. 2). The range of temperatures which can be used goes from 5 to 40°C but the majority of cultures are maintained at temperatures between 10 and 30°C. Fig. 2 illustrates how minimum, optimum and maximum (T_1, T_2, T_3) temperatures induce different rates of growth when other factors are equal. Unfortunately it is not possible to regulate precisely the temperature of outdoor cultures and day/night differences may sometimes be large. In spite of this, the growth of microalgae is more often limited by light than by temperature and some species may grow even at low temperatures.

1.1.4 Movement of the culture medium and the depth of the rearing vessel

Light is absorbed through the whole depth of an algal culture and thus the cells at the bottom receive less light than those near the surface. The absorption of light is rapid and only 2% of the incident light reaches a depth of 7 cm in a culture with an optical density of 0.1 and a wavelength of 560 nm (Richmond *et al.* 1980).

The mean quantity of light received by an algal cell is more important than the amount arriving at the surface of the water as the position of the cell, and its illumination, will vary. This depends, in a given volume of water, on the movement and the number of algal cells (i.e. biomass) and also the depth. This is entirely different from the situation in the natural environment where the algal density is millions of times lower than in culture. The depth of the culture vessel is one of the determining factors, even though it is infinitely less than that of the natural environment.

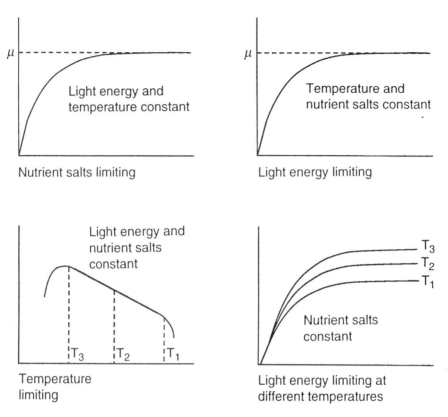

Fig. 2. Relationship between algal growth rate (μ) and environmental parameters (after Goldman 1979).

Induced movement evens out the exposure of algal cells to the light but also has other roles: it prevents cells from settling on the bottom of the culture vessel and keeps them in suspension; prevents thermal stratification from developing; and also stops the development of gradients of salts or dissolved gases. Strong agitation also facilitates the transfer of carbon dioxide gas between the atmosphere and the water. However, it has been demonstrated that increased agitation cannot overcome the loss of production with depth, although there is disagreement as to the extent: Richmond *et al.* (1980) obtained improved growth by increasing the rotation of a paddle wheel to 30 rpm. In the ocean, variations in light intensity brought about by waves are known to increase photosynthesis.

1.1.5 Rate of renewal of the culture medium
The growth of the algal population increases the biomass in the volume of water used for culture. It is essential to begin to harvest from it at the optimum time (the start of the plateau phase) to avoid slowing down the growth rate, especially through lack of light. Goldman (1979) showed that the percentage dilution should be equal to the daily growth rate which is characteristic for each species. There are two ways of treating the culture:

either the whole culture is harvested at its optimum concentration (the batch culture technique) or part of the culture is harvested every day and replaced by the equivalent quantity of water enriched with nutrient salts; this is continuous culture and is continued over a few weeks or months.

The outdoor intensive culture of algae is limited by temperature and the available light. It can be optimized by finding the best combination of mixing rate, turnover and depth to maintain the highest biomass and obtain the best possible production. These are the only factors which can be altered by man, as long as the supply of mineral salts is guaranteed. To maximize the conversion of solar energy it is necessary to adapt the concentration of algae to the available light; this involves varying the rate at which algae are harvested and new medium added (turnover rate). De Pauw & de Leenheer (1985) harvested 4% of the culture volume every day in winter and 60% in summer.

1.2 Production of microalgae for feeding humans and other livestock

1.2.1 Aims
This aspect of the use of algae is almost unknown in Europe but is highly developed elsewhere. There are several objectives:

— Providing nutritional supplements for humans in developed nations. This production is a well-developed activity in warm waters (many farms in South-East Asia produce around 2000 tonnes of spirulines and chlorellas destined for use). The composition of the microalgae compares favourably with soya, a well-known food source (Table 1). Salt-water species are the only ones to synthesize certain polyunsaturated fatty acids; their consumption (0.5–1 g day^{-1}) reduces the risk of death from cardiovascular disease by 40% and also decreases the likelihood of death from cancer (Anon. 1990 and Part IV, Chapter 2).
— Producing food for livestock. Because of their richness in various substances, there is a large market for microalgae. This aspect has been reviewed by Soeder (1980) who demonstrated the relationship between the cost of production of this material and its market potential, which, if prices are low, could exceed several million tonnes. It is worth noting that the production of microalgae per unit area and time is much higher

Table 1. Dry matter composition of microalgae in comparison to the digestible component of soya beans (after Soeder 1980)

Components	Scenedesmus	Spirulina	Soya beans
Protein	50–60	56–62	34–40
Lipids	4–8	2–3	16–20
Glucides	12–14	16–18	19–35
Fibre	3–10		3–5
Ash	6–10		4–5
Residual water	4–8	10	7–10

than that in traditional agriculture (Table 2), especially in hot countries, and that these cultures may be set up in low fertility agricultural regions or in deserts close to the sea.

Table 2. Comparison of production from conventional agriculture and micro-algae (adapted from Grobbelaar 1979)

Cereals	Dry weight (t ha^{-1} yr^{-1})	Proteins (t ha^{-1} yr^{-1})
Wheat	3–6	0.4–0.8
Maize	7–18	0.8–2
Soya	6–7	1.8–2.5
Algae	Dry weight (t ha^{-1} yr^{-1})	Proteins (t ha^{-1} yr^{-1})
Cultured in fresh water		
Europe	25–30	14–16
Warm countries	70	35–40
From water purification systems		
California	50	22
Tropics	160	80

Microalgae also play a part in extensive aquaculture as they form the starting point of the food chain which ends with the production of fish. There is a positive benefit to the inclusion of fresh microalgae (spirulines) in diet; 70% algae in the diet of tilapia gives a growth rate identical to that when herring meal is included (Korn 1989).

— Creating particular products (phycobiliproteins, isotopic compounds, beta-carotene, xanthophyll, polysaccharides, amino acids, polyunsaturated fatty acids). Many other products which have applications in biotechnology are still confidential: these include hydrocarbons, alcohols, hydrogen and energy-releasing pathways. Research is being carried out in Europe into the production of hydrocarbons using the species *Bottryococcus*. Photosynthetic bacteria are being studied at the same time as the algae.
— Producing medicines (anti-cancer drugs and antibodies), vitamins (C and E) or products which can be used in research.

The biological basis of this type of production has been reviewed by Goldman (1979), Shelef & Soeder (1980) and Becker (1985). Becker also gives details of several examples of applications.

1.2.2 Methods of production
The systems most frequently used in large-scale culture are longitudinal tanks (raceways) linked to each other to form racecourse-shaped structures (Fig. 3) with a paddle wheel or an air-lift structure ensuring circulation and aeration. Aeration induces photosynthesis which helps to meet the oxygen requirements of respiration. These tanks can be up to

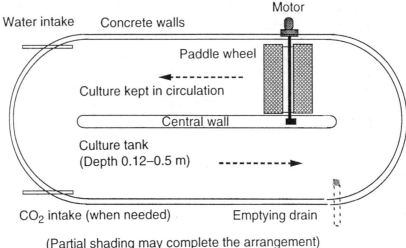

Fig. 3. Plan of a tank for microalgal production.

500 m^2 in area and can be grouped together to cover areas of tens of hectares. The smallest (20–100 m^3) are under glass, the largest out of doors (usually in South-East Asia). Depth varies from 0.3 to 0.5 m. The mean species cultured are the Chlorellas (*Chlorella ellipsoidea* and *C. pyrenoidosa*) in fresh water, the spirulines (*Spirulina maxima*) in carbonated fresh water and *Dunaliella salina* in salt or hypersaline water. These systems have been described by Kawaguchi (1980), together with details of the economics of culture. One of the largest systems for the production of algae is a giant solar evaporator (3200 m diameter, 900 hectare area) installed by the French petroleum institute in Mexico (Durand-Chastel 1980).

These few examples of monospecies culture depend on the ability to select a species for rearing in large volumes out of doors: the spirulines and *Dunaliella* are species suited to particular environments (carbonated and hypersaline) and the chemical composition of the water does not allow the development of other species. The fresh water in which the *Chlorella* species are cultured is fertilized with acetates or glucose to provide a source of carbon for the algae and this favours the mixotrophic nutrition (combined autotrophic and heterotrophic) which is peculiar to these algae. However, in most circumstances, man cannot choose the species of algae which will develop in a given water body; even with an inoculum of the favoured species another, unwanted species is likely to become dominant within a few days. This is therefore a very different situation from agriculture where not only a species, but a variety of that species, can be selected for culture. This problem is a serious constraint on the development of the mass culture of plankton where a particular algal species is required for the production of a chemical or to feed larvae.

1.3 Production of microalgae for aquaculture

There is an extensive scientific literature on the subject of this type of aquaculture. The culture methods which have now become routine are based on research carried out mainly in Japan and Britain, but there was a resurgence of interest in the culture of microalgae following the 1974 energy crisis. It is one of the most important of all aquaculture operations because it is the starting point for the culture of many species of fish and crustaceans.

The two types of rearing described below are carried out in volumes varying from a test tube to a tank of several cubic metres capacity to produce monocultures of different species of unicellular algae in culture rooms which are protected from contamination. Several species are cultured to meet the nutritional requirements of rotifers, molluscs, etc. (Table 3).

Table 3. Characteristics and use of microalgae in aquaculture (from Maitre-Alain, pers. comm.)

Species	Normal size	Size range	$T°C$ optimal	Salinity S‰	Concent. millions/m	Species to which this is fed
Unicellular green algae						
Tetraselmis suesica	6–7	4–11	<22	25	2–5	B-R-A-P
Dunaliella tertiolecta	4–5	3–11				C
Dunaliella salina	4–5					C
Chlorella	2–4		<28		50	low
Nannochloris	1–3		<28	5–30	50	R
Unicellular brown algae						B
Isochrysis galbana	3	2–7	16–20		5–15	B
Monochrysis lutheri	4	2–9	16–20		10–20	C-P_e
Pseudoisochrysis	5–6		16–20		idem	
Unicellular diatoms						
Phaeodactylum tricornutum	5	3–20	18–26			B-P_e
Diatom chains						
Skeletonema costatum	50	3–9				B-C
Chaetoceros calcitrans (rendered unicellular by ultrasound)	100+	3–5	16	20–30		B-H

A: Artemia, B: Bivalves (larvae + spat), C: Cupped oyster, H: All oyster species, P_e: Penaeid shrimps, R: Rotifers, P: Clams.

The starting point for the culture of all species is the same: a small fraction of a pure culture is used as an inoculum; as the cells multiply the culture is transferred to increasingly larger vessels.

1. All manipulations take place in aseptic conditions, the same as those in a microbiological laboratory. The water used for incubation is filtered through a filter with a 1 μm or smaller pore size. Artificial sea water, made by adding salts to fresh water, is sometimes used. In spite of this the culture may become contaminated with bacteria and

ciliates. These precautions—the use of mineral salts and vitamins in making up the special medium (Table 4), the necessity of permanent artificial illumination (intensity 4000–5000 lux) and a temperature of around 20°C—makes this an expensive form of culture: Audineau & Blancheton estimated the cost of production to be 1.27 French francs per litre in 1986. However, these small volumes (often around one cubic metre) are an essential part of any hatchery producing molluscs, crustaceans or marine fish. Even smaller volumes may be used: the above authors describe the system used in the IFREMER Laboratory at Palavas:

2. A pure culture, maintained in a test tube is subcultured every 15 days into a fresh test tube to which is added enriched sea water (Table 4). This culture serves as an inoculum (20 ml for a 250 or 500 ml Erlenmayer flask). The developing culture is then transferred to a two litre flask of still water. The next stage uses a round-bottomed 20 l flask through which air, enriched so that the carbon dioxide content reaches 2%, is bubbled. These round flasks, in their turn, are used to inoculate transparent plastic tubes (80–120 l volume); these are the actual production units. These tubes (double-layered plastic bags) are thrown away after use, avoiding cleaning and contamination. The bags are illuminated from the side; the diameter of the bag thus represents the 'depth' and rarely exceeds 20 cm. This is continuous culture and a fraction is regularly removed. Its advantage is that labour costs are reduced but, in spite of all precautions taken, the system is in unstable equilibrium. It takes 20 days, starting with the test tube, to obtain the first production. Fig. 4 (adapted from Audineau & Blancheton 1986) shows this production diagrammatically.

Table 4. Conwy medium (Helm *et al.* 1979; after Walne 1966)

Nutrient solution		Composition of trace metal solution	
$FeCl_3.6H_2O$	2.60 g	$ZnCl_2$	2.1 g
$MnCl_2.4H_2O$	0.72 g	$CoCl_2.6H_2O$	2.0 g
H_3BO_3	67.20 g	$(NH_4)6Mo_7O_{24}.4H_2$	0.9 g
EDTA (Na salt)	90.00 g	$CuSO_45H_2O$	2.0 g
$NaH_2PO_4.2H_2O$	40.00 g	Distilled water:	adjusted to 100 ml
$NaNO_3$	200.00 g		
Trace metals		HCl: enough to clear the solution	
(solution)	2.0 ml		
Distilled water:	adjusted to 2000 ml		

1 ml is added to each litre of water

Composition of the vitamin solution

B_{12}	10 mg
B_1 (thiamine)	200 mg
Distilled water	adjusted to 200 ml

0.1 ml is added to each litre of water

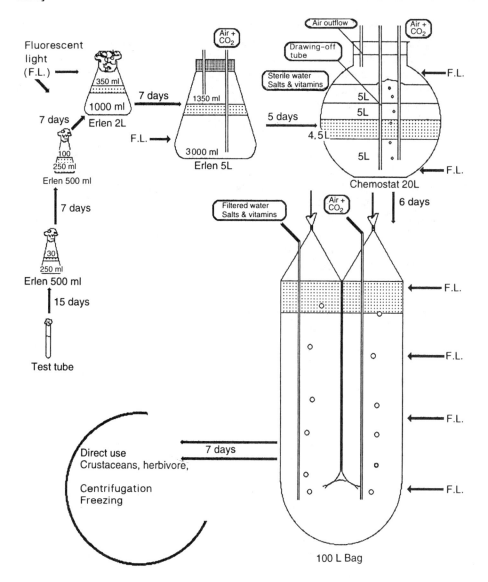

Fig. 4. Diagram of the process of microalgae production in a hatchery, from the mother culture in a test tube to the production in a transparent bag (adapted from Audineau & Blancheton 1986).

A similar system has been developed at the Conwy Laboratory in Wales (Helm *et al.* 1979) based on the use of 200 l capacity cylindrical tanks (150 cm high, 40 cm diameter). In the centre of each cylinder is a transparent tube, 19 cm in diameter; this contains four fluorescent tubes giving a light of similar composition to daylight, each of 80 watt output. Two aeration tubes provide 15 litres of air per minute, bubbled through the culture. This air is enriched with 0.25% CO_2 to maintain a pH of between 7.8 and 8.2. The culture is maintained at temperatures between 15 and 24°C. *Tetraselmis* is reared at a salinity of

25‰ and the water used is passed through a 0.4 μm filter. The inoculum is 5% of the final volume. These culture vessels are cleaned *in situ* with detergent and sterilized (with bleach). The maintenance and operation of ten units like this requires 3 man-hours daily. At the Lowestoft Laboratory of the same organization (MAFF), transparent plastic tubes (480 l capacity) are used. Each group of tubes is illuminated by five 80 W fluorescent tubes. One-quarter of the culture is removed every three days.

3. The alternative is to harvest the entire culture when it has reached its optimum growth phase: this is used either to start new cultures (depending on the density of the cells, the volume used to inoculate a new culture is between 1/50 and 1/10 of the final volume) or is fed to the farmed molluscs, rotifers or crustaceans. Production units increase in size from a 200 ml Erlenmayer flask to 500 then 1500 then 10 000–29 000 l rectangular and cylindrical tanks made from rigid plastic (opaque or translucent) indoors under artificial light (2000–5000 lux). This technique, used in mollusc hatcheries, has been described by Le Borgne (1989). In warm countries, installed tanks can be sheltered from direct sunlight or placed under special filters; the techniques are similar to those used for the production of algae for direct consumption.

In some instances, progress has allowed the production of monospecific cultures out of doors. Two examples of this are given below, one adapted to the algal species being cultured, the other adapted to the type of culture unit and the favourable light intensity.

1. *Nannochloris* sp. (2–6 μm diameter), a species of chlorella, have been isolated from sea water near Kiel (where they are the dominant species) by successive dilutions without sterilization. Water is fertilized with agricultural fertilizer (N/P = 11). Tanks are between 4 and 27 m^2 in area and 25–50 cm deep. The tanks are aerated by a lateral channel which distributes air through a network of submerged pipes perforated with 5 mm diameter holes. Baltic sea water (15–25‰ salinity), filtered through a 1.5 m deep gravel bed is warmed slightly so that the temperature never drops below 4–6°C. Witt *et al.* (1981) showed that *Nannochloris* has an almost identical growth rate between 5 and 30‰ salinity (optimum 10–20‰). This species is also extremely eurythermal: the doubling time for the culture is 2–3 days, even at a temperature of 1°C. The rate at which fertilizer is applied is calculated from a count of the algal cells and a measurement of the biomass which allows a calculation of the biomass in culture (Table 5): a dry weight of 13–20 mg algae contains 1 mg nitrogen and 60–125 mg dry algae contains 1 mg phosphorus; the higher values are used for the calculation of the amount of fertilizer to be added. The culture is tapped when the density has reached between 8 and 20 million cells ml^{-1}. In summer 20–30% are removed daily; in winter there may be no removal as a regular second harvest may be impossible because of the limitations of light. The proliferation of herbivores in the culture makes it necessary to clean the tanks and to start new cultures every 2–4 weeks. In the absence of CO_2, pH sometimes reaches 10.5. Mean production goes from 1 g m^{-2} day^{-1} in January to 15 g m^{-2} day^{-1} in June (maximum 25 g). This has shown that, at the latitude of Kiel, production of algae without artificial light is only possible between April and October.

2. In Israel, Boussiba *et al.* (1988) ran a pilot production plant for *Isochrysis galbana* in outdoor tanks of 2.5–100 m^2 from stocks reared in laboratories according to the techniques described above. Culture tanks of 2.5 m^2 area are oval in shape with a central

Table 5. Relationship between cell diameter and dry weight in *Nannochloris*
sp. (from Witt *et al.* 1981)

Diameter (μm)	Volume (μm^3)	Conversion—dry weight	
		millions cell mg^{-1}	μg (million cell)$^{-1}$
3	14.1	283.7	3.5
3.5	22.4	178.6	5.6
4	33.5	119.4	8.4
5	65.4	61.2	16.3
6	113	35.4	28.2

partition. The culture volume is 300 l: the depth of the water is 12 cm (Fig. 3). Artificial
sea water is used and CO_2 is injected to maintain the pH between 6.5 and 7.5. The water
is kept in motion with a paddle wheel; six tanks provide the inoculum for 100 m^2 of
rearing units. There are two channels, 2 m wide and 25 m long. Water is again 12 cm
deep and a paddle wheel 1 m wide turns at a rate of 15 rpm. The inoculation culture has a
density of 3×10^6 cells ml^{-1}. Injection of CO_2 maintains the pH between 6.5 and 8.
Temperatures go from 23–24°C in the morning to 32–34°C in the afternoon and 18–21°C
at night. The proliferation of zooplankton is restricted by maintaining a concentration of
2 mM ammonium sulphate [$(NH_4)_2SO_4$]. In order to maintain the culture at a constant
density part is removed every day and replaced by fresh medium. The final harvest is by
precipitation with ferric chloride and flotation. A doubling time of 9.5 h has been recorded in the laboratory; this corresponds to an hourly growth rate (μm) of 0.073, the
highest recorded for this species. In 100 m^2 tanks, the stationary phase occurs 12 days
after the start of culture when the density of cells has reached 43 million cells ml^{-1}. The
mean production is 23 g m^{-3} day^{-1}. This species has a high lipid content (24%), a large
proportion of which is in the form of polyunsaturated fatty acids which are essential for
the larvae of cultured species. The authors had problems in scaling up from 3 m^2 tanks to
100 m^2 tanks: this is not uncommon. Monospecies cultures have been maintained for four
months without contamination; the authors attribute this success mainly to the constant
temperatures (>28°C), the biomass in the tanks and to the addition of the ammonium
sulphate which has eliminated the predators. Under European climatic conditions De
Pauw & de Leenheer (1985) found that the proliferation of protozoan predators (ciliates,
flagellates), which most researchers know only too well, leads to the destruction of
cultures.

These different approaches show that there is constant development and overturning of
previous ideas in the culture of microalgae. Several other approaches are being investigated: Wong *et al.* (1977) showed that diluted, sterilized waste water was more effective
as a culture medium than complex solutions of minerals and vitamins (e.g. the Conwy
medium, Table 4). De Pauw & de Leenheer (1985) showed that small additions (0.1%) of
organic fertilizer such as pig manure or inorganic agricultural fertilizer increased production by 20%. Complex mixtures of mineral salts can also be replaced by agricultural

fertilizers with no adverse effects. These fertilizers may, in fact, perform better and cost 100 times less (Gonzalez-Rodriguez & Maestrini 1984). Microalgae can be used in the purification of waste water: *Nannochloris*, for example has an affinity for NH_4.

2. PURIFICATION OF WASTE WATER AND THE MASS PRODUCTION OF PLANKTON

Water purification lagoons produce algae as a by-product; the biomass may be huge (Table 2). Unlike the culture of monospecific stocks of algae, bacteria are also present: this limits the use of the algae produced in water purification lagoons.

The algae and also the zooplankton produced can be used directly—for example, by introducing planktonophagic fish into the last purification ponds or ponds alongside—or indirectly by harvesting the algae and preparing them for use. Such systems are very similar to those for integrated aquaculture units in which wastes produced by domesticated animals are recycled by fish (Part IV, Chapter 5).

Lagoons are systems of ponds exposed to air (Fig. 5) or artificially aerated. Conditions are similar to those in ponds (but amplified), with similar trophic webs. However, environmental conditions are often incompatible with the survival of sensitive species such as fish. One member of the human population of an industrialized country produces between 150 and 200 l of waste water per day and this effluent, which is rich in organic matter, is broken down in the water, or on the bed of the pond, where heavy metals sediment out. In mid-water aerobic bacteria mineralize organic matter, at least partially, while on the bed of the pond the sediments are colonized by anaerobic bacteria. The water remains in the first of the series of ponds for around ten days and then passes into the second pond. Here, active photosynthesis takes place because of the richness of the elements available for the phytoplankton and the natural light; the lagoon is very similar to a fertilized pond. Lagoons are 1.2–2 m deep to avoid invasion by reeds. The biggest (non-aerated) lagoons exceed several hectares in area. The lagoon systems usually comprise three ponds, in the last of which the zooplankton (rotifers, daphnia and, sometimes, copepods) are most abundant. Fresh-water zooplankton are harvested at the outflow of lagoons where bacteria are cleansed from the water. This harvest is carried out by pumping water through filter bags or by draining (Barnabé 1979, 1980). Edeline (1979) has described the operation of lagoon systems in water purification. Barnabé & Périgault (1983) have described the use of algae from such systems to rear filter-feeding prey.

3. THE CULTURE OF ZOOPLANKTON

In the wild the larvae of marine planktonic species mainly consume copepods (which make up around 90% of the zooplankton in the marine environment) and rotifers, copepods and cladocerans in fresh water. Man, although able to go to the moon, does not really know how to rear planktonic species and, in spite of numerous attempts, only one species has been routinely cultured in the marine environment. This is the rotifer, *Brachionus plicatilis*. Although Daphnia can be reared on fresh water it has not been the object of systematic culture as many tonnes can be harvested from lagoons (Barnabé 1979). Note

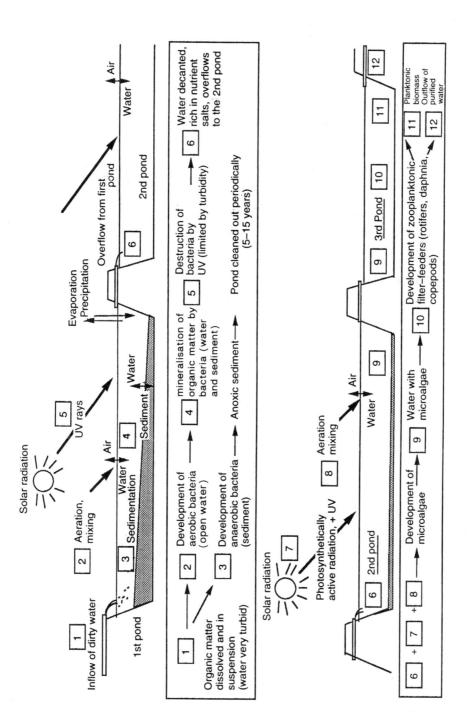

Fig. 5. Lagoon systems, purification system and aquatic production.

that both these species can reproduce rapidly by parthenogenesis when environmental conditions are favourable: this means that they are suitable for culture.

3.1 Culturing the rotifer, *Brachionus plicatilis*

In Japan, Ito adapted the brackish water rotifer *Brachionus plicatilis* to salt water in the 1960s. It is the only species in the world which is raised on a large scale but there are often many difficulties. The essential biological information on this species has been given by Pourriot (1990). Fig. 6 shows the anatomy of *Brachionus plicatilis* diagrammatically; Fig. 7 illustrates the life history patterns of this species.

There are two stocks of different sizes: the small (S) stock measures 120–160 μm, the large (L) 250–300 μm. There is a degree of polymorphism, the origin of which is not yet understood. The small animals are of more interest than the large as they are more suited to feeding small larvae (e.g. sea bream). In favourable conditions (high temperature, abundant food) this rotifer reproduces parthenogenetically. A female produces eggs asexually (amictic); the offspring are amictic females while the conditions persist. Growth and reproduction are at their optimum rate at 27–30°C and the generation time is less than four days. *Brachionus* is a filter feeder which normally consumes microalgae; cultured species such as *Dunaliella*, *Isochrysis*, *Monochrysis* and *Chlorella* are suited because of their size (<20 μm). The volume of a rotifer is 6×10^5 μm^3; Pourriot calculated

C : Ciliary
      crown
E : Stomach
M : Mastax
O : Egg
Or : Pedal gland
P : Foot
Pe : Penis
V : Bladder
Vi : Ovary

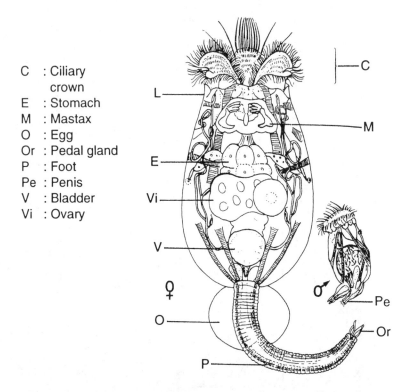

Fig. 6. *Brachionus plicatilis*, female and male (diagram by Schach Y., after Koste 1980).

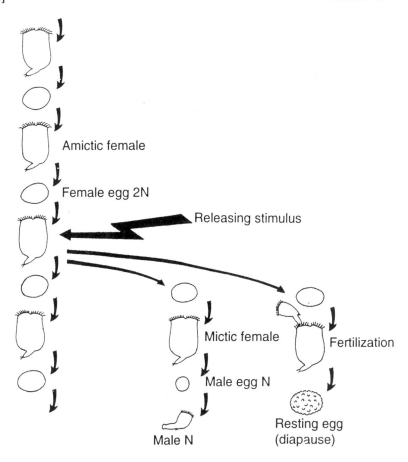

Fig. 7. Parthenogenetic and sexual cycle in monogonont rotifers (after Pourriot & Clement, unpublished).

that one of these females ingests between five and ten times its own volume of algae each day. The complications and cost of production of such quantities of algae have led to their replacement by baker's yeast, but cultures of rotifers fed on this nutrient alone do not survive for more than a few days; there are problems with pollution and nutrition. Because of this, a mixture is used (10% cultivated algae, 90% baker's yeast). More importantly, the polyunsaturated fatty acid requirements of marine larval species can only be met by rotifers fed on marine microalgae which synthesize these substances themselves.

In intensive production, rotifers are reared in volumes of less than 10 m^3 in Europe, although the Japanese and others work in cylindrical or rectangular flat-bottomed tanks of $5–30 \text{ m}^3$ in volume.

In Japan the water (with no flow-through) in a rearing tank is strongly aerated to keep the algal and yeast cells in suspension. At the start the density of algal and yeast cells is between 5 and 10 ml^{-1} and algae are given as the only food. After this, 2–3 g of yeast per million rotifers per day are given in two distributions. The maximum biomass is attained after eight days at 25°C; after this one-third to one-quarter of the volume is removed each day, and is replaced by 'green water' (cultured algae in the bloom phase) or sea water. The density at the end of the rearing stage may reach 1000 ml^{-1} but is more often between 70 and 200 ml^{-1}. The Japanese culture plant has four 100 m^3 units for the production of algae and 180 m^3 units for rotifers; these are used in the production of one million larval sea bream each year.

Many variants on the Japanese technique have been developed and there are 241 references in one bibliography produced at the end of the 1970s. The use of microparticles, sludge from water purification works and marine yeasts has done little to advance rearing techniques; algae are irreplaceable even as a small (10%) component of the diet for routine culture, and yeasts remain a major part of the diet.

In France rotifers are reared in cylindro-conical tanks with a volume between 0.5 and 0.8 m^3 (Fig. 8). Rearing water (salinity 25‰, $T = 25–27$°C) is filtered and agitated strongly by aeration. Rearing is carried out in small tanks which are used to inoculate bigger ones. When the concentration has reached around 50 individuals per litre, the volume is progressively increased by adding culture medium and nutrients. Rotifers are counted in a 1 ml pipette by naked eye or under a binocular microscope using a haemocytometer. When the concentration reaches 150–200 individuals ml^{-1} in the production units, 25% of the volume is taken off each day and replaced by the addition of medium. Rotifers are harvested in a 40–50 μm mesh sieve (Fig. 8, after Coves et al. 1989).

In spite of the use of artificial sea water to prevent contamination with ciliates, problems still arise. Pollution, caused by the use of the yeast, means that tanks must be emptied and completely cleaned as frequently as once a week. The maintenance of five tanks of 2 m^3 volume requires 3.5–4 h work every day (Audineau et al. 1984). Because of all the constraints, the cost of the production of rotifers is over £130 kg^{-1} (net weight).

An enrichment technique for rotifers (known as 'doping') allows an improvement in their nutritional value and in the quality of the larvae which feed on them. This has been developed in Japan and entails harvesting the rotifers and placing them in an algal culture for 6 h. This enriches them in the essential unsaturated fatty acids which are essential to larval marine fish. Direct enrichment with cod liver oil (rich in fatty acids, Part IV, Chapter 2) emulsified in egg yolk is also sometimes used (Iisawa 1983), as well as more sophisticated formulations, including vitamins and antibiotics (Robin et al. 1984). The rate of production of rotifers is improved by additives containing lactic bacteria and cereal derivatives (Gatesoupe et al. 1989). These products also improve the survival rate of larval fish fed on these prey.

The nutritional value of these cultured rotifers and their composition has been the subject of a detailed study (Watanabe et al. 1983).

3.2 Producing Artemia salina nauplii

Artemia salina is a small branchiopod crustacean which is found in brackish and salt waters. It can even live in hypersaline waters (up to 300‰) where it is abundant as other species do not have such a high salinity tolerance. The female releases eggs with a hard outer coating which accumulate around the banks of lakes. These eggs or cysts contain a

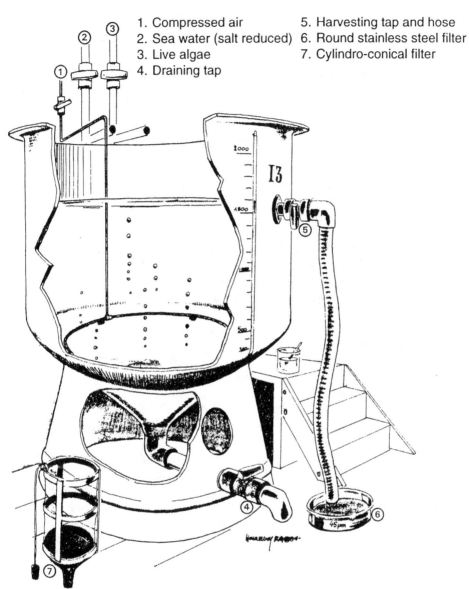

1. Compressed air
2. Sea water (salt reduced)
3. Live algae
4. Draining tap
5. Harvesting tap and hose
6. Round stainless steel filter
7. Cylindro-conical filter

Fig. 8. Vessel for rotifer production.

dehydrated embryo which is in the gastrula stage and is extremely robust; it can remain dehydrated for a long period of time (diapause). The cysts are harvested from the banks either mechanically or manually (10–100 kg ha^{-1} yr^{-1}). The cysts are separated from sand by filtration and washing in hypersaline water (300‰), washing in fresh water, drying on paper and packing, dry, in tins, preferably under an atmosphere of nitrogen. They can be transported easily and stored for several years. In 1983, 60 tonnes of cysts were used in aquaculture; since then the total has doubled. Artemia are not farmed; however, there are special techniques for obtaining the nauplii from cysts.

After hydration, cysts are incubated in fresh water (1–2 h at a biomass of 2–5 g l⁻¹). They are then placed in tanks of brackish or salt water ($S = 5$–35‰) which is strongly aerated and illuminated (1000 lux). Nauplii hatch 18–24 h later at 27°C. Turning off the aerator allows the nauplii (which are in mid-water) to be separated from the empty shells (on the surface) and unhatched cysts (on the bottom). The nauplii are strongly phototrophic, which means that they can be concentrated by illumination through the side of translucent tanks (cylindro-conical tanks of several hundred litres volume). The hatching rate varies, depending on the stock (35–83%); from 1 g of cysts 105 000–300 000 nauplii can be obtained (Canler 1983). There are around 62 000 nauplii g⁻¹; they have a mean length of 420–430 μm (Fig. 9).

The cysts can be decapsulated; their shell (chorion), which is composed of lipoproteins, can be dissolved by hypochlorite without affecting viability.

The nauplii (termed A0) should be distributed as soon as they hatch to benefit from their high energy content. However, they cannot be 'doped' at this stage as they do not begin to feed until 15 h after hatching (after their first moult).

The metanauplii (Fig. 9) are obtained from nauplii held at 25°C in the hatching tank or in a new tank for one day (A1) or two days (A2). These are used as large prey items for older larvae. They are given a specific diet: cultured algae, baker's yeast, dried spiruline algae or specially manufactured diets (e.g. 'Topal' made by Artemia Systems in Belgium or A09 Bis made by INRA in France). This latter diet is composed of 87.4% brewer's yeast, 4% cod liver oil, 3.6% vitamin complex, 1% methionine and 4% choline. Other diets are based on fish protein (80%). The cost of one million Artemia nauplii is around 4FF (£0.40) and one million metanauplii 6FF (£0.60). Details of all the techniques for the production of Artemia from cysts are given by Ardineau *et al.* (1984) and Versichelle *et al.* (1990).

As with the use of rotifers fed solely on yeast, the use of Artemia nauplii (or freshwater plankton) can lead to deficiencies in polyunsaturated fatty acids in the larvae. When live prey are used at the beginning of the larval stage they should be enriched. Many dietary supplements are now sold ('Fippak' by France Aquaculture, 'Selco' by Artemia Systems). The supplements are added in baths (see above) but for a shorter duration (1–3 h).

3.3 Collecting zooplankton

We have seen in Chapter 2 the part played by zooplankton in the aquatic trophic web and their relationship with larval fish and other groups of importance to aquaculture (see also Part IV, Chapter 4).

The mean density of copepod larvae in oceanic water is between 17 and 40 per litre and copepodites between 1 and 7 per litre according to Hunter (1980). This density is too low to supply larval fish cultures, and from various information it has been calculated that the copepod prey should be concentrated by 1.3 in relation to the mean natural density so that the average fish larva has a chance of survival.

From this it might be assumed that collecting and concentrating zooplankton from the wild is a useful way of supplying hatcheries with their requirements. Barnabé has been

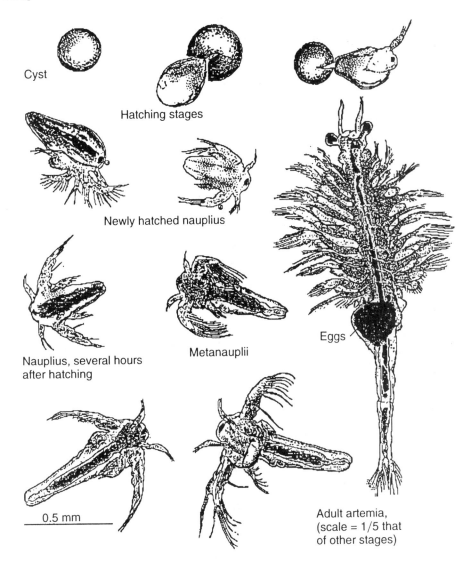

Cyst

Hatching stages

Newly hatched nauplius

Nauplius, several hours
after hatching

Metanauplii

Eggs

0.5 mm

Adult artemia,
(scale = 1/5 that
of other stages)

Fig. 9. Different stages in the Artemia life cycle (from Audineau *et al*. 1984).

investigating the problem since 1976 but unfortunately, in the Mediterranean, the density of copepods is only $0.5–10 \, l^{-1}$ (mean $1 \, l^{-1}$) while the littoral zones are also poor. Some wetlands have high numbers seasonally but, without doubt, the most productive systems are water purification lagoons (Barnabé 1979). The value of harvesting plankton as prey for larval fresh- or salt-water fish and several other industrial applications has been demonstrated (Barnabé 1980, 1981); tonnes of frozen copepods from salt water are sold and rotifers and daphnids from lagoons are used to feed tropical ornamental fish. Many hatcheries use them as either part or all of their food; plankton harvested from various sources are fed to fresh-water or marine fish.

The technology of harvesting has been reviewed in the works cited above. Simple, cheap harvesting devices (Fig. 10), capable of continuous operation, are an easy way of collecting zooplankton. This technology has been developed so that the smaller forms of zooplankton (<40 μm) can be harvested.

3.4 Collecting phytoplankton

Although tested experimentally as a means of providing food for rotifers (Barnabé & Périgault 1983), collecting phytoplankton is a lot less easy than zooplankton because of the small size of the algal cells. These authors used centrifugation, but this is expensive. There are other methods (reviewed in Barnabé 1990) including flocculation but it is very difficult to re-separate the microalgae once they have been flocculated; they form a mat which traps larval molluscs or rotifers which are attracted to feed there. In addition to this, the flocculants (iron or aluminium salts) have their own effects; aluminium is toxic. At present the collection of phytoplankton is at the experimental stage.

4. CONCLUSION

There are two differing systems for the production of algae:

— Purification of productive waste waters to remove algal biomass (Table 2) is a biological process requiring no human intervention. Of all the microbial methods for producing proteins, the use of algae is the least sophisticated (Grobbelaar 1979).
— In contrast to this, the production of predetermined species of microalgae in controlled conditions is always unstable and expensive. De Pauw *et al.* (1984) reviewed the findings of various authors and found that the cost of production of dry microalgae from waste water is around US$0.25, that of cultivated spirulines $1, that of chlorellas between $9 and $11 compared with $120–200 for monospecific cultures of marine species grown indoors or under glass.

The commercial culture of single species of algae is a scaled-up version of culture in the microbiology laboratory, and for this reason it is always likely to be costly. There is therefore a continued search for improved methods of outdoor culture under natural light conditions. Boussiba *et al.* (1988) kept their cultures at depths of 12 cm; De Pauw & de Leenheer (1985) tried depths from a few centimetres to 2 metres, underlining the stability of their cultures in deep water. In addition to this there is the potential for adapting species such as *Nannochloris* to higher temperatures and salinities (Barclay *et al.* 1987). There is a vast potential for development of methods and simplification of culture procedures.

The move towards the simplification of the production of microalgae for aquaculture is also likely to result in the simplification of the culture of rotifers and open up the way for the culture of other species of zooplankton which, at present, are limited by the availability of food.

The major development may come from the use of microalgae for purification of waters conveying human wastes. In a time when environmental problems are well to the

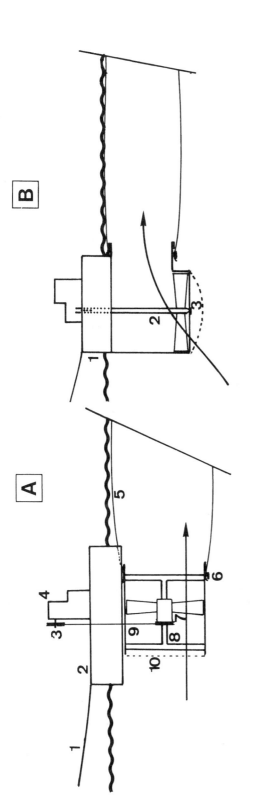

A: Harvester for plankton, stationary and autonomous, horizontal axis

1: Electricity supply from a battery on the bank and mooring rope.
2: Float (PVC tubes or expanded polystyrene).
3: Motor shaft and pulley.
4: Motor (car windscreen wiper motor).
5: Filter bag (bolting silk, 70–200 μm mesh, S-20 m^2 area).
6: Tightening collar for the filter bag on the tube (diameter 30 cm).

7: Ventilation fan.
8: Support and axis for the fan.
9: Drive belt.
10: Large mesh grid.

B: Harvester for Plankton, autonomous and stationary vertical axis

1: Float.
2: Vertical axis.
3: End wall of the axis and protective grid.

Fig. 10.

fore, this system of purification has two purposes and is simple and cheap to operate. Producing more microalgae in this way will be of great benefit to the future development of aquaculture.

REFERENCES

Anon., 1990. Omega-3 'very beneficial'. *World Aquaculture*, **21** (2): 91.

Audineau P., Coves D., Hameury P., 1984. Production de proies vivantes: *Brachionus plicatilis et Artémia salina*. IFREMER, Equipe MEREA, Palavas. Doc ronéoté, 26 pp.

Audineau P., Blancheton J., 1986. Production d'algues unicellulaires. IFREMER, Equipe MEREA, Palavas. Duplicated document (not paginated).

Barclay W., Terry K., Nagle U., Weisman J., Goebel R., 1987. New strains of algae for aquaculture. *J. World Aquac. Soc.*, **18** (4): 218–228.

Barnabé G., 1979. Utilisation des chaînes alimentaires naturelles et du recyclage des eaux usées dans la production à grande échelle de juvéniles pour l'aquaculture. *Actes de Colloque*, No. 7. CNEXO, Paris: 221–238.

Barnabé G., 1980. Système de collecte du zooplancton à l'aide de dispositifs autonomes et stationnaires. In: *La pisciculture en étang*, R. Billard (ed.). INRA Publ., Paris: 215–220.

Barnabé G., 1981. Collecte et utilisation de plancton. In: *Compte rendu des Activités Scientifiques du Centre de Recherches du Lagunage de Mèze*. Rapport 1980–1981: 112–132.

Barnabé G., 1990. La collecte des micro-algues. In: *Aquaculture*, G. Barnabé (ed.). Technique & Documention (Lavoisier) Publ., Paris: 193–199.

Barnabé G., Périgault C., 1983. Collecte et utilisation du phytoplancton produit dans les étangs de lagunage de Mèze: Données Préliminaires. In: *Recherches biologiques en Aquaculture*. C. R. Travaux GIS-ARM, CNEXO Ed., **1**: 161–175.

Becker E. W. (ed.), 1985. Production and use of microalgae. *Arch. Hydrobiol. Beih. Ergebn. Limnol.*, **20**: 198 pp.

Boussiba S., Sandbank E., Shelef G., Cohen Z., Vonshak A., Ben-Amotz A., Arad S., Richmond A., 1988. Outdoor mass cultivation of the marine microalga *Isochrysis galbana* in open reactors. *Aquaculture*, **72**: 247–253.

Canler A., 1983. Améliortion de l'utilisation d'Artémia salina en écloserie de Loups et en écloserie d'ecrevisses. Rapport Stage DESS Cultures marines, Univ. Caen: 89 pp.

Coves D., Audineau P., Nicolas J. L., 1990. Les Rotifères—Technologie d'elevage. In: *Aquaculture*, G. Barnabé (ed.). Technique & Documentation (Lavoisier) Publ., Paris: 225–240.

De Pauw N., de Leenheer L., 1985. Outdoor mass production of marine mciroalgae for nursery culturing of bivalve molluscs. *Arch. Hydrobiol. Beih. Ergebn. Limnol.*, **20**: 139–145.

De Pauw N., Morales J., Persoone G., 1984. Mass culture of microalgae in aquaculture systems: Progress and constraints. *Hydrobiologia*, **116/117**: 121–134.

De Pauw N., 1981. Use and production of microalgae as food for nursery bivalves. In: *Nursery culturing of bivalve molluscs*, C. Claus, N. De Pauw, E. Jaspers (eds). EAS Publ., Bredene, Belgique: 35–69.

Durand-Chastel H., 1980. Production and use of Spirulina in Mexico. In: *Algae biomass, production and use*, G. Shelef & C. Soeder (eds). Elsevier Publ., Amsterdam: 51–63.

Edeline F., 1979. *L'épuration des eaux usées résiduaires*. Ed. CEBEDOC, Liège (Belgique): 243–255.

Gatesoupe J., Fukusho K., Watanabé T., 1989. Improvement of the production rate of rotifers with food and bacterial additives. *Aquaculture Europe 89*: Short communications. EAS Ed., Sp. Publ. No. 10: 109–110.

Goldman J., 1979. Outdoor algal mass culture. I. Applications. *Water Research*, **13**: 1–19; II. Photosynthetic yield limitations. *Water Research*, **13**: 119–136.

Gonzalez-Rodriguez E., Maestrini S., 1984. The use of some agricultural fertilizer for the mass production of marine algae. *Aquaculture*, **36**: 245–256.

Grobbelaar J. U., 1979. Observations on the mass culture of algae as a potential source of food. *South Africa Journal of Science*, **75**: 133–136.

Helm M., Laing I., Jones E., 1979. The dévelopment of a 200 l algal culture vessel at Conwy. *Fish. Res. Techn. Rep.*, **53**: 1–12.

Hunter J., 1980. The feeding behavior and ecology of marine fish larvae. In: *Fish behavior and its use in the capture and culture of fish*. ICLARM Conference Proceedings, Manilla: 287–330.

Iisawa M., 1983. Ecologie trophique des larves du Loup *Dicentrarchus labrax* (L.) en élevage. Thèse Doct. 3° Cycle. Univ. Sci. Techn. Languedoc, Montpellier: 141 pp.

Kawaguchi K., 1980. Microalgae production system in Asia. In: *Algae biomass, production and use*, G. Shelef, & C. Soeder (eds). Elsevier Publ., Amsterdam: 25–33.

Korn M., 1989. Microalgae as fish feed. *Aquaculture Europe 89*: Short communications EAS Ed., Sp. Publ. No. 10: 135–136.

Le Borgne Y., 1990. La culture des micro-algues. In: *Aquaculture*, G. Barnabé (ed.). Technique & Documentation (Lavoisier) Publ., Paris: 183–194.

Pourriot R., 1990. Les rotifères—Biologie. In: *Aquaculture*. G. Barnabé (ed.). Technique & Documentation (Lavoisier), Publ., Paris: 203–223.

Richmond A., Vonshalk A., Arad S., 1980. Environmental limitations in outdoor production of algal biomass. In: *Algae biomass, production and use*, G. Shelef & C. Soeder (eds). Elsevier Publ., Amsterdam: 65–72.

Robin J. H., Gatesoupe J., Stephan G., Le Delliou H., Salaun G., 1984. Méthodes de reproduction de filtreurs proies et amélioration de leur qualité nutritive. *Océanis*, **10** (5): 497–504.

Shelef C., Soeder C. J. (eds), 1980. *Algae biomass, production and use*. Elsevier, Amsterdam: 852 pp.

Soeder C. J., 1980. Massive cultivation of microalgae: results and prospects. *Hydrobiologia*, **172**: 197–209.

Versichelle D., Léger P., Lavens P., Sorgeloos P., 1990. L'utilisation d'artémia. In: *Aquaculture*, G. Barnabé (ed.). Technique & Documentation (Lavoisier) Publ., Paris: 241–259.

Watanabe T., Kitajima C., Fujita S., 1983. Nutritional values of live organisms used in Japan for mass propagation of fish: a review. *Aquaculture*, **34**: 115–143.

Witt U., Koske P. H., Khulman D., Lenz J., Nellen W., 1981. Production of *anochloris* (Chlorophycae) in large-scale outdoor tanks and its use as a food organism in marine aquaculture. *Aquaculture*, **23**: 171–181.

Wong M. H., Yip S. W., Fan K. Y., 1977. Chlorella cultivation in sludge extracts. *Environ. Pollut.*, **12** (3): 205–209.

Part II
Mollusc culture

P. Lubet

Introduction

BACKGROUND TO MOLLUSC CULTURE

Mollusc farming or the culture of marine bivalves is the oldest form of aquaculture. However, it was not until the end of the nineteenth century that the reliable techniques which allowed the rapid growth of the industry were developed. A new leap has been made, around a century later: this has been the development of the hatchery for the production of spat. On top of this, overexploitation of wild stocks has caused the fishermen to consider management of the resource and to look to restocking and culture in deep water using juveniles originating in the hatchery.

The practices involved in marine farming, such as technological improvements, the prospect of new methods, particularly relating to biotechnology suggests that there will be a growth in the number of problems relating to the biology of the cultured species. In this section, the most recent information concerning nutrition, growth and reproduction is presented.

This chapter is not restricted to widely cultured species: other species which are fished intensively and which are therefore suitable candidates for culture are included. These include the king scallop (*Pecten maximus*) and the cuttlefish (*Sepia officinalis*). This is not the place to review the classical information on anatomy and physiology of molluscs. The object here is to provide a fundamental basis for the better understanding of the biological problems linked to the culture of these species. In addition, we have omitted any information on diseases and the response of these species to pollution.

1

Environment and nutrition

INTRODUCTION

Very few species are as yet cultured. These are, essentially, bivalves and one genus of gastropods.

Bivalves

— Mussels: *Mytilus edulis-galloprovincialis*, Europe; *M. edulis*, Europe, Korea; *Perna* sp., Asia and New Zealand.
— Oysters: *Ostrea edulis*, Europe; *Crassostrea gigas*, North Pacific, Europe; *Crassostrea virginica* (Atlantic coast of the USA); *Saxostrea commercialis*, Australia; *S. forskali*, Mauritius.
— Scallops: *Patinopecten yesoensis*, Japan.
— Pterids: *Pteria fucata*, Japan (pearl oyster).
— Venerids (clams): *Ruditapes philippinarum*, North Pacific, Europe, North Africa; *R. decussatus*, Europe.

Archeogastropods

— Haliotids: Several species of *Haliotis*, North Pacific; *H. tuberculata*, Europe.

Spat from many other bivalve species has been produced on an experimental scale in hatcheries in order to test the feasibility of techniques and the possibility of producing juveniles for restocking and eventually for culture. In addition, cuttlefish have been reared through all stages from the egg.

1. ENVIRONMENTAL CONSTRAINTS

All species are subject to the constraints imposed by the climatic, hydrographic, and biological characteristics of the environment. These parameters limit a species geographical distribution. Knowledge of these factors is particularly important for implementing aquaculture operations and evaluating the maximum capacity of an oysterbed.

Limiting factors should be determined precisely when the operator wishes to attempt to culture new species. Unfortunately, we have only fragmentary information in this area; results which have been obtained from laboratory experiments do not necessarily reflect those that would have been obtained in the wild under natural conditions. The most significant results are given in Table 1 and Figs 1 and 2. Problems associated with the capacity of aquaculture are described by Héral (1990).

2. NUTRITION

2.1 Bivalves

With the exception of species which are carnivorous, saprophytic or living symbiotically with algae or bacteria, most bivalves are microphagous and feed on particles which are deposited on the surface of the sea bed (deposit feeders) or in suspension in sea water (suspension feeders). Most of the species which are commercially exploited belong to this latter category.

2.1.1 Morphogenesis, organization and cytology of the alimentary system

A. Larvae

The ontogeny and structure of the digestive system are well known through the work of Ansell (1962), *Venus striatula*; Creek (1960), *Cardium edule*; Sastry (1965), *Aequipecten irradians*; Hickmann & Gruffydd (1970), *Ostrea edulis*; and Bayne (1970) and Masson (1975), *Mytilus edulis*.

Differences between species are slight: the description given here is for the mussel (Fig. 3).

In the trochophore stage, the mass of endodermic cells hollows out, caves inwards making a junction with the invagination of the blastopore. Proctodeal piercing takes place next. The digestive tract appears as a U shape and is not differentiated into distinct regions: at this stage it is not functional.

In D stage larvae (at the straight hinge stage), from 48 hours after fertilization, the different regions of the digestive tract begin to differentiate: the oesophagus is short and straight; the stomach is a large bag with a functioning crystalline style, the distal extremity of which continually wears out against a gastric shield; the primitive intestine is rectilinear and extends rapidly, forming a curve. The digestive gland is made up of several large, highly pigmented cells which form a simple outgrowth of the stomach wall above the diverticulum (caecum), containing the crystalline style. The bolus of food is in direct contact with the apical cells of the digestive gland.

The digestive tube is functioning by now: water and particulate food is taken in through the combination of movements of the velum and the preoral palp, the directing of the food bolus in the tract, the rotation of the crystalline style, the inclusion of algal pigments in the cells of the digestive gland, and faecal evacuation.

In the veliger larva the stomach is completed by the formation of an intestinal groove originating from its ventral aspect and delimited by two folds (*typhlosoles*) which isolate this area of the diverticulum containing the crystalline style. The intestine bends and elongates in two branches, one recurrent and one terminal, opening above the posterior

Table 1

Species	Temperatures LT$_{50}$(°C)		Salinities SL$_{50}$(‰)	Substrate	Turbidity (mg l^{-1})	Depth (m)	
Bivalves							
Mytilus edulis	0°	22°	(7).15	38	D.G.Sc.Sv	++++	ML	−40
Mytilus galloprovincialis	5°	26–27°	17	39	D.G.Sc.Sv	+++	ML	−40
Ostrea edulis	2°	20–25°	15	38	D.G.Sc	+	IL	−30
Crassostrea gigas	0°	27°	12	39	D.G.Sc.Sv	++++	ML	−10
Ruditapes decussatus	1–2°	27°	18	38	Sc.Svmc	+++	ML	
Ruditapes philippinarum	2–3°	28°	18	38	Sc.Svmc	+++	ML	
Mercenaria mercenaria	2–3°	28°	20	38	Sc.Svmc	+++	ML	
Pecten maximus	6°	23°	23	38	Sc.	++	IL	−120
Chlamys varia	6°	27°	23	38	D.Sc.	++	IL	−30
Patinopecten yesoensis	0°	23°	23	38	Sc.Sv	++	IL	−60
Gasteropods							
Haliotis tuberculata	5°	23°	25	38	D-algae	+	IL	−10
Cephalopods							
Sepia officinalis	6°	27°	25	38	Pelagic demersal	+	IL	−50

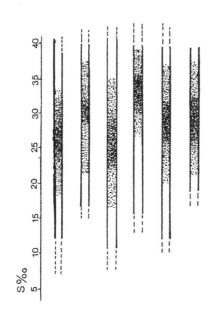

T°C

Fig. 1. Lethal temperature limits (TL$_{50}$) and reproductive cycles for several species of bivalves.

sexual dormancy

gametogenesis

Dotted: sexual cycle; hatched: sexual dormancy.

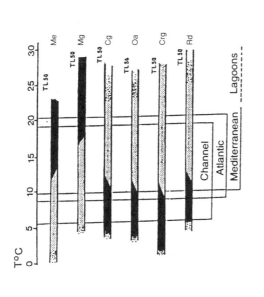

S‰

Preferred salinity

Fig. 2. Lethal salinity limits (SL$_{50}$) for several species of bivalve.

Me = *Mytilus edulis*, Mg = *Mytilus galloprovincialis*
Cg = *Cerastoderma glaucum*, Oe = *Ostrea edulis*
Crg = *Crassostrea gigas*, Pm = *Pecten maximus*
Rd = *Ruditapes decussatus*

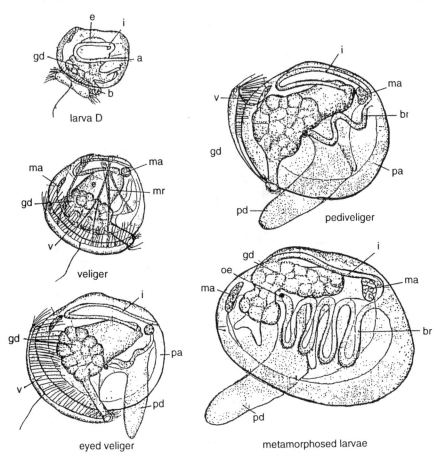

Fig. 3. Morphogenesis of the digestive tract of *Mytilus edulis* (after Masson 1975). a = anus, b = mouth, br = gills, es = stomach, gd = digestive gland, i = intestine, ma = adductor muscles, mr = retractor muscles of the velum, oe = ocelle, p = mantle, pl = palps, pd = foot, v = velum.

adductor muscle. The digestive gland increases in size but always remains confined by the left side of the stomach with which it is always linked by a large canal. At the end of the larval phase (pediveliger stage), the gland appears as a highly developed left lobe and a right lobe where growth begins: these are the beginnings of the two caecae of the adult gland.

Metamorphosis involves few changes in the anatomy of the digestive system. In the juvenile (spat), the morphology of the digestive gland becomes progressively completed and changes from a sac-like structure to an acinar structure in the form of a bunch of grapes. This post-larval morphogenesis begins with the burgeoning of the lobes of the gland with primary channels of ciliated epithelium which, in their turn, ramify to form secondary ducts with brush borders, ending in acinar cul de sacs made up of the digestive and secretary cells (Owen 1955).

Observations made by Masson (1975) and Masson & Herlin (1976) have shown that the differentiation of the digestive gland takes place between 24 and 48 hours after fertilization. There are already flagellated 'mother' cells from which digestive cells that are rich in vacuoles and secretary cells (low in number in larvae) are likely to arise.

B. Juveniles and adults

a. Anatomy

a1. The two gills, in a W formation, are formed from folds in the mantle and are symmetrical in the sagittal plane. They are made up (Fig. 4) of a layer of external and internal filaments. Each sheet comprises a direct (descending) and recurrent (ascending) part. In the most simple type of system (filibranchs, mussels), coherence between adjacent filaments is ensured by ciliary junctions; the ciliary zones are opposite each other. The layers can also be connected by transverse septa (Fig. 4a).

These can be numerous in pseudo-lamellibranchs (scallops, pterids, oysters) where the gill layers are folded over and where the filaments are fused at their extremities, having the appearance of a ploughed field with lines of furrows. For the most highly evolved species (e.g. lamellibranchs—clams), the extremities of the ascending gill filaments are fused respectively to the mantle (external layer) and to the visceral mass (internal layer). The mantle cavity is therefore separated into two distinct regions. Finally, many septa change the appearance of the gills into a grid-like structure.

Each filament is hollow, filled with haemolymph and formed by a ciliated epithelium around mucus cells which secrete glycoproteins. Haemocytes move by diapedesis, gliding on the exterior along the furrows.

The cilia form tracts on the four surfaces of the filament. These were studied by Atkins (1939) and then by several other authors (reviewed in Morton (1983)) (Fig. 4b). On the external surface there are cilia which are directed forwards, and on the opposite surface the cilia face away from the front (Fig. 4b). The lateral surfaces are covered by lateral cilia which are responsible for pumping water across the branchial grid. The frontal tract is defined by the lateral-frontal cilia, which is large in size and combines to form a moving triangular structure which can retract the surface of the channel in order to transport particulate matter.

a2. The labial palps are situated on both sides of the mouth and are formed by two triangular foliated flaps. These ensure the continuity of the transport of particles from the gills to the mouth. This transport is carried out by the internal side which is highly ciliated and formed into alternating crests and troughs.

a3. The digestive tube (Fig. 5). The mouth has foliar 'lips' which may be folded and meshed (scallops). A short, curved oesophagus opens into a large stomach which is irregular and complex in form with highly folded ciliated walls. This extends posteriorly to a cylindrical caecum, the sac for the crystalline style which may be separate from the stomach groove (cockles) or fused to it (mussels, oysters, clams). In this latter case the intestinal groove of the posterior stomach region is separated from the stylet sac by two folds (major and minor typhlosoles), coming from the stomach. At the posterior extremity

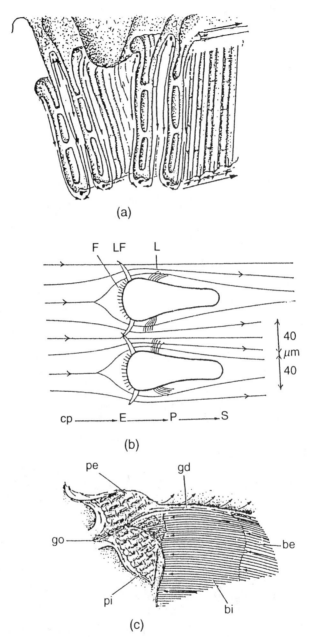

Fig 4. (a) Arrangement of the gills. (b) Section through the filaments of a mussel (after Silvester & Sleigh 1984 modified). (c) Lab-al palps (after Morton 1978). cp = palleal cavity, E = filtration, F = frontal cilia, L = lateral cilia, Lf = latero-frontal cilia, P = pumping, Sc = bronchial cavity. bi–be = half-internal and external gill; gd–go = distal and proximal food groove; pe–pi = external and internal palp.

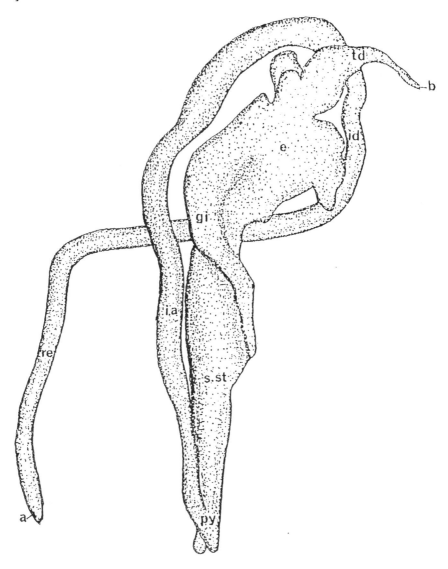

Fig. 5. Reconstruction of the digestive tract of *Crassostrea gigas* (right view). The digestive gland is not shown (after Lebesnerais 1985). a = anus; b = mouth; c.d. = digestive coecum; g.i. = intestinal groove; i.a. ascending intestine; py = pylorus; re = rectum; s.st = stylet sac; t.d. = oesophagus; e = stomach.

the groove opens into the recurrent intestine (ascending branch) while the extremity of the stylet sac becomes detached and slightly elongated. Depending on the species the intestine may form several loops around the digestive gland into which it may become incorporated, and then directed towards the exterior of the animal (descending branch). It is extended by a rectum bordering the dorsal surface of the adductor muscle and may (mussels) or may not (oysters) penetrate into the ventricle. It has an anus which opens into the mantle cavity close to the ostium or exhalant syphon (clams).

The stomach is surrounded by a voluminous brownish mass, the digestive gland. It is tubular-acinar in structure and made up of a large number of blind-ended tubules in communication with the stomach through ramified ducts. Several glandular tubules open into a short secondary canal; these secondary canals lead into main canals, which join into ducts of increasing diameter, opening out into the stomach.

b. Cytology

With the exception of the gastric shield, the oesophagus, stomach and intestine have a uniform histological structure. These are made up of a ciliated epithelium containing many glandular cells for the secretion of glycoprotein; only varying, according to region, in the density of the glandular cells, the height of the ciliated cells and the morphology of the cilia. The gastric shield is made up of a layer of vertically elongated cells coated with a thick chitinous, sclerotized cuticle. The muscular wall of the digestive tract is never very well developed.

The digestive gland (Fig. 6) has a characteristic structure, similar in all species. The principal ducts have a region with a ciliated epithelium and a region with a brush border epithelium: the epithelium contains glandular cells; these are more numerous in the region with the brush border. The elongated ciliated cells have strong ciliary roots with a highly developed system of cilia and microvilli; their apical region is rich in mitochondria and glandular cells. The brush border cells are also rich in mitochondria and have many tall and serrated microvilli. Finally, the glandular cells are cup shaped with a nucleus in the basal position surrounded by cytoplasm which is rich in glycoprotein granules.

The secondary ducts are made up of an epithelium containing cells with a brush border, identical to the cells of the principal ducts.

The glandular tubules (acini) are covered with several types of cell.

— The *digestive cells* (10–20 μm). The most common ones are tall and have microvilli in their apical region. The nucleus is in the basal position; the cytoplasm contains mitochondria, lipid globules and lysosomes, as well as many heterophagic vacuoles (= digestive) in different stages of development.
— The *basophilic cells* (10–15 μm) also have microvilli and, sometimes, cilia in the apical position. The nucleolus is always well developed and the cytoplasm contains numerous vesicles of granular endoplasmic reticulum.
— The *smallest cells* may be the originators of the other two types of cell. Henry (1987) suggested a progression passing through intermediate flagellar forms which ensures that the digestive gland is always being repaired.

Finally, there are many haemocytes coming from blood spaces. In mussels and oysters tubules and ducts are often surrounded by tissue which is rich in reserves of lipids and glycogen. In venerids (clams), the cells of the principal ducts accumulate glycogen (Henry 1987).

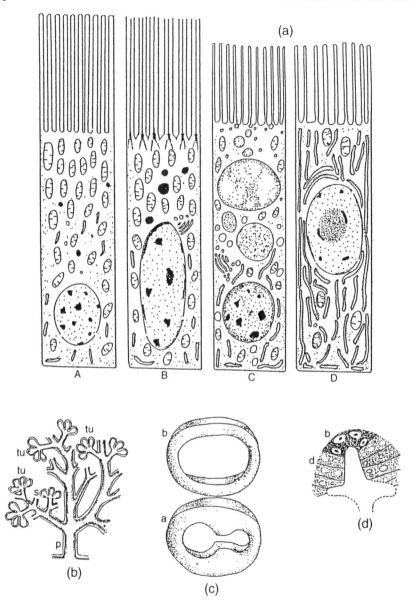

Fig. 6. (a) Diagrammatic representation of the principal cell types in the digestive gland of *Crassostrea gigas* (after Boucaud *et al.* 1985). (b) Structure of the digestive diverticules (after Owen 1955). (c) Section through principal duct. (d) Arrangement of the digestive cells ('d') in a digestive tubule of *C. gigas* (after Lebesnerais 1985), ('b', basophilic cells). A: Cells with a brush border from the principal and secondary ducts, inhalent passage. B: Ciliated cell from the principal duct, exhalent passage. C: Digestive cell. D: Basophilic cell. cp = principal ducts, ciliated over part of the surface, cs = secondary ducts, non-ciliated, tu = digestive tubules.
a = after Owen (1955); b = after Lebesnerais (1985)—*Crassostrea gigas*.

2.1.2 The intake of food and its transit through the digestive system

A. Larvae

The intake of food is through the currents induced by the cilia of the velum and the labial palps. The particles adhere to a mucous film produced or even rejected by the oral palp according to the state of fullness of the tract. Capture appears therefore to be essentially a mechanical process, leaving open the possibility of ingesting various types of particles (algae or aggregates of organic matter). The limiting factor appears to be the size of the particle, which is limited by the size of the mouth (Masson 1975). This author has also shown the existence of pinocytosis by the epidermal cells through the microvilli.

B. Juveniles and adults (Fig. 4)

The intake of food takes place during filtration, through the frontal ciliary tract which takes in small-sized particles, below 10 μm (Tammes and Dral 1956). Water is pumped towards the interior of the gill filaments by the movements of the lateral cilia; particles of appropriate size are retained by the outer surface of the filaments (frontal tract). They are guided by the tract towards the ciliated grooves in which they are propelled towards the labial palps. The size of the frontal tract depends on the position of the laterofrontal cilia which determines the size of the material that can be transported. In certain species (Arcides) a first grading may take place at the level of the gills, but in most bivalves (mussels, scallops, oysters and venerids) selection is made by the labial palps (Morton 1983). Big particles are expelled through the furrows while the smallest cross the crests and are guided towards the oral groove. This general scheme does not hold for scallops (Beninger et al. 1988; Beninger 1989) where the selection is essentially negative; only a small number of 'undesirable' particles are likely to be ingested. Particles of very different sizes are found in the stomach (Shumway et al. 1987), the size is often very much bigger than that of the ciliary tract. The sites of selection are still not well known. Colloidal suspensions may also play a role in nutrition.

C. Dynamics of the transit through the digestive system

For the different species of commercially important bivalves which have been studied, the structure of the digestive tract appears to be similar: the musculature surrounding the stomach and the intestine is poorly developed while the cilia are well developed, the transit of the alimentary bolus will be ensured by the cilia covering the various cavities.

D. Dynamics of the digestive system

Experiments feeding controlled diets carried out by Lebesnerais (1985) and Boucaud-Camou et al. (1985) for *Crassostrea gigas* have allowed the precise definition of the different stages of the movement of food through the digestive system (Fig. 7).

d1. Filling up the digestive system. When the animal has been starved or when the digestive system contains very little (e.g. oysters that have been out of the water for a long period) all of the cavities become completely empty. Whole algae have been observed in the stomach, the main ducts of the digestive gland, the intestine and the rectum. Without having been subject to enzyme action, this material begins to be expelled through

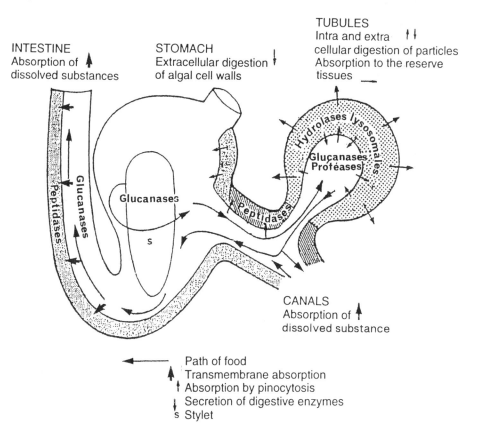

INTESTINE
Absorption of
dissolved substances

STOMACH
Extracellular digestion
of algal cell walls

TUBULES
Intra and extra
cellular digestion of particles
Absorption to the reserve
tissues

Hydrolases lysosomales

Glucanases
Protéases

Glucanases

Peptidases

Peptidases

Glucanases

s

CANALS
Absorption of
dissolved substance

Path of food
Transmembrane absorption
Absorption by pinocytosis
Secretion of digestive enzymes
s Stylet

Fig. 7. Schematic representation of the different sites of digestion (after Boucaud *et al.* 1985).
s = stylet.

the anus, surrounded by a mass of mucus. The duration of this stage is strongly influenced by temperature: 6 h at 10°C and 3 h at 20°C. The animal continuously compensates for this loss by taking in new food.

d2. Digestion. From the beginning of their entry into the ducts of the digestive gland (1–2 h), the nutrient particles are attacked, although living cells (algae) have been observed for around 6 h after food has reached the stomach and 8–16 h in the intestine.

Morton (1983) has shown that, in the digestive gland, regions with a brush border lining the ducts will be 'inhalant', allowing the transfer of particles from the stomach in the tubules, while the expulsion of wastes in the stomach will be carried out in the reverse direction by the ciliated zones (exhalant). In the Japanese oyster (Fig. 6), these zones are not separated by crests (typhlosoles) and it is difficult to see the existence of countercurrents as suggested by Owen (1955). It is possible (Morton 1983) that the currents alternate with time: Mathers (1972) has shown the existence of an interval between the time when the food arrives in the ducts and the return to the intestine of residues from the digestive process.

These residues appear in the faeces 3–6 h after the first intake of food, mixed with live algae whose proportion decreases with time. The texture of the faeces changes: at first they have the consistency of a mucus-rich sausage, they then become enriched with tightly packed residues, forming a folded ribbon moulded by the rectal groove. After around 10 h, the faeces become homogeneous, hard and rolled into twists; they are ejected in a continuous fashion. If a starved animal is given a single meal (3 h), it takes around 50 h at 20°C and 70 h at 10°C for the digestive tube to empty itself completely.

d3. Influence of external factors.   These experiments have shown that the animal fills the digestive tube completely and then progressively ejects a part of its contents.

— *Temperature* has a significant influence on the speed of transit by accelerating the frequency and intensity of the movements of the cilia of the tract (Lebesnerais 1985). The shortening of the length of digestion at 20°C affects mainly the start of digestion (transit of the food bolus) and the end (elimination of faeces); the phase of true digestion is less dependent on temperature.
— *The length of the feeding period*—short (3 h) or long (22 h) does not appear to have a significant modifying effect on the chronology described. Experimental results are in accord with the existence of a feeding rhythm linked to tides for intertidal animals (Morton 1977, 1983). Examination of the stomach contents (Lubet 1978) shows clearly the existence, during a major part of digestion, of intact algae in the stomach and intestine while those which have reached the digestive gland are immediately attacked by enzymes.

2.1.3 Enzyme action and absorption
Much cytochemical and biochemical research (reviewed by Morton 1983) has allowed the definition of the nature, location and activity of enzymes of the digestive tract. More recent work has been carried out on *Mytilus edulis* (Janssen 1981), *Crassostrea gigas* (Lebesnerais 1985; Boucaud-Camou *et al.* 1985), *Ruditapes philippinarum* (Henry 1987) and *Pecten maximus* (Boucaud, in press); this shows great similarity between the species in the location of the enzymes within the cells.

A. The enzymes and their location
Table 2 shows the range of enzyme activity found in the tubules and ducts of the digestive gland.

The activities of the carbohydrases (amylase, cellulase and laminriase) are located throughout the digestive tract in the epithelia, such as the lysosomal enzyme, N-acetyl glucosaminidase, which is particularly strong around the gastric shield and in the cells of the tubules of the digestive gland.

Non-specific esterases are present in the canal of the digestive gland and the apical part of the stomach and intestinal epithelium. Acid phosphatase, a lysosomal enzyme, is very active in the digestive gland while alkaline phosphatase, which indicates active transport, is localized in the canals of the digestive gland and the base of the tubules.

Table 2. Localization of enzymes in the digestive tract of *Crassostrea gigas*
(after Lebesnarais 1985)

Enzymes	Digestive gland		Stomach		Intestine	
	Canals	Tubules	Ciliated cells	Gastric shield	Ascending	Descending
Amylase	+++	+++	+	+	+	+
Cellulase	++	++	++		++	++
Laminarinase	+	+	+		+	+
Alginase	–	–	–	–	–	–
B. N-acetyl gluco- saminidase	++	++	++	+++	++	++
B. glucuronidase	–	–	–	–	–	–
Non specific esterases	+++	++	+++		+	
Lipase	–	–	–	–	–	–
Acid phosphatase	+	+++	++		++	
Alkaline phosphatase	++	+	–	–	+ weak	+ weak
Proteases	–	++	–	–	–	–
Chymotrypsin	++	weak +	+		weak +	weak +
Cathepsin B	+		++		++	++
Cathepsin D	–	–	–	–	–	–
Cathepsin C or DAP I	+++	–	++	–	+	–
DAP II	+++	–	++	–	++	++
DAP III	–	–	–	–	++	++
DAP IV	+	–	++	–	–	–
Aminopeptidase M	++	++	++	+	+++	+++

Peptidase activity is weak: Cathepsins and dipeptidyl-peptidases are found around the brush border of the canals of the digestive gland and in the apical region of the ciliated epithelium of the stomach and the intestine.

Lipase activity has been shown around the crystalline style (Palmer 1979).

B. Enzyme action

The pH of the digestive tube is between 6.2 and 6.5 (Morton 1983).

a. Carbohydrates. The cell walls and reserves (starch, laminarin) of unicellular algae are digested by glucanases. While there is enzyme activity through all the epithelium, the most intensive is found in the digestive diverticulae where the algae are instantly attacked by enzymes. At the start of the digestive process it is probable that the digestive gland secretes amylase towards the stomach. The crystalline style is not, at this stage, charged with enzymes and exerts a mechanical action, grinding the particles against the gastric shield (Yonge 1923). The crystalline style progressively incorporates the glucanases which are secreted by the digestive gland and the stomach (Mathers 1972), which explains the long delay in the breakdown of algae in the stomach. The mechanical action of the style is aided by a chemical action, favouring extracellular digestion (Morton 1983); the incorporation of particles to the interior of the style has been observed (Prieur 1981). The digestion of carbohydrates is intracellular under the action of lysosomal enzymes produced by the digestive gland and probably the stomach also. The stylet appears to operate in a similar manner to a perpetual screw (Bernard 1973), transporting fine particles towards the front part of the stomach. These are then absorbed by the epithelium of the stomach.

b. Proteins.   Protease activity, in the tubules of the digestive gland and some parts of the intestine, is low at the start of digestion but builds up towards the end. These enzymes have an optimum pH of 5.5, as with lysosomal enzymes; this is associated with intracellular digestion. However, taking into account their localization on the apical zones of brush borders, it would appear that these enzymes were membrane-bound (Boucaud-Camou *et al.* 1985).

c. Enzyme activity.   Variations in the activity of acid phosphatase during the course of the digestive cycle reflect the passage and the intensity of intracellular digestion by digestive cells (digestive gland) and the epithelium of the stomach. The alkaline phosphatases found in the canals of the digestive gland and in the intestine indicate active transfer zones where the products of extracellular digestion are absorbed.

d. Utilization of bacteria.   Bivalves are capable of using bacteria in the feeding process (Zobell & Felthman 1938; McHenery *et al.* 1979). Prieur (1981) showed that there were numerous stocks of bacteria within the bivalve digestive tube. The animal is able to remove these selectively from the microflora of sea water. Some species multiply within the digestive cavities. It is probable that the bacteria secrete extracellular enzymes (proteases) which aid digestion. Many bacteria are incorporated within the crystalline style. The selection of particular stocks of bacteria within the digestive tract can be explained by the presence of lysozyme which has been found in the gills, the mantle, the crystalline style and the digestive gland of many species of bivalve (McHenery *et al.* 1979). Some strains of bacteria are resistant to lysozyme, the others are digested. The use of bacterial strains which have been marked with radioactive tracers has shown that while nucleotides (DNA) are rejected, the animal retains a major fraction, contributing significantly to the energy and carbohydrate requirements (Birbeck & McHenery 1982). Conway (1987) demonstrated in the cockle (*Cerastoderma edule*) the existence of a rhythm of lysozyme activity in the gills, the mantle and the visceral mass similar to that of the tides. Most extracellular digestion and the absorption of bacteria takes place during low water.

It is probable that lysozyme is also present in larval bivalve molluscs. Larval *Mytilus edulis* fed on filtered cultured algae (*Platymonas*), which eliminated the algae but retained many bacteria, grew more rapidly and had a higher survival rate after metamorphosis than larvae which had been fed on an unfiltered algal culture (Faveris 1982).

C. *Conclusion* (Fig. 7)

The bivalves which inhabit the intertidal zone fill up the cavities of their digestive systems as the tide rises. Particles enter the stomach, and the canals of the digestive gland (inhalant zones) and the intestine. Substances which are directly assimilable are absorbed with the aid of the enzymes located on the membranes of the brush borders. The destruction of the algal cell walls by the glucanases takes place in the channels of the digestive gland and then in the stomach through the action of the crystalline style. It is completed progressively through the biochemical action of the enzymes coming from the digestive gland and the stomach which are attached to the style. The nutrients which are being thus broken down may either continue to enter the digestive gland (extra- and intracellular digestion, absorption) or be absorbed by the epithelium of the stomach. The wastes eliminated by the ciliated zones (exhalant) of the digestive gland and found in the stomach are expelled to the intestine by means of the intestinal groove. Absorption and intracellular digestion may continue in the intestine, especially in the recurrent branch (= ascendant). In the rectum, fine residues accumulate in the main groove. The bacterial flora which is found in the digestive tract may play a major role in digestion through extracellular enzymes. Some strains of bacteria are digested by the lysozyme which is present in the digestive tract.

2.1.4 Feed

The food requirements of commercially important bivalves are still poorly understood (reviewed in Raimbault 1966; Lubet 1978; Morton 1983). The following statement, made by Galtsoff (1942) relating to oysters, still holds true today: 'At present we must confess our ignorance and consequently our inability to suggest a practical solution of forced feeding and production of fat oysters'.

A. *Qualitative aspects*

Most of the commercially important species of bivalve molluscs feed on plankton. However, the examination of the stomach contents does not give precise information on the nature of the diet. The material retrieved is heterogeneous and varies seasonally and with location for the same species. It contains organic or mineral (clay, silicaceous or calcareous silt) colloidal substances, particles of organic detritus and living organic particles (bacteria, planktonic eukaryotic cells).

In the littoral zones which are strongly influenced by tides, bivalves often grow rapidly even though phytoplankton are relatively scarce; bacteria and organic wastes are abundant. Tidal currents pick up and carry particles which have been deposited on the substrate (phytobenthos, bacterial detritus) which make up most of the contents of the stomach. The distinction between a detritivore and a filter feeder of suspended particles can therefore be subjective if based on a consideration of the particles found in the stomach.

Tammes & Dral (1956) showed that particles retained by bivalves should be less than 10 μm. However, these results cannot be applied to scallops (Beninger 1989) which are opportunist detritivores and do not select particles in the same way as other bivalves.

Dissolved organic substances (monosaccharides, amino acids) or lipid droplets may be absorbed directly by the epithelium of larvae or adults. Many studies (reviewed by

Amouroux 1982) have demonstrated the existence of active transfer in the mantle and the gills, while certain carbohydrates present in sea water stimulate the rate of pumping and particle retention.

Dissolved organic matter definitely plays an important role in nutrition, possibly as a growth factor, particularly in littoral or estuarine zones. Héral *et al.* (1983) have established a correlation between the production of flesh in cupped oysters in the Marennes-Oléron basin and the concentrations of dissolved organic matter, expressed as carbon or nitrogen.

Finally, the rate of particle retention depends on the density of the particles in sea water; this varies as a function of the current speed, tides and runoff from the land, and results in significant deposition (faeces, pseudo-faeces). Oyster farmers claim that oysters feed during the flood tide. In fact, if the density of particles is very high, the animal may reject a large quantity as pseudo-faeces and the best rate of retention occurs when the density of particles diminishes.

B. Estimation of feeding rate

It is possible to assess quantitatively the rate at which water is pumped and purified by bivalves; this gives an estimate of the rate at which particles are totally or partially removed from suspension. If particles are homogeneous, less than 10 μm and do not exceed a certain concentration, the water pumped is completely purified.

a. Methods of measurement (reviewed in Blanchard 1989).

Direct methods for measuring the quantity of water pumped consist of estimating the volume exhaled (ostium-syphon) either in isolation (Lubet & Chappuis 1966) or by physical methods such as a temperature-sensitive probe (Amouroux 1982). The probe, heated by an electrical current, is placed near to the syphon; the cooling depends on the flux of water through the exhalant syphon and can be estimated mathematically.

Indirect methods allow an estimation of the rate of purification of water by the bivalves (= rate of retention, E) by determining, in relation to time, the reduction in the concentration of particles in filtered water.

A simple calculation, similar to that proposed by Willemsen (1972), allows the calculation of E. If V is the volume of water in a tank, C the concentration in numbers of particles at time t, the quantity of material retained between t and $t + dt$ is $V.dC = -E.C.dt$. If at time t_1 the concentration is C_1 and at t_2 it is C_2, E can be calculated as:

$$E = V \times \frac{1}{t_2 - t_1} \times \log \frac{C_1}{C_2} = V \times \frac{1}{t_1 - t_2} \times \log \frac{C_2}{C_1}$$

Measurements of concentration are usually made by photometry; the optical densities D_1 and D_2 are proportional to C_1 and C_2.

$$X = V \times \frac{1}{t_1 - t_2} \times \log \frac{D_2}{D_1}$$

The concentration is estimated either by nephelometry or by fluorometry. Particle counters are also used; these allow a more precise estimation of the quantity of water purified and the volume of material retained. Winter (1969, 1974) used a chemostat which recorded the quantity of unicellular algae retained by bivalves and the quantity retained in the rearing tank continuously for several months (Fig. 9).

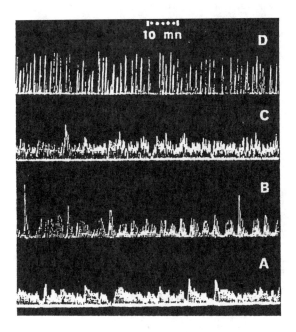

Fig. 8. Patterns of filtration rhythm of *Mytilus edulis* by the direct method (after Chappuis & Lubet 1966). A–B–C: Animals showing pseudo-rhythms of filtration. D: An animal which has been subjected to bilateral ablation of cerebroid ganglia.

b. Results. It is difficult to compare rates of retention obtained by different authors for the same species because of the mixture of different experimental methods and the way in which the results are expressed.

Values obtained from different authors are:

Ostrea edulis (2 years) $1–3 \, l \, h^{-1} \, g^{-1}$ dry weight
Crassostrea gigas (2 years) $2–6 \, l \, h^{-1} \, g^{-1}$ dry weight
Mytilus edulis (18 months) $0.3–2 \, l \, h^{-1} \, g$ dry weight

Variations in flow rate and 'rhythms' : Lubet & Chappuis (1966) have shown the existence of short pseudo-rhythms in mussels (Fig. 8) which are independent of external factors. However, filtration can be considered as a continuous process if measured over a sufficiently long period of time (1 h).

b_1. Action of external factors. (1) Temperature acts on the rate of retention which, in mussels, doubles between 5 and 20°C (Lubet & Chappuis 1966) and may even quadruple, as in flat oysters. While a rapid drop in temperature causes a decrease in the rate of filtration, there is a progressive adaptation to low temperatures and an increase in the

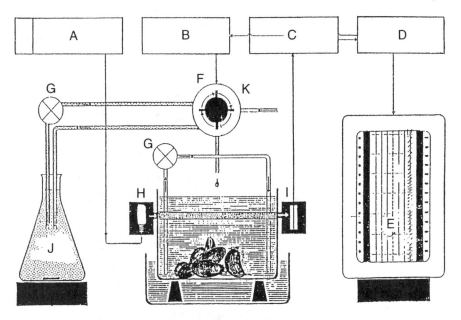

Fig. 9. Chemostat (after Winter 1974). A = voltage regulator, B = electrofilter activating relay, C = amplifier and 3-way switch, D = alternative connector, E = recorder; each point corresponds to an addition of algae, F = electrofilter, G = pump, H = light source, I = photoelectric cell, J = algal culture, K = compressed air.

volume filtered. Similar observations have been made concerning the effects of changes in salinity. In mussels, Cole & Hepper (1954) and Lubet & Chappuis (1966) have shown that low salinities (7–13‰) or high salinities (38–40‰) cause filtration to slow down and eventually cease. However, the mussels are able to adapt to low salinities (Kruger 1960); this explains why there are populations of mussels in estuaries and low-salinity seas such as the Baltic and the North Sea.

(2) While the size of particles plays an important role (Fig. 10a) (optimum capture efficiency between 6 and 10 μm), their concentration also has an influence on the rate of retention. Winter (1974) showed that below a certain level (C), which varies according to the nature of the particle, filtration activity is greatly reduced. However, the bivalve can progressively increase this activity, but this is accompanied by a significant loss of weight because the energy used in filtration exceeds that supplied by the food.

For a series of concentrations between C_1 and a new level C_2 ($C_1 > C_2$), the filtration activity is stimulated, whatever the nature of the particles, but the quantity retained is almost independent of the concentration (Fig. 10b).

When the value C_2 is exceeded, filtration slows down and eventually stops. This point is marked by the appearance of pseudo-faeces. The determination of C_1 and C_2 is very important for rearing broodstock in a hatchery (Table 3).

The production of pseudo-faeces in the natural environment is a function of the density of particles in the water. It varies over the course of a day in relation to tidal currents,

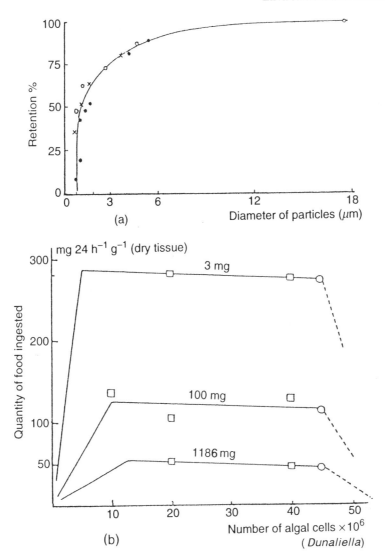

Fig. 10 (a) Retention of particles as a function of their diameter in the mussel (*Mytilus edulis*) (after Silvester & Sleigh 1984). Values obtained with different particles by Jorgensen (1975) and Mohlenberg & Risgard (1978). (b) Relation between the quantity of food ingested and algal density (3, 100, 1186 mg representing the live weight of the mussels). The appearance of pseudo-faeces is shown by the circles (after Winter 1976).

although faeces are produced continuously. According to Héral *et al.* (1983), this biodeposition is the best indication of seston present in sea water. It has been estimated to be 47.5% of the energy available to the animal.

b₂. Internal factors. The rate of retention varies with age; it is greater for young animals than for older ones (Lubet & Chappuis 1966; Wahl, 1973). This is a consequence of the

Table 3. Effects of algal concentration and algal size on filtration experimental temperature, around 20°C (after Winter 1969)

Species	Food	Size in μm	Appearance of pseudo-faeces $\times 10^6$ cells/l.	Critical cell density $\times 10^6$ cells/l.
Crassostrea virginica	*Chlorella* sp.	5	450	2000
Crassostrea virginica	*Chlorella* sp.	5	450	–
Modiolus modiolus	*Dunaliella* sp.	7.5×5	60	–
Arctica islandica	*Dunaliella* sp.	7.5×5	60	–
Mytilus edulis	*Platymonas suecica*	7.2×9.4	35–40	70–80
Modiolus modiolus	*Chlamydomonas* sp.	10×7.5	50	–
Mya arenaria	*Dicrateria* sp.		50	–
Mytilus edulis	*Phaeodactylum tricornutum*	$3.2–4.2 \times 19.2–27.2$	30–40	–
Crassostrea virginica	*Nitzschia closterium*	40–50	20–30	70
Crassostrea virginica	*Euglena viridis*	60	2	3–5

decrease in metabolic rate as the animal gets older (Kruger 1960; Thompson & Bayne 1972, 1974; Wahl 1972) (Fig. 11).

Finally, it must not be forgotten that the rate of filtration is a physiological function which constantly adapts to the requirements of the organism; it depends on a number of organs such as the adductor muscles of the gills, the mantle, the ciliary tract, etc.

The nervous system (Lubet & Chappuis 1966) plays a fundamental part in the regulation of filtration; the cerebral ganglia increase filtration rhythms in an irregular manner (Fig. 8); visceral ganglia decrease the rhythm. Serotonin, which is particularly abundant in the visceral ganglia, plays an important role in the coordination of the movements of the ciliary tract.

C. Estimation of ration size

Many feeding regimes using cultured phytoplankton have been tested but their use in hatcheries is often empirical. Mixtures of a variety of species of algae give the best results. Quantitative aspects have been studied by Winter (1969) and Winter and Langton (1976) which have given the following information:

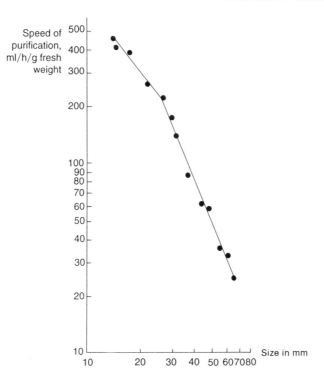

Fig. 11. Variations in the speed of purification of *Mytilus galloprovincialis* as a function of size
(after Lubet & Chappuis 1966).

(a) The increase in dry weight of the flesh of bivalve molluscs is directly proportional
to the weight of algae supplied. This has been demonstrated for the mussel fed a mixture
of *Isochrysis* and *Dunaliella* as a maintenance ration (1 g dry weight of algae corresponds
to 4% dry weight of flesh) and for maximum growth (3 g algae corresponds to 15%). The
best results with the food supplied show a high coefficient of digestibility (Fig. 12). The
optimum ration decreases with the weight of the animal.

(b) *Gross growth efficiency* (K1) (Fig. 12). This is an expression of the relationship
between the increase in dry weight of flesh of the bivalve for the duration of the experiment and the ration ingested over the same period. This could also be expressed in terms
of energy. The optimal ration will be obtained when the maximum value of K1 is
achieved. If the ration is increased beyond this point K1 will fall because the concentration of particles becomes too high, filtration decreases and the animal eliminates pseudo-
faeces. If K1 = 0, there is no growth; this is the maintenance ration and covers only the
basic essentials of routine metabolism (Fig. 12b). If K1 is less than 0, the ration is
insufficient. The maintenance ration varies with the age of the individual. In mussels it is
5.56% of the dry weight of flesh for an animal of 9 mg and only 2.23% when it reaches
35 mg.

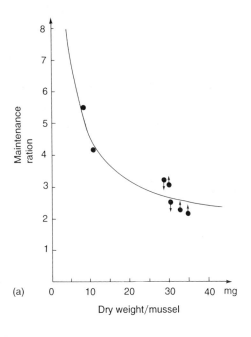

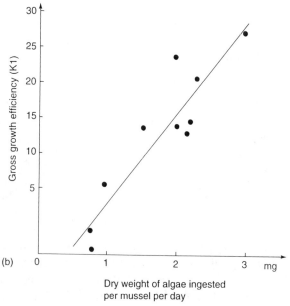

Fig. 12. (a) Variation in the maintenance ration as a function of age (*Mytilus edulis*). (b) Gross growth efficiency (*Mytilus edulis*) (after Winter 1976).

2.2 Gastropods

(Abalone: *Haliotis tuberculata* L.)

Although several species of abalone (ormer) have been raised from the fertilized egg, the current state of knowledge on the organization of the digestive tract and the mode of digestion remains fragmentary.

2.2.1 Morphogenesis and organization of the digestive organs

A. Morphogenesis of the larval abalone

This is well understood (Fig. 13). The larva, which is lecithotrophic and pelagic, metamorphoses rapidly. The juvenile feeds on the algal film which covers the substrate to which it becomes attached. The morphogenesis of the digestive tract is no different from that of bivalves, apart from an invagination of ectodermal origin in the buccal region giving rise to a blind sac in which the radula forms.

B. Anatomy of the digestive tract (Fig. 14)

This has been clearly described by Crofts (1929), *Haliotis tuberculata*; Campbell (1965), *H. cracherodii*; MacLean (1970), *H. rufescens*.

The general organization of the digestive tract is similar to that of bivalves. The two salivary glands open into the buccal cavity which contains the fan-shaped radula (Rhipidoglossa) (Fig. 14). The oesophagus has two lateral appendages in the proximal region.

After a short, narrow initial length, the oesophagus enlarges for most of its length and forms a zone (the 'crop') where food accumulates. The walls of the crop are thin walled and elastic.

The stomach is relatively small, bent back on itself and equipped with a spiral caecum. The bilobed digestive gland is linked to the stomach cavity through several channels. The intestine has a recurrent branch (the anterior), followed by a descending branch (posterior) which opens into the rectum. The anus opens into the pallial cavity.

C. Cytology

Although there is little known on the cytology of abalone (MacLean 1970), there appear to be no major differences from that of bivalves or other archeo-gastropods. The digestive cavities are covered by epithelium which is made up of ciliated and secretory cells, among which are numerous mucocytes. The flexible walls of the crop are very thin with a poorly developed musculature and connective tissue envelope.

The stomach wall has well-developed cilia but also a secretary zone, the gastric shield. Inside the stomach cavity is a cylindrical, elongated, mucus-rich structure, the protostylet, resembling the crystalline style. This is formed from mucus filaments which aggregate particles of food and particulate material from the digestive gland. This organ has the same organization and ultrastructure as that described for bivalves 'secretory' and 'digestive' cells are found in the acinar tubules.

The region at the junction of the stomach and the intestine is specialized for grading particles and is made up of a zone of alternating parallel grooves and crests (Fig. 15).

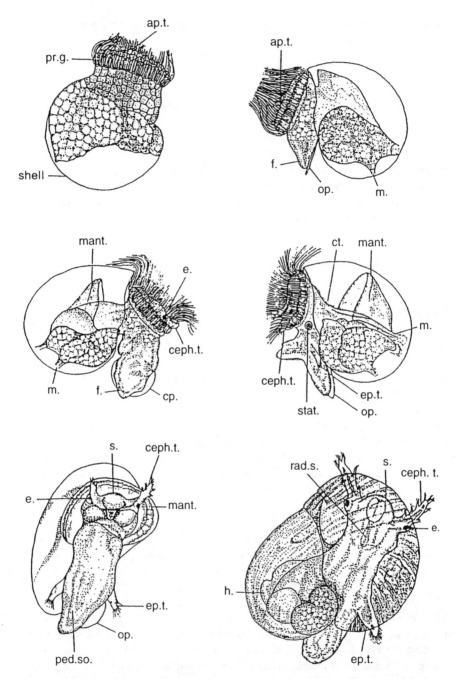

Fig. 13. Larval development of *Haliotis discus-hannai* (after Koike 1978). ap.t = apical tuft; ceph.t = cephalic tentacle; e_1 = ciliary lobe; e = eye; ep.t = epipodial tentacle; f = foot; h = heart; L = digestive gland; L.sh = larval shell; m = muscle; mant - mantle; op = operculum; ped.so = foot; rad.s = radula sac; s = snout.

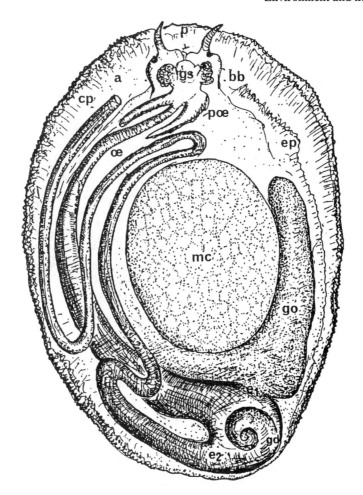

Fig. 14. Morphology of the digestive tract of *Haliotis tuberculata* (after Crofts 1929). a = anus, b
= mouth, bb = buccal bulb, cp = palleal cavity, ep = epipodium, e_1–e_2 = stomach, go = gonad, gs
= saliva gland, gr = radular gland, gd = digestive gland, m = muscle, oe = oesophagus, poe =
oesophageal pockets.

This leads to an intestinal furrow which is ciliated and has two crests on the ventral
surface (the major and minor typhlosoles). The intestine is enclosed by connective tissue
and many blood lacunae, rich in haemocytes.

2.2.2 Transit through the digestive system, digestion and absorption

A. Algal food
This is attacked by the radula which sticks out from below the buccal cavity and is moved

(a)

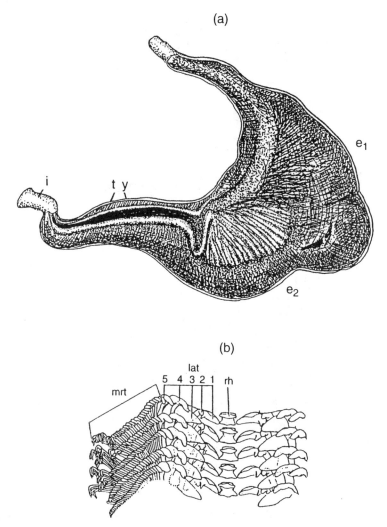

(b)

Fig. 15. (a) *Haliotis tuberculata* stomach. (b) *Haliotis tuberculata* radula (after Fretter & Graham 1962). e = stomach, i = intestine, ty = typhlosoles. b = lateral teeth, mrt = marginal teeth, rh = median tooth.

forwards and backwards by a strong musculature. The algae are scraped off and shredded by the lateral teeth (Crofts 1929).

Saliva is added to the ingested bolus; this accumulates in the extensible region of the oesophagus (the crop) and then enters the stomach progressively. Particles become attached to the protostylet while part of the digestive fluid enters the digestive gland. This process can be detected 45 min after food has been taken in, but reaches its maximum after 6–10 h at 13°C (MacLean 1970). The undigested material in the stomach (residues from the digestive gland, residual fluid) is expelled through the grading zone which acts selectively on wastes which enter into the intestine. The cilia attached to the crests reject the large particles and orientate the small ones into the grooves where they are transported by cilia and enclosed in mucus, forming filamentous bodies. The intestine contains a mass of such filaments, small pieces of undigested algae and digestive fluid. Faeces are relatively solid and emitted almost continuously during the digestive process. The progression of the digestive fluid is probably due to the combined action of the ciliature and peristalsis of the muscular walls. The total transit time depends on temperature and, from our observations, is comparable to that of bivalves (oysters, mussels).

B. Enzyme activity
Enzyme activity remains poorly understood. MacLean (1970) detected the presence of glucanases throughout the digestive system; these have an amylase activity. They are present in high concentrations in the crop area and the digestive gland (maximum activity at pH 7.75). The presence of glucose has been demonstrated in the crop fluid and in the haemolymph of the adjacent lacunae.

Proteolytic activity has been found in the salivary gland (maximal activity at pH 7.72), in the contents of the crop, the stomach and the digestive gland. Lipase activity is present in the crop.

C. The process of extracellular digestion
As with bivalves, extracellular and intracellular digestion operate together, the latter being particularly important in the digestive gland. Extracellular digestion begins in the crop and continues in the stomach and even in the intestine. MacLean (1970) used markers and radioactive tracers to demonstrate the existence of several absorption zones. The first metabolites liberated (monosaccharides, amino acids, alcohols and fatty acids) may be transferred by the epithelial cells of the crop into the haemolymph, but most of the absorption takes place in the digestive gland while digestion and absorption continue into the intestine.

2.2.3 Diet

A. In the natural environment
In the wild, abalone feed on microscopic algae during the post-larval phase, changing progressively to macroscopic algae (seaweeds) either alive or as detritus. They are active at night, hiding from light and spending the day in a sheltered refuge. At night they leave the refuge to feed, moving over distances not exceeding a few metres. 50% of the population moves between 19.00 and 04.00 in the morning with a maximum of 80% between 23.00 and 24.00 (Cox 1962; Newman 1966; Poore 1972).

The lecithotrophic (i.e. feeding off its yolk) and pelagic larva become fixed in zones which are rich in the red seaweeds (lithothamnia); here they complete metamorphosis. Many juveniles congregate in these nurseries, eventually moving to nearby areas. Morse et al. (1979) showed that the GABA (gamma amino butyric acid) liberated by the lithothamnia favours the implantation of the larvae on the substrate and their metamorphosis.

B. In hatcheries

The above information has been used in the development of the hatchery rearing of abalone. The percentage survival after metamorphosis and synchronization has been greatly improved by the addition of GABA to the water in the tanks containing 4-to-5-hour-old veligers.

Post-larvae are fed on microalgae, either attached to collectors (the Japanese technique) or covering the walls of the rearing vessel (French technique). The transition to feeding on seaweeds takes place after a period where granulated artificial diet is fed; this allows rapid growth. The daily consumption of algae is around 10% of body weight.

Food tables for IFREMER hatcheries have been established by Flassch & Aveline (1984).

2.3 Cephalopods (cuttlefish, *Sepia officinalis* L.)

There have been many studies which have increased our knowledge of the digestive system of cephalopods (reviewed in Boucaud-Camou & Boucher-Rodoni 1983; Boucher-Rodoni *et al.* 1987). In this section we shall review information on the cuttlefish, a species of major commercial importance where there are attempts to manage stocks and which can be reared from the egg.

2.3.1 Morphogenesis and organization of the digestive organs

A. Embryonic phase

The germinal disc takes the form of a crown on top of the mass of glycolipoprotein yolk. During embryogenesis, the yolk divides into two regions, one outside the embryo (Fig. 16) and the other inside, made up of an anterior sac adjacent to the outline of the two lobes of the digestive gland and a smaller posterior sac in the area of the adult gonad. The vitelline reserves are resorbed through the surrounding epithelium through the activity of enzymes. The released metabolites are carried in the haemolymph to all parts of the embryo.

As with all molluscs, the digestive tract develops from endoderm, apart from the buccal cavity which is ectodermal in origin. By the time of hatching the external vitelline sac has disappeared, the internal sacs are resorbed within 20 to 30 days while the digestive gland becomes progressively functional (Boucaud & Yim 1980).

B. Juveniles and adults (Fig. 17)

a. Organization of the digestive tract.   This is very similar to that of other molluscs. In the middle of the circle of tentacles there is a buccal mass which can project out or cause rotational movements during the capture of prey. The mouth, enclosed between two lips, has a hardened beak, in two pieces, which can be moved by the strong musculature which forms the buccal mass (bulb). Three organs are present, from the ventral zone to the dorsal zone of the buccal cavity: the salivary papillae; the entrance for the rough tongue, which closes the orifice of the buccal cavity (Fig. 18) (the canal leading from the posterior salivary glands opens at the extremity); the radula and the lateral lobes, connected to

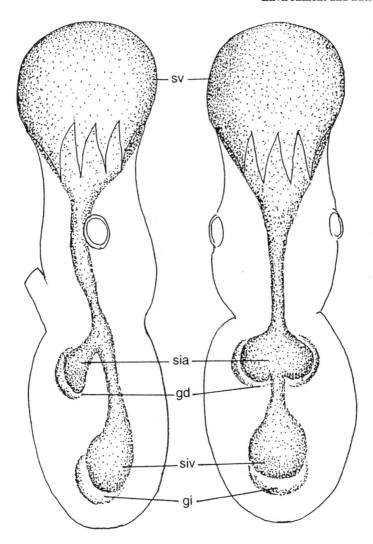

Fig. 16. Cuttlefish embryo. gd = digestive gland, se = external vitelline sac, sia = internal anterior vitelline sac, siv = internal posterior vitelline sac.

the anterior salivary glands. The buccal epithelium encloses glandular areas around the lips and the salivary papillae.

The elongated, rectilinear oesophagus leads to a vast stomach, coupled with a caecum. The digestive gland (liver) of the cuttlefish is made up of two huge lobes which communicate with the caecum through two ducts which have numerous diverticulae (= 'pancreas'). A complex system of sphincters and grooves separates the stomach from the caecum; in the cuttlefish this controls the movement of the fluid exhaled and the fluid inhaled by the digestive gland. The intestine and the rectum are short and rectilinear, the anus is situated in the anterior region of the pallial cavity.

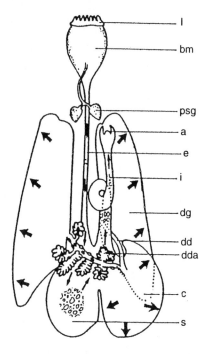

Fig. 17. Morphology of the digestive tract of *Sepia officinalis*. a = anus, bm = buccal bulb, c = caecum, dd = digestive ducts, dda = 'pancreas' or diverticules of the ducts, dg = digestive gland, e = oesophagus, i = intestine, l = lips, psg = posterior salivary glands, s = stomach. Thin arrows represent the track of particles in the digestive tract, thick arrows, absorption sites.

b. Cytology. The oesophagus and the stomach are covered by an epithelium which secretes a strong protective cuticle, while the ducts of the digestive gland, the caecum and the intestine have a ciliated epithelium. There is a well-developed musculature around the digestive tract which allows movement of food by peristalsis.

The digestive gland has the same structure as other molluscs. The tubules contain a single type of cell, in different stages of development, from juveniles with a granulose reticulum to the adult form shown in Fig. 19a. These mature cells which have microvillae and heterophagic vacuoles are characterized by 'boules', large protein inclusions, and by the presence of 'brown bodies', large vacuoles containing crystalline masses (Boucaud-Camou 1973). This author described the ultracellular development of these cells during the course of the digestive cycle; the number and size of 'boules' relates to the stage of digestion; the rhythm of their secretion and their biochemical character suggest that they contain digestive enzymes which are liberated to the exterior when the 'boules' break up.

2.3.2 Intake of food and transit of the digestive system

A. Capture
Details of the predatory behaviour of cuttlefish are well known (reviewed by Messenger

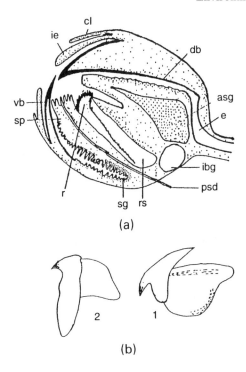

Fig. 18. (a) Section through the buccal bulb of *Sepia officinalis*. (b) Beak of *Sepia officinalis* (after Boucaud-Camou & Boucher-Rodini 1983). 1: dorsal beak; 2: ventral beak. asg = anterior salivary glands, db = dorsal beak, ibg = inferior buccal ganglion, ie = internal lip, e = oesophagus, cl = external hp, psd = posterior salivary gland canal, r = radula, rs = radula sac lip, sg = sub-mandibular gland, sp = salivary papilla, vb = ventral beak.

1977). The prey captured by the retractable arms is held by other tentacles and directed to the mouth. Before being ingested, the prey is paralysed by a cephalotoxin, a glycoprotein manufactured by the posterior salivary glands. The beak breaks up the prey (crustaceans) mechanically, the radula only plays a minor role. A mucus secretion accompanies the food bolus through the oesophagus; the juice released by the posterior salivary glands has a proteolytic function.

The capture and ingestion of prey involves a series of complex processes, requiring the coordination of the mechanical actions of the arms, the buccal mass and secretion from the posterior salivary glands. This is under the control of the cephalopod nerve centres (superior and inferior buccal ganglia, subradular ganglia) (Young 1971). The processes are initiated by visual signals (Best 1981).

B. Transit through the digestive system (Fig. 17)
The food bolus is moved through the oesophagus by peristalsis. The digestive fluid begins to act through the digestive enzymes secreted by the salivary glands; in the stomach they are joined by enzymes from the digestive gland. The fluid enters the caecum and, in Sepia and Octopus, a large proportion is drawn into the digestive gland where extra- and intracellular digestion takes place. The fluid which remains in the

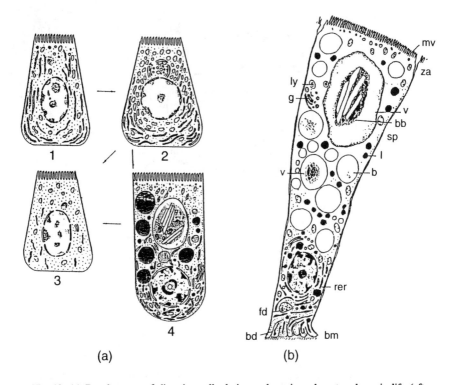

Fig. 19. (a) Development of digestive cells during embryonic and post embryonic life (after Boucaud-Camou & Yim 1980). 1: embryonic cell, 2: cell at the end of the embryo stage, carrying out synthesis, 3: post-embryonic cell, resting, 4: functional, post embryonic cell. (b) Vacuolated cells (after Boucaud-Camou 1973). bb = brown bodies, b = vacuoles, bd = fold of the basal membrane, bm = basement membrane, fd = ferric deposits, g = golgi apparatus, l = lipids, ly = lysosomes, mv = microvill, rer = granular endoplasmic reticulum, sp = tight junction, v = vacuole, za = zomila adherens.

caecum is itself digested and the metabolites are absorbed—a process that is particularly important in squid. The undigested material is evacuated from the stomach to the intestine together with the residues from the digestive gland and the caecum. They are coated by mucus filaments secreted by the caecum and the intestine. These faeces contain small amounts of residue (pieces of carapace) and are evidence of the excellent conversion of prey. The time taken for food to transit the digestive system (reviewed in Boucher-Rodoni *et al.* 1987) varies as a function of temperature, for the squid, 20 h at 15°C and 15 h at 18°C but only 4 h to 6 h at the same temperature for cuttlefish, probably because the digestive fluid does not transit through the 'liver' but remains mainly in the caecum.

2.3.3 Enzyme activity and absorption
Enzyme activity is very low at the time of hatching; it develops as the animal begins to feed and shows no difference from that in adults.

A. Enzyme activity
A very strong proteolytic (chymotrypsin-like) activity has been shown in the digestive

gland ('boules' cellules). This is also found in the posterior salivary glands and the caecum. Lipase and amylase activity can be detected histochemically in the digestive gland and the caecum (Boucaud-Camou 1973).

There is no enzyme activity in the oesophagus and the stomach. Digestion is extracellular and also intracellular in the 'boules' cellules; the brown bodies constitute the remains of this digestive process and are evacuated in the faeces.

B. Absorption sites

The absorption sites shown by the use of markers or radioactive tracers are, in cuttlefish, found mainly in the digestive gland, the diverticulae ('pancreas'). The ducts of this gland can absorb small organic molecules (amino acids) or minerals. While the caecum is also a major site of absorption, the role of the intestine appears to be negligible.

2.3.4 Diet

A. Live prey

Cephalopods are active predators and feed on live prey (molluscs, crustaceans) and even fish (Teuthöids).

Juvenile squid feed mainly on small crustaceans (amphipods, mysids, euphausiids, young decapods). The weight of prey, over 24 h, is equivalent to 40% of the weight of the animal. Hunting for prey is at its most intensive at night, juveniles take several meals over a 24 h period; adults take one meal only but this meal represents 40–60% of the weight of the adult. The weight of prey captured appears to be correlated with the activity of the animal and its metabolic demands; it is lowest in benthic octopuses which do not move very far (20–40% of live weight) while it is 250–100% for pelagic squid which require a large amount of energy for their movements.

B. Coefficients of conversion

Cephalopods grow very rapidly, some Teuthöids can reach large sizes and weights, making them the largest invertebrates. Because of their movements which can be very rapid, they consume a large amount of energy; this requires not only regular intake of food but also efficient use of the prey. Conversion coefficients are particularly high in these animals.

BIBLIOGRAPHY

Bivalves

Amouroux J. M., 1982. Ethologie, filtration, nutrition, bilan énergétique de *Venus verrucosa*. Th. Doc. Sc. Nat., Univ. Paris VI: 1–99.

Ansell A. B., 1962. The functional morphology of the larva and the post larval development of *Venus striatula*. *J. Mar. Biol. Ass. U.K.*, **42**: 419–443.

Bayne B. L., 1970. Some morphological changes that occur at the metamorphosis of larva of *Mytilus edulis*. In: *IVth Europe Mar. Biol. Symp.*, 1971. Cambridge Univ. Press: 259–280.

Beninger P. G., 1989. Structures and mechanisms of feeding in scallops: paradigms and paradoxes—VIth Pectinid workshop—Portland (U.S.A.). *J. Shellfish. Res.* (in press).

Birbeck J. H., McHenery J. G., 1982. Degradation of bacteria by *Mytilus edulis*. *Marine Biology*, **72**: 7–15.

Bernard F. R., 1973. Crystalline style formation and function in the oyster *Crassostrea gigas*. Th. *Ophelia*, **12**: 159–170.

Boucaud-Camou E., Lebesnerais C., Lubet P., Lihrmann A., 1985. Dynamique et enzymologie de la digestion chez l'huître *Crassostrea gigas* (Thbg). *Bases Biologiques de l'Aquaculture*, Colloque IFREMER (1): 75–96.

Chappuis J. G., Lubet P., 1966. Etude du débit palléal et de la filtration de l'eau par une méthode directe chez *Mytilus edulis* et *M. galloprovincialis*. *Bull. Soc. Linnéenne Normandie*, **7**: 210–216.

Cole H. A., Hepper B. T., 1954. The use of neutral red solution for the comparative study of filtration rates by Lamellibranchs. *J. Cons. Int. Exp. Mer. Copenhagen*, **20**: 117–203.

Collier A., 1953. Effect of dissolved organic substances on oyster. *Fish. Bull. U.S.*, **54**: 167–185.

Conway N., 1987. Occurrence of lysozyme in the common cockle *Cerastoderma edule* and the effect of the tidal cycle on lysozyme activity. *Marine Biology*, **95**: 231–235.

Creek G. A., 1960. The development of *Cardium edule*. *Proc. Zool. Soc., London*, **135**: 243–260.

Faveris R., 1982. Recherches sur les possibilités de réalisation de cultures de phytoplancton à partir d'effluents domestiques et de leur utilisation pour la nourriture de Mollusques lamellibranches. Th. Doct. Sp., Caen: 1–152.

Galtsoff P. S., 1942. Problems of productivity of oysters bottons. *Address Nat. Shell. Fish Ass'nt Philadelphia*: 1–7.

Henry M., 1987. Glande digestive de la palourde (*Ruditapes decussatus* L.). *Vie Marine, H.S.* **9**: 1–439 (Thèse Doc. Sc. Nat. Marseille).

Héral M., Deslous-Paoli J. M., Sornin J. M., 1983. Transferts énergétiques entre l'huître *Crassostrea gigas* et la nourriture potentielle disponible dans un bassin ostréicole. *Océanis*, **9** (3): 169–194.

Héral M., 1990. L'ostréiculture traditionnelle. *Aquaculture*, **1**: 348–398.

Hickmann R. W., Gruffydd L. L. D., 1970. The histology of the larva of *Ostrea edulis* during metamorphosis. In: *IVth Europ. Mar. Biol. Symp. 1971.* Cambridge Univ. Press: 281–294.

Kruger F., 1960. Zur Frage Grössenabhängigkeit des Sauers. *Helgolânder wiss. Meeresunters*, **7**: 125–148.

Lebesnerais C., 1985. Etude expérimental de la digestion chez l'huître japonaise *Crassostrea gigas* (Thbg). Thèses Doct. Sp. Univ., Caen: 1–102.

Lucas A., 1982. Remarques sur les rendements de production chez les Bivalves marins. *Haliotis*, **12**: 47–60.

Lubet P. E., 1978. Nutrition des Lamellibranches. *Océanis*, **4** (1): 23–54.

Lubet P. E., Chappuis J. G., 1966. Etude de la filtration de l'eau chez *Mytilus galloprovincialis*; influence de la taille et de la salinité. *C. R. Soc. Biol. Paris*, **158** (11): 2125–2128.

McHenery J. G., Birbeck J. H., Allen J. A., 1979. The occurrence of lysozyme in marine bivalves. *Comp. Biochem. Physiol.*, **63** (B): 25–28.

Masson M., 1975. Etude expérimentale de la croissance et de la nutrition de la larve de *Mytilus galloprovincialis* (LMK) Mollusque pélécypode. Thèse Doct. 3e cycle, Caen, 1975: 1–125.

Masson M., Herlin P., 1976. Grande digestive de la moule (*Mytilus edulis* L.). Organogenèse et structure. *Bull. Soc. Zool. France*, **101** (5): 889.

Mathers N. F., 1972. The tracing of natural algae food labelled with a ^{14}C isotope through the digestive tract of *Ostrea edulis*. *Proc. Malacol. Soc. London*, **40**: 155–124.

Morton B. S., 1977. The tidal rythm of feeding and digestion in the Pacific oyster *Crassostrea gigas* (Thbg). *J. Exp. Mar. Biol. Ecol.*, **26**: 135–151.

Morton B. S., 1983. Feeding and digestion in Bivalvia. In: *The Mollusca, Vol. 5: Physiology (2)*, A. S. N. Saleuddin & K. M. Wilburg (eds). Academic Press: 65–147.

Owen G., 1955. Observations on the stomach and the digestive diverticule of the Lamellibranchia. Part I: The anisomyaria and Eulamellibranchia. *Q. J. Microsc. Sci.*, **96**: 517–537.

Owen G., 1978. Classification and the bivalve gills. *Phil. Trans. R. Soc. London (B)*, **284**: 377–385.

Palmer R. E., 1979. An histological and histochemical study of digestion in the bivalve *Artica islandica* (L.). *Biol. Bull.*, **156**: 115–129.

Prieur D., 1981. Experimental studies of trophic relationship between marine bacteria and bivalve molluscs. (Proc. 15th. Eur. Mar. Biol. Symp.), *Kieler Meeresforsch.*, **5**: 376–383.

Raimbault R., 1966. Plancton et coquillages. In: *Fléments de planctonologie appliquée. Rev. Trav. I.S.T.P.M.*, Paris.

Sastry A. N., 1965 The development and external morphology of pelagic larval and post-larval stages of the bay scallop *Aequipecten irradiens concentricus* Say, reared in the laboratory. *Bull. Mar. Sci.*, **15** (2): 417–435.

Shumway S. F., Selvin R. Schick D. F., 1987. Food resources related to habitat in the scallop *Placopecten magellanicus* (GmL), a quantitative study. *J. Shellfish. Res.*, **62**: 89–95.

Stefano G. B., Aiello E., 1975. Histo fluorescent localization of 5 H.T. and dopamine in the nervous system and gill of *Mytilus edulis*. *Biol. Bull.*, **148** (1): 141–156.

Tammes P. M. L., Dral A. D. G., 1956. Observations on the straining of suspensions by mussels. *Arch. Neerl. Zool*, **11**: 87–112.

Thompson R. J., Bayne B. L., 1972. Active relationships between growth, metabolism and food in the mussel *Mytilus edulis. Mar. Biol.*, **27**: 317–326.

Thompson R. J., Bayne B. L., 1974. Effects of starvation on the structure and the function in the digestive gland of the mussel (*Mytilus edulis* L.). *J. Mar. Biol. Ass. U.K.*, **54** (3): 699–712.

Wahl O., 1972. Particle retention and relation between water transport and oxygen uptake in *Chlamys opercularis* L. (Bivalvia). *Ophélia*, **10**: 67–74.

Wahl O., 1973. Efficiency of particle retention in *Mytilus edulis* L. *Ophélia*, **10**: 17–25.

Willemsen J., 1972. Quantities of water pumped by mussels (*Mytilus edulis*) and cockles (*cardium edule*). *Arch. Neerl. Zool.*, **10**: 152–160.

Widdows J., Bayne B. L., 1971. Temperature acclimatation of *Mytilus edulis* with reference to its energetic budget. *J. Mar. Biol. Ass. U.K.*, **51**: 827–834.

Winter J. E., 1969. Uber den Einfluss der Nahrungskonzentration und anderer Faktoren auf Filtreirleistung und Nahrung sausnutzung der Mucjeln *Artica slandica* L. und *Modiolus modiolus. Mar. Biol.*, **4**: 87–135.

Winter J. E., 1974. Growth in *Mytilus edulis* using different types of food. *Ber. dt. wiss. Kom. Meeresforch.*, **23**: 360–375.

Winter J. E., Langton R. W., 1976. Feeding experiments with *Mytilus edulis* L. at small Laboratory scale. I. Influence of the total amount of food ingested and food concentration on growth. *Ibid.*, **1**: 565–581.

Yonge C. M., 1923. Studies on the comparative physiology of digestion. I—The mechanism of feeding, digestion and assimilation in the lamellibranch. *Mya. J. Exp. Biol.*, **I**: 15–63.

Yonge C. M., 1926. The digestive diverticule in the lamellibranchs. *Trans. R. Soc. Edinburgh*, **54**: 703–718.

Zobell C. E., Felthman C. B., 1938. Bacteria as food for certain marine invertebrates. *J. Mar. Res.*, **I**: 312–327.

Gastropods (abalone)

Akashige S., Seki T., Kan-no H., Nomura T., 1981. Effects of aminoburic acid and certain neuro transmitters on the settlement and metamorphosis of the larvae of *Haliotis discus hannai* IND (Gastropoda). *Bul. Tohoku Reg. Fish. Res. Lab.*, **43**: 37–45.

Campbell J. L., 1965. The structure and function of the alimentary canal of the black abalone, *Haliotis cracherodii* leach. *Trans. Amer. Micr. Soc.*, **84**: 376–395.

Cox K. W., 1962. California abalone, family Haliotidae. *Fish. Bull.*, **118**: 1–113.

Crofts D., 1929. *Haliotis. Liverpool Marine Biology Committee Memoirs*. Univ. Press of Liverpool, **29**: 1–174.

Flassch J. P., Aveline C., 1984. Production de jeunes ormeaux à la station expérimentale d'Argenton. C.N.E.X.O., Rapp. Sc. & Tech., **50**: 1–68.

Fretter V., Graham A., 1962. British prosobranchs Molluscs—their functional anatomy and ecology. *The Ray Society London*, **114**: 1–755.

MacLean N., 1970. Digestion in *Haliotis rufescens* (*Gastropoda prosobranchia*). *J. Exp. Zool.*, **173**: 303–318.

Koike Y., 1978. Biological and ecological studies on the propagation of the ormer, *Haliotis tuberculata*, Linnaeus. I. Larval development and growth of juveniles. *Bull. Soc. Franco-Jap. Oceano.*, **16** (3): 124–136.

Morse, D. E., Hodker, N., Duncan H., Jensen L., 1979. Amino butiric acid, a neurotransmitter, induces planktonic abalone larvaes to settle and begin metamorphosis. *Science*, **204**: 407–410.

Morse D. E., Tegner M., Duncan H., Hodker N., Trevel G., Cameron A., 1980. Induction of settling and metamorphosis of planktonic molluscan (*Haliotis*) larvae: signaling by metabolites of intact algae is dependent on contact. *Chemical*: 67–86.

Newman G. G., 1966. Growth of the South Africa abalone *Haliotis midae. Div. Sea. Fish. Union South Africa*, Tech. Rept. 67: 124.

Poore G. C. B., 1972. Ecology of New Zealand abalones *Haliotis* species. *N.Z. J. Mar. Freshwater Res.*, **6** (4): 534–559.

Seki T., Kan-no H., 1981. Observations of the settlement and metamorphosis of the Veliger of the Japanese abalone. *Haliotis discus hannai, Ino, Haliotidae, Gastropoda. Bull. Tomoku Reg. Fish. Res. Lab.*, **42**: 31–39.

Seki T., Kan-no, H., 1981. Induced settlement of the Japanese abalone: *Haliotis discus hannai*, veliger by the mucous trails of the juvenile and adult abalones. *Bull. Tohoku Reg. Fish. Res. Lab.*, **43**: 29–36.

Uki N., 1981. Feeding behaviour of experimental populations of the abalone *Haliotis discus hannai. Bull. Tohoku Reg. Fish Res. Lab.*, **43**: 53–58.

Cephalopods

Best, E. H. M., 1981. Aspects of the digestive system and its control in *Octopus vulgaris*. Ph.D. Thesis, Cambridge.

Boismery J., 1988. Structure et développement des glandes annexes de l'appareil génital de la seiche femelle *Sepia officinalis* L. *Bull. Soc. Zool. France*, **113** (3): 321.

Boletzky S. V., 1983. *Sepia officinalis. Cephalopod life cycles*, Vol. 1. Academic Press, London: 31–52.

Boucaud-Camou E., 1973. Etude de l'appareil digestif de la seiche—*Sepia officinalis* L. Essai d'analyse expérimentale des phénomènes digestifs. Th. Doct. Sc. Nat., Univ. Caen.

Boucaud-Camou E., Yim M., 1980. Fine structure and function of the digestive cell of *Sepia officinalis* (*Mollusca cephalopoda*). *J. Zool.*, **179**: 261–271.

Boucaud-Camou E., Boucher-Rodoni R., 1983. Feeding and digestion in Cephalopods. *The Mollusca*, Vol. 5, *Physiology*, Part 2. Academic Press, London: 149–187.

Boucher-Rodoni R., Boucaud-Camou E., Mangold K., 1987. *Feeding and digestion—Cephalopod life cycles*, Vol. II. Academic Press, London: 85–108.

Messenger J. B., 1977. Prey, capture and learning in the cuttlefish. *Sepia. Symp. Zool. Soc. London*, **38**: 347–376.

Young J. Z., 1971. The anatomy of nervous sytem of *Octopus vulgaris*. Oxford Univ. Press (Clarendon), Oxford.

2

Growth and reserves

During the progress of sexual maturation a highly variable competition for energy develops between the somatic processes (growth, maintenance, metabolism) and the reproductive processes of the mollusc. When food supplies fail to meet all demands the animal must call on reserves and make adjustments to the distribution of energy between the different processes. The situation may become critical when growth rate is increasing at the same time as the gonads are developing. It may result in a change in reproductive strategy which will affect fecundity and recruitment. This chapter will focus on studies of somatic growth to give an understanding of cultured or wild mollusc stocks. Emphasis will be placed on new methods which show processes at the tissue or cellular level and open up the way for the selection of broodstock or biotechnological methods.

1. GROWTH

There are two types of technique for the study of growth:

1. *'Retrospective methods'*. These allow a determination of the growth made by individuals in a population in relation to elapsed time, i.e. their age.
2. *'Prospective methods'*. These are based on the measurement of instantaneous growth at the level of the whole organism or tissues or cells.

1.1 Retrospective methods

These are based on the measurement of linear growth or growth in weight over a period of time or of making precise measurements of selected anatomical features such as the shell, beak (of cuttlefish), ridges on the shell or the ligament in bivalves, and the operculum in gastropods.

1.1.1 Methods

The methods used most frequently are described below.

A. Overall measurements such as length and weight

When there are no anatomical features which can be used for the determination of age (e.g. fish scales), the age structure of the population must be estimated indirectly by studying the statistical distribution of a measurable character (e.g. length) from a sample.

If we assume that the population has a constant recruitment and a regular mortality pattern throughout the year and that the size of the individuals in a year class is distributed normally, the polymodal distribution obtained means that the population can be broken down into K age classes.

The component age classes making up the overall distribution can be separated graphically by identifying their modes using methods of varying degrees of complexity:

— the Petersen graphic method where each mode is assumed to be the mid-point of an age class (Fig. 20a);
— the successive maximum method of Le Guen using the lowest half of the distribution of the last age class, eliminating each successive age class in the sample (Fig. 20b);
— the logarithmic difference method (Bhattacharya) based on a transformation leading to a linearization of the difference between the logarithms of adjacent frequencies as a function of the length ($\log y_i + 1 - \log y_i / x_i$);
— the Harding & Cassie method where each normal distribution is linearized by probit transformation (according to Henry's Law);
— the Hasselblad best fit method established by computer.

Whenever possible, in order to check results obtained using these methods, it is a good idea to carry out research on animals which have been grown in the same environmental conditions. These animals should be from the same age class and be the same size at the start of the experiment. It is therefore possible to measure the monthly and even daily growth rate. It is also possible to mark sedentary or migratory molluscs (abalone, cuttlefish).

B. Mathematical models of growth

Growth in length or weight cannot be reduced to a simple equation; because of this, theoretical curves best approximating to experimental data are used.

The Von Bertalanffy model (1938): reviewed by Beverton & Holt (1957) (Fig. 21). This is the most frequently used model for the study of growth of various species of fish and molluscs. Von Bertalanffy considered that the growth in weight of an animal is the result of constant antagonism between two physiological processes:

— the coefficient of catabolism, *D*, which is estimated in proportion to the weight, *W*, of the animal, *S*;
— the coefficient of anabolism, *H*, which is estimated from the absorptive surfaces of the animal.

The integration of these metabolic parameters leads to the equation:

$$dW/dt = HS - DW \qquad (1)$$

where dW/dt represents the rate of variation of weight as a function of time. If growth is isometric, i.e. the animal retains the same overall shape, the surface area is proportional

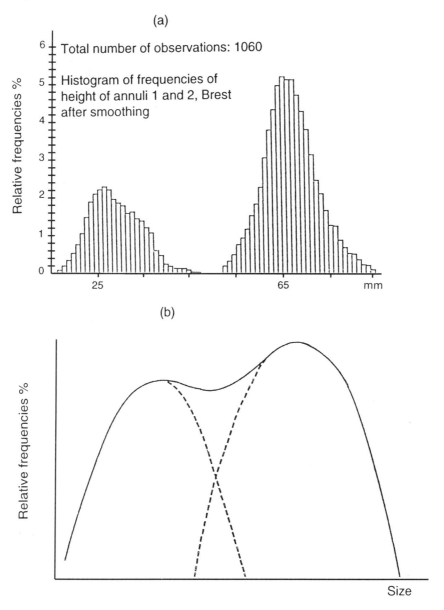

Fig. 20. (a) Histogram of frequencies of the heights of the annuli on the flat shell or the scallop
(Rade de Brest after Antoine 1979). (b) Successive maxima.

to the square of the length, $S = pL^2$ and the weight is proportional to the cube of the
length ($W = qL^3$), p and q being constants.

A relationship is obtained which depends on the length of the animal; this can be
expressed in the form:

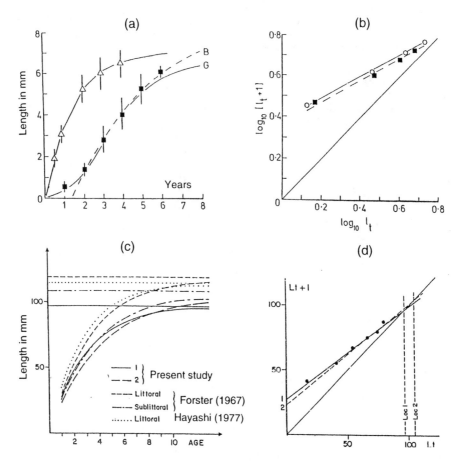

Fig. 21. (a) Growth of *Mytilus edulis*; B = Von Bertalanffy model, G = Gompertz model; Population from Calvados rocks (Lubet); Population Lynker, UK (after Bayne & Worral 1980). (b) Growth of *Mytilus edulis*—Ford-Walford plot. Populations from Lynker & Cattewater UK (after Bayne & Worral 1980). (c) Growth of abalone *Haliotis tuberculata* from the rade de Brest (after Cochard 1982). Von Bertalanffy model, influence of zonation. (d) Growth of abalone *Haliotis tuberculata* (Ford–Walford diagram) (after Cochard 1982).

$$L_t = L.\left[1 - e^{-k(t-t_0)}\right] \tag{2}$$

where L_t is the length at age t, L is the maximum length to which the animal can grow; k is a special constant which describes the growth rate; and t_0 is the age at which L is zero (this has no biological reality).

For practical purposes, the use of the Von Bertalanffy model depends on the determination the parameters in the above equation. This determination is not always strictly carried out so computers or simple graphic techniques are used. Of these latter, the methods developed by Gulland & Holt and Ford-Walford are easiest to use.

Ford-Walford Diagram (Fig. 21).   Using the time $t + 1$, where the year is taken as the unit of time, equation (2) can be written:

$$L_{t+1} = L(L - L_0) e^{-k(t+1)}$$

and thus :

$$L_{t-1} = L(1 - e^{-k}) + L_t e^{-k} \tag{3}$$

which is the expression of the Ford-Walford plot, with the slope e^{-k}, representing the relationship between the length L_{t+1} and the length, L_t, of animals which are one year younger.

Some experimental observations allow the establishment of this line, and thus the determination of k, L, L_0 and t_0.

The same technique can also be used to determine the parameters of the equation for growth in weight ($W^{1/3}$ replaces L):

$$W_t = \left[W^{1/3} - \left(W^{1/3} - W_0^{1/3} \right) e^{-kt} \right]^3 \tag{4}$$

The Gulland diagram.   This method consists of showing graphically the linear increase per unit of time (annual growth $L_{t-1} - L_t$) as a function of the length L_t.

$$L_{t-1} - L_t = L(1 - e^{-k}) - L_t (1 - e^{-k}) \tag{5}$$

This equation is the expression of a line with the slope $-(1 - e^{-k})$ and enables the parameters to be determined graphically.

Gompertz model.   This is a much older model which is now seldom used. The equation is:

$$\log L_t = \log L \left[1 - e^{-k(t-t_1)} \right]$$

where t_1 is the age at length L_1. To estimate the parameters of this equation one proceeds as for the Von Bertalanffy model, inserting the logarithms of the length.

C. Study of anatomical parts

In practice, size is often assessed by measuring the parts of the body which are simplest to assess: shell length, length of the beak in cuttlefish, position of winter bands indicating slow growth on shells, bivalve ligament or opercula of prosobranch gastropods. Growth is then assessed using the models described above.

The microscopical examination of the shells of intertidal molluscs shows daily growth bands (Fig. 22). These can be studied from cross-sections of the shells or from impressions taken of the shell's external surface with a fine layer of plastic or gelatine. It is clear that temperature is not the only factor which influences daily growth rates and that the bands may be an indication of stress from various causes.

The most accurate method for determining daily growth uses isotopic dating. The profile of the values of the isotope $d^{16}O$ (the ratio of $^{18}O/^{16}O$), in the calcareous part of the shell reveals temperature related changes: the highest values correspond to the coldest periods. This method has been successfully applied to *Placopecten magellinacus* (Tan *et al.* 1988) and *Pecten maximus*.

1.1.2 Results
The overall growth as assessed by the Von Bertalanffy or Gompertz growth models is similar (Fig. 21); this has been applied with success to several species of edible bivalves and allows comparison between the growth of different stocks throughout the range of the species. However, estimating the maximum size attained by a species remains inaccurate. Growth of a species varies greatly in relation to environmental conditions and to latitude (Fig. 21).

The method of studying growth from the position of winter bands leads, for some populations, to results which agree with ageing using isotopes. The scallop *Pecten maximus* in Scotland can be aged in this way. For other populations (e.g. in the western part of the English Channel) the classical method gives an underestimate of the age of the animals in relation to results from isotope dating.

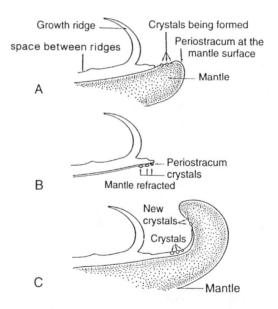

Fig. 22. Formation of daily growth ridges (after Clark 1974).

1.2 Methods for the future
These are based on the determination of growth either at the whole body level, which is difficult, or in tissue or isolated target cells.

1.2.1 Determination of overall growth activity
The quantification of synthesis activity associated with growth is possible through the use of marked precursors which become incorporated in macromolecules which are insoluble in trichloroacetic or perchloric acid.

— Protein synthesis (cellular growth) is measured by using marked amino acids. Leucine is most frequently used; phenylalanine has the advantage of being the most soluble, and is also part of a smaller pool of free amino acids than leucine. Tyrosine is also used to measure the rate of synthesis of certain proteins such as those of the periostracum where this amino acid is particularly abundant.

— RNA synthesis (cell synthesis—cellular growth) and DNA (cell division) is measured by the incorporation of radioactive uridine and thymidine respectively.

— The RNA/DNA relationship is used as an index of growth in fish and many invertebrates. The variations in the concentrations of RNA and DNA allow a very rapid estimate of the changes in the growth rate of the animals.

— Enzyme activity: the measure of the activity of ornithine decarboxylase (ODC) is considered to be an index of growth rate. This enzyme catalyses a stage of the biosynthesis of polyamines (spermidine and spermine) considered to be factors in cellular growth.

Finally, results are expressed in percentage activation according to the formula $X - T/T \times 100$, where X is the incorporation in the presence of extract and T the incorporation without (in cpm).

1.2.2 Determination of growth activity using target cells

As research on whole animals requires radioactive precursors and as these are not distributed homogeneously through the animal because of the presence of vessels and sinuses, studies are carried out *in vitro* either on tissue cultures or, preferably, on target cells after they have been parted from tissues and separated on a Percoll density gradient.

Bioassays for the assessment of neuroendocrine factors affecting growth have been developed (Toullec *et al.* 1988). Results obtained with these techniques are given below.

1.3 Factors affecting growth

1.3.1 External factors

The action of external factors on growth is still poorly understood.

A. Temperature

Temperature acts either directly by speeding up metabolic rate and the rate of synthesis of substances used for growth, or indirectly by affecting the primary production of the natural environment and therefore the food of the molluscs. The relative roles of both is difficult to estimate. The use of a growth hormone allows a study of their action on different targets to be made *in vitro*.

1. *Linear growth*: Studies of daily growth rings show a slowing down of growth in winter (winter rings in scallops). However, Dare & Deith (1989) have shown that this method is not always valid: in Scotland scallops were subject to major, regular annual temperature fluctuations. However, elsewhere, where temperature fluctuations are smaller and irregular (for example when there are abrupt changes in the stratification of water masses off Cornwall), only the method of isotopic dating is valid.

Finally, the excellent ability of molluscs to adapt to temperature leads to a wide geographic distribution.

2. *Growth in weight*:   Temperature has a kinetic action on gametogenesis. There is competition between somatic and germinal growth but both of these forms of growth lead to an increase in weight.

Mytilus edulis-galloprovincialis (Lubet 1983): When temperatures exceed 15–17°C, the reproductive cycle slows down or stops (see Chapter 3, this part). The weight of flesh increases after winter and spring spawning through the build up of tissue in which glycogen reserves accumulate. This situation extends through to the autumn. Linear growth is also excellent during this phase of sexual dormancy while growth in weight is modulated by gonad activity.

Oysters (Mann 1979a): The action of temperature is different. This author raised Japanese oysters and flat oysters over five months at different, but constant, temperatures (12, 15, 18 and 21°C).

At the end of the experiment, the wet weight of *Crassostrea gigas* was shown to have increased significantly with temperature but dry weight remained the same. The quantity of glycogen decreased in all batches as a function of gametogenetic activity which is increased as temperature rises (see Chapter 3).

In *Ostrea edulis*, where sexual activity is less intense (low fecundity), a rise in temperature significantly increases both dry and wet weight while the level of glycogen remains almost unchanged.

Identical experiments with the Japanese clam (Mann 1979b) gave results which were comparable with those from the flat oyster, although in this case a decrease in the level of glycogen accompanied the progress of gametogenesis which is activated by the increase in temperature.

B. Population density

In populations which form large population groups (mussels, oysters, cockles), increase in density brings about a decrease in the growth rate (Lubet 1959) because of the decrease in the supply of nutrients to the individual. The problem of the high density of individuals in a mollusc culture basin is fundamental.

C. Nutrition

This is *the most important factor*, governing the rate of growth in bivalves which are subjected to the same climatic conditions. Animals in the intertidal zone show reduced growth compared with those which are constantly submerged in areas where the primary production is high (Rodhouse *et al.* 1984) (Fig. 23). The study of daily growth lines clearly shows the influence of strong tides on bivalves growing under conditions of intermittent immersion. Thus, growth varies markedly according to the energy available in the intertidal zone. Attempts have been made to quantify the energy available in the environment and optimize the production of bivalves such as oysters in a culture bed.

However, little is known of the qualitative requirements, except for the larvae of bivalves where the addition of a range of vitamins to the algal regime has reduced mortality and accelerated growth.

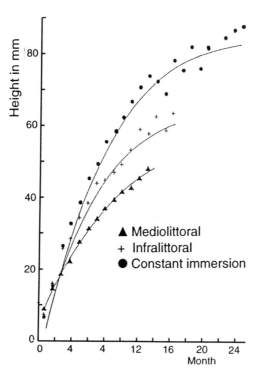

Fig. 23. Growth of *Mytilus edulis* in Killary Harbour (Ireland) (after Rodhouse *et al.* 1984).

1.3.2 Internal factors

A. Endocrine factors

This is a completely new area of study for marine molluscs. Research carried out by Toullec *et al.* (1988) has shown that there is a protein which is immunologically similar to vertebrate growth hormone with a molecular weight of around 22 kDa, but which does not have a direct effect on the protein synthesis of the somatic target cells; the role has still to be determined.

A low molecular weight peptide (1 kDa) which stimulates the synthesis of proteins, DNA and RNA, has been separated by molecular exclusion chromatography and bioassay from extracts of cerebral ganglia or haemolymph serum. This endocrine factor, coming from mussels, has a similar action on targets (cells isolated from the mantle) taken from other species (*Pecten maximus, Crassostrea gigas, Ruditapes philippinatum*) (Fig. 24). These results suggest the existence of a growth factor in bivalves, secreted at the level of the neurosecretory cells of the cerebral ganglia and transported in the haemolymph. This factor is of low molecular weight, a peptide, hydrophillic, non-thermolabile and non-acid soluble.

Purification of this factor will allow specific antibodies to be obtained. These can be used in immunological procedures (RIA-Elisa) to quantify the levels present. This will allow an assessment of the growth potential of individuals and, eventually, the selection of broodstock which have an above-average growth potential.

B. Genetic factors

a. Heterozygosity. The increase in heterozygosity in broodstock appears in hatcheries to favour the subsequent growth of the offspring (reviewed in Allen 1987).

b. Triploidy. The development of techniques for inducing triploidy has allowed the production of sterile animals. This is of considerable benefit to aquaculture as the energy normally converted to reproductive products can be diverted to the production of flesh. This technique is now being developed for application in the culture of molluscs (Table 4).

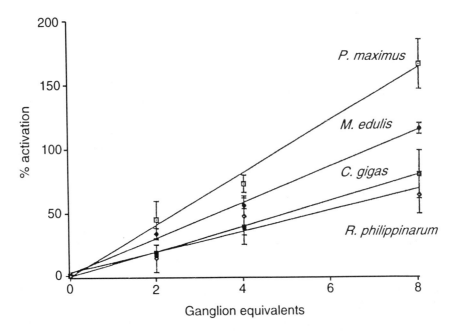

Fig. 24. Incorporation of ³H Leucine in the protein fraction of suspension of cells detached from the edge of the mantle of bivalves. Effects of extracts of cerebral ganglia of *M. edulis* expressed as a percentage of activation in relation to controls. Incubation time: 3 hours (after Toullec 1988).

METHODS

The principal method consists of inhibiting either the first or second meiotic division so that the oocyte remains in the diploid state. When fertilized by a haploid spermatozoa, a triploid egg results. Three methods of altering meiotic divisions have been developed: two physical (pressure and temperature) and one chemical (cytochalasin B).

Table 4. Triploids—Main species, studied

Species	Authors
BIVALVES	
Scallops	
Chlamys varia	Baron *et al.* 1989
Chlamys nobilis	Komaru *et al.* 1988
Pecten maximus	Beaumont 1986
Argopecten irradians	Tabarini 1984
Oysters	
Crassostrea virginica	Stanley *et al.* 1981
Crassostrea gigas	Chaiton & Allen 1985
Pearl oyster	
Pinctada fucata martensii	Wada *et al.* 1988
Clams	Beaumont & Contaris 1988
Tapes semidecussatus	Gosling & Noalan 1989
Mya	
Mya arenaria	Allen *et al.* 1982
GASTROPODS	
Abalone	
Haliotis discus hannai	Katshutoshi *et al.* 1988

Pressure: Chaiton & Allen (1985) subjected the eggs of the Pacific oyster, 10 min after fertilization, to a pressure of 6000–8000 atmospheres for 10 min (57% triploid).

Temperature: Eggs, 10 min after fertilization, were subjected to a temperature of 30–38°C for 10–20 min. The best results (50–60% triploid) were obtained at temperatures of 32°C (*Ruditapes philippinarum*) or 35–38°C (*Crassostrea gigas*) (Quillet & Panelay 1986).

For abalone, a drop in temperature of 6°C for 12–32 min, 15 min after fertilization, appears to be most effective.

Chemical: Cytochalasin B inhibits the formation of the mitotic spindle and therefore cell division. However, this compound is hydrophobic and must therefore be dissolved in DMSO (dimethyl sulphoxide) at a concentration of 1.0 mg of cytochalasin B for 0.1% DMSO. The most effective concentrations tested vary between 0.1 and 1 mg cytochalasin per litre. The eggs are treated 15–20 min after fertilization and then placed in a solution of 0.1% DMSO for 20 min (50–100% triploids). This technique is now in commercial use for the production of triploids in hatcheries.

Cytological control of triploids: It is a good idea to determine which larvae or juveniles are triploids. These can be identified either by the establishment of a karyotype (Arai *et al.* 1986) or by the use of cytophotometric methods which allow an assessment of the amount of DNA in the nucleus (Allen 1983; Allen & Dowing 1986) or by the measurement of the diameter of the nuclei of diploids and triploids (cytomorphometry).

RESULTS

Mortality: During the few days following treatment, the mortality of the triploid larvae is higher than that of diploids although this is only of minor significance when the high fecundity of certain species (*Crassostrea, Pecten*) is taken into account. During the end of the larval stage, metamorphosis and post-larvae (juveniles), the mortality of the two types is identical.

Growth and reproduction: Few results suitable for analysis have been obtained. Comparison of immature and mature individuals (diploid) and triploids of *Crassostrea gigas* has shown a significant increase in the growth of flesh and glycogen stores in triploids. Similar results have been obtained for the scallop *Chlamys nobilis* (Komura & Wada 1989), the bodyweight doubled in the triploids in comparison to the diploids while the gonad decreased by 50%. Triploidization has a considerable effect on spermatogenesis (blocking meiosis) and oogenesis. Gonads develop but remain small in size. The development of germ cells is abnormal and fails to result in the production of viable gametes (Komura & Wada 1989). Finally, the reduction of the gonad in triploid pearl oysters (*Pinctada fucata martensii*) has a considerable benefit to the technology of the production of pearl mussels. The pearl sac is grafted into the gonad; this operation is only possible in diploids outside the period of sexual maturation which extends over much of the year. Grafting is possible throughout the year in triploids (Wada *et al.* 1988).

2. RESERVES

A significant proportion of the energy coming from food is used to cover the metabolic requirements, particularly those associated with reproduction. The reserves are stored either in tissues which are specific to this function or in tissues which have a different physiological function (muscle, digestive gland, haemocytes). Variations in the weight of these tissues can affect the overall weight of the animal; a good physiological equilibrium between the somatic and germinal parts is demonstrated by a study of the condition factor (CF) for the bivalves.

$$CF = \frac{\text{Weight of the flesh (g)} \times 100}{\text{Intervalve volume (ml)}}$$

2.1 Reserve tissues

2.1.1 Cytology
This is described for mussels and oysters of the genus *Ostrea.*

A. Mytilus edulis L. (Fig. 25)

These tissues are present in the mantle and in the visceral mass and are made up of two types of cell in juxtaposition: adipogranulous cells (ADG); vesiculous cells with glycogen (vesicular connective tissue (VCT)) (Lubet 1959; Houtteville 1974; Pipe 1987). The ADG cells (6–20 μm) are spherical and occupy the spaces between the VCT. Glycogen is very abundant and accumulates in a voluminous vacuole.

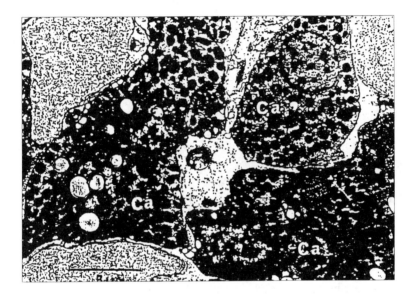

Fig. 25. Reserve tissues (*Mytilus edulis*); (after Lubet *et al.* 1976). ca = adipo-granulous cells,
cv = vesicular cells.

B. Ostrea

In the flat oyster there are only vesiculous cells with glycogen. The study of their infrastructure has shown that there are differences from *Mytilus* cells; there are neutral lipid cells in the cytoplasm and there is also an abundant reserve of glycogen. The vesicular connective tissue forms a bed which may be highly developed, adhering to the mantle by the external surface and enveloping the visceral mass (gonad and digestive gland).

2.1.2 Ultrastructural and biochemical development

A. Mussels

A correlation between the activity of the reserve tissue and that of the gonad was established several years ago. A study of the annual pattern of the glycogen level (Gabbott 1976) showed that glycogen disappeared progressively during the maturation of the gonad and at the same time the reserve tissue regressed in terms of both volume and number of cells (Peek *et al.* 1989) (Fig. 26).

The study of the ultrastructure and the cytochemistry (Houtteville 1974; Pipe 1987) revealed the existence of complex processes of autophagy linked with the action of lysosomes which destroy the ADG. Metabolites are thus released and these products of

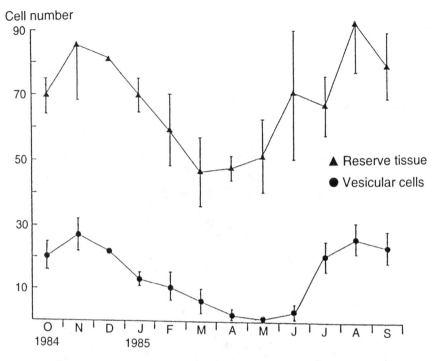

Fig. 26. Seasonal variations in the cells in reserve tissues (*Mytilus edulis*) (after Peek *et al.* 1989).

lysis become distributed around the tubules of the gonad and are absorbed by the young oocytes (pinocytosis). Most of the ADGs are destroyed. The VTCs also lose their glycogen through lyposomal activity; this may also entail a total lysis of cells. The activity of glycogen phosphorylase is intense.

Regeneration of reserve tissue begins at the end of the annual period of reproduction (Houtteville 1974) and is brought about by haemocytes which accumulate between the gonad tubules and then form a loose mesh tissue. The synthesis begins from spring onwards and lasts to the autumn in mussels around the French coast, which corresponds both to the annual cessation of spawning (see Chapter 3, this part) and the period when food is at its most abundant. There is a significant increase in somatic growth following the drop due to the emission of gametes; this appears as a marked increase in the rate of linear growth.

B. Oysters (genus Ostrea)

The reproductive strategy is different from that of mussels. There is no relationship between the glycogen reserves and gonad maturation (reviewed in Walne 1970; Gabbott 1976).

In *Ostrea edulis* gametogenesis and the accumulation of glycogen take place at the same time. The increase in the availability of food in spring meets the energy requirements of the oyster and the increase in temperature at this time accelerates gametogenesis. An identical pattern occurs in *Crassostrea virginica* and *C. gigas*. In all species, workers have found the highest level of glycogen before the resumption of sexual activity, the period where growth in oysters is particularly fast. Glycogen levels drop dramatically after the emission of gametes in summer in species where fecundity is high, but this drop is less marked in the flat oyster which has a lower fecundity.

2.1.3 Factors affecting reserves

A. External factors

Below 20°C, temperature appears to have little effect on the build-up of reserves in *Mytilus edulis* as long as there is enough food available (Widdows & Bayne 1971). Above 20°C an increase in temperature and an insufficient supply of food stresses the animal and reserves are reduced: this explains why the distribution of *Mytilus edulis* is limited at its southern end while *M. galloprovincialis* is more resistant (Lubet *et al.* 1987).

Nutrition plays the principal role in the equilibrium between tissue containing reserves and gonad tissue in mussels. When these bivalves are growing in very favourable conditions (high primary productivity, low temperature, constant immersion) these two 'compartments' develop at an equal rate. Growth is continuous, as is sexual activity, and reserve tissues are always present between the gonad tubules, while there is a slight reduction in the volume at the time when the gonad is at its fullest (Lubet *et al.* 1987).

In contrasting circumstances, fasting entails a breakdown of the ADG while the VCT does not appear to be affected. The breakdown of the glycogen stored in the latter occurs when gametogenesis is reactivated; products liberated by the ADG meet the somatic requirements in cases of stress. According to Gabbott & Bayne (1973), the energy lost during fasting is compensated for in summer by the use of glycogen, in autumn by that of glycogen and lipids and in winter by that of proteins. Finally, the breakdown of glycogen at the same time as gametogenesis allows the synthesis of the lipids which are incorporated in the gametes (triacylglycerides, phospholipids).

B. Internal factors

Organ culture techniques have demonstrated that cerebral ganglia are essential for the maintenance of activity in storage tissue. Isolation and culture of vesicular cells has allowed the study of factors controlling the synthesis and breakdown of glycogen. The incorporation of radioactive precursors, particularly $U^{14}C$ glucose has demonstrated that the major constituents are glycogen and amino acids, very small quantities of ^{14}C being present in lipids and proteins.

The incorporation of glucose in glycogen takes place in a linear fashion and obeys the kinetics of saturation as a function of the concentrations of exogenous glucose, i.e. glycaemia of the haemolymph.

Lenoir (1989), using isolated vesicular cells, was able to demonstrate the existence of two factors which could be separated by *molecular exclusion chromatography*. The first, with a molecular weight of between 20 and 30 kDa has an inhibitory effect on the uptake of glucose and activates the breakdown of glycogen (FADG). The second, which has a low molecular weight (1.5 kDa), appears to have the reverse effect and favours the synthesis of glycogen (FADG) from glucose as well as organic acids and lipids.

2.2 Other reserves

2.2.1 Clams
Medhioub & Lubet (1988) demonstrated two types of reserves in the Japanese clam. The first, formed by vesicle-type cells (VCT), is rich in glycogen and localized in the lumen of the gonad tubules. The cells lyse at the start of the gonad reconstruction phase and the products released contribute to the build-up of new germinal tissue. A major part of the energy requirements for growth and reproduction is liberated through cytolysis of the body wall during gametogenesis. After the annual reproductive cycle, the muscle masses are restored through a 'stock' of muscle cells which have atrophied (are differentiated myocytes) and that of vesicular cells from haemocytes. (Fig. 27).

2.2.2 Scallops
Here there are no vesicular cells containing glycogen. The energy required to sustain the reproductive effort comes from the adductor muscle, the weight of which decreases during the annual spawning cycle through the liberation of fatty acids and amino acids. Glycogen appears to be extremely unimportant (Lubet *et al*. 1987; Faveris & Lubet 1989).

2.2.3 Cephalopods
These animals do not have significant amounts of reserves. Their very high growth rate is sustained only by their feeding, as is the development of the gonad. At the time of spawning cephalopods are no longer feeding and utilize amino acids and fatty acids mainly coming from the muscles of the mantle (O'Dor & Wells 1978).

BIBLIOGRAPHY

Growth
Allen S. K., 1983. Flow cytometry: essaying experimental polypoïd fish and shellfish. *Aquaculture*, **33**: 317–328.
Allen S. K., 1987. Genetic manipulations: critical review of methods and performances for shellfish. In: *Proc. World Symp. on Selection, Hybridisation and Genetic Engineering in Aquaculture*, Vol. II, K. Tiews (ed.). Heinemann, Berlin: 128–143.
Allen S. K., Dowing S. L., 1986. Performance of triploïd Pacific oyster (*Crassostrea gigas*). *J. Exp. Mar. Biol. Ecol.*, **102**: 197–208.
Allen S. K., Gagnon P. S., Hidu H., 1982. Induced triploidy in the soft shell clam: cytogenetic and allozy confirmation. *J. Hered.*, **73**: 421–428.
Allen S. K. Jr., Hidu H., Stanley J. G., 1986. Abnormal gametogenesis and sex ratio in triploïd soft-shell clams (*Mya arenaria*). *Biol. Bull.*, **170**: 198–210.
Antoine L., 1979. La croissance de la coquille St-Jacques (*Pecten maximus* L.) et ses variations en mer Celtique et en Manche. Th. Doct. Sp. Univ. Brest: 1–105.
Arai K., Naito F., Fujino K., 1986. Triploidization of the Pacific abalone with temperature and pressure treatments. *Nippon Suisan Gakkaishi* (*Bull. Jpn. Soc. Sci. Fish*), **52**: 417–422.
Baron J., Ditter A., Bodoy A., 1989. Triploidy induction in the Black Scallop (*Chlamys varia* L.) and its effect on larval growth and survival. *Aquaculture*, **77**: 103–111.
Bayne B. L., 1976. *Marine mussels—their ecology and physiology*. Cambridge University Press: 1–495.
Bayne B. L., Worral C. M., 1980. Growth and production of mussels *Mytilus edulis* from two populations. *Mar. Ecol. Prog. Ser.*, **3**: 317–328.
Beaumont A. R., 1986. Genetic aspects of hatchery rearing of the scallop, *Pecten maximus* (L.). *Aquaculture*, **57**: 99–110.

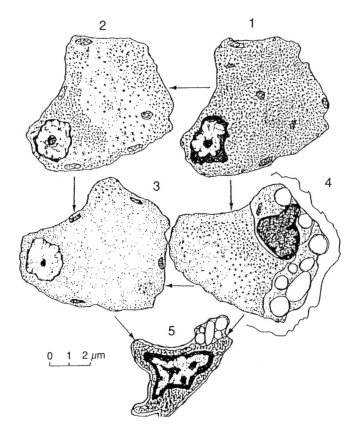

Fig. 27. *Ruditapes philippinarum*: diagram showing the hypothetical development of a muscle cell (after Medhioub & Lubet 1988). 1: Intact myocyte; 2: atrophy of myofibrills; 3: degeneration, with accumulation of glycogen; 4: degeneration with expulsion of hyaline body and residual body; 5: myocyte in breakdown.

Beaumont A. R., Contaris M. H., 1988. Production of triploid embryos of *Tapes semidecussatus* by the use of cytochalasin B. *Aquaculture*, **73**: 37–42.

Bertalanffy, L. Von, 1938. A quantitative theory of organic growth. *Human Biology*, **10** (2): 181–213.

Beverton R. J. H., Holt S. J., 1957. On the dynamics of exploited fish populations. *Fisheries Investigations*, **2** (19): 533 pp, MAAF, London.

Tan F. C., Cai D., Roddick D. L., 1988. Oxygen isotope studies on sea scallops, *Placopecten magellanicus*, from Browns Bank, Nova Scotia. *Can. J. Fish. Aquat. Sci.*, **45**: 1378–1386.

Toullec J. Y., Lenoir F., van Wormhoudt A., Mathieu M., 1988. Non-specific growth factor from the cerebral ganglia of *Mytilus edulis*. *J. Exp. Mar. Biol. Ecol.*, **119**: 11–127.

Wada K. T., Komura A., Uchimura Y., 1988. Triploïd production in the Japanese pearl oyster, *Pinctada fucata martensii*. *Aquaculture*, **76**: 11–19.

Reserves

Bayne B. L., 1976. *Marine mussels: their ecology and physiology.* Cambridge Univ. Press: 1–495.

Faveris R., Lubet P., 1989. Energetic requirements of the reproductive cycle in the scallop *Pecten maximus* L. *J. Shellfish. Res. U.S.A.* (in press).

Gabbott P. A., 1976. Energy metabolism. In: *Marine mussels: ecology and physiology*, B. L. Bayne (ed.). Cambridge Univ. Press: 213–355.

Gabbott P. A., Bayne B. L., 1973. Biological effects of temperature and nutritive stress on *Mytilus edulis*. *J. Mar. Biol. Ass. U.K.*, **53**: 269–286.

Herlin-Houtteville P., Lubet P., 1974. Analyse expérimentale, en culture organotypique de l'action des ganglions cérébro pleuraux et viscéraux sur le manteau de la moule mâle, *Mytilus edulis* L. *C.R. Acad. Sci., Paris, D*, **278**: 2469–2472.

Holland D. L., Spencer B. E., 1973. Biochemical changes in fed and starved oysters *Ostrea edulis* during larval development, metamorphoses and early spat growth. *J. Mar. Biol. Ass. U.K.*, **53**: 287–298.

Houtteville P., 1974. Contribution à l'étude cytologique et expérimentale du cycle annuel du tissu de réserve chez la moule, *Mytilus edulis* L. Th. Doct. Sp., Univ. Caen: 1–125.

Lenoir F., 1989. Mise au point de techniques de purification et de cultures cellulaires chez la moule *Mytilus edulis*. Applications à l'étude des régulations du métabolisme du glucose et du glycogène. Th. Doc. Sp., Univ. Caen: 1–130.

Lenoir F., Robbins I., Mathieu M., Lubet P., Gabbott P. A., 1989. Isolation, characterization and glucose metabolism of glycogen cells (= vesicular connective tissue cells) from the labial palps of the marine mussels. *Marine Biology*, **101**: 495–501.

Lubet P., 1959. Recherches sur le cycle sexuel et l'émission des gamètes chez les *Mytilidés* et les *Pectinidés*. *Rev. Fr. I.S.T.P.M.*, **23**: 396–545.

Lubet P., Herlin P., Mathieu M., Collin F., 1976. Tissu de réserve et cycle sexuel chez les Lamellibranches. *Haliotis*: 59–62.

Chaiton J. A., Allen S. K., Jr., 1985. Early detection of triploidy in the larvae of Pacific oysters, *Crassostrea gigas*, by flow cytometry. *Aquaculture*, **48**: 35–43.

Clark G. R., 1974. Calcification on an unstable substrate: marginal growth in the mollusc *Pecten diegensis*. *Science*, **183**: 968–970.

Cochard J. C., 1982. La croissance de l'Ormeau *Haliotis tuberculata* en rade de Brest. *Haliotis*, **12**: 61–69.

Komura A., Uchimura Y., Leyama H., Wada K. T., 1988. Detection of induced triploid scallop, *Chlamys nobilis*, by DNA microfluorometry with DAPI staining. *Aquaculture*, **69**: 201–209.

Komura A., Wada K., 1989. Gametogenesis and growth of induced triploid scallops *Chlamys nobilis*. *Nippon Suisan Gakkaiski*, **55** (3): 447–452.

Le Gall P., 1970. Etude des moulières Normandes, renouvellement, croissance. *Vie et Milieu*, **21** (B): 545–589.

Lubet P., 1959. Recherches sur le cycle sexuel et l'émission des gamètes chez les *Mytilidés* et les *Pectinidés* (Moll. Lamellibranches). *Rev. Trav. I.S.T.P.M., Paris*, **23** (3): 387–458.

Lubet P., 1983. Experimental studies on the action of temperature on the reproductive activity of the mussel (*Mytilus edulis*). *J. Malacol. Soc. London*, **12**A: 100–705.

Mann R., 1979a. Some biochemical and physiological aspects of growth and gametogenesis in *Crassostrea gigas* and *Ostrea edulis* grown at sustained elevated temperatures. *J. Mar. Biol. Ass. U.K.*, **59**: 95–110.

Mann R., 1979b. The effect of temperature on growth, physiology and gametogenesis in the Manila clam *Tapes philippinarum* (Adams & Reeve, 1858). *J. Exp. Mar. Biol. Ecol.*, **38**: 121–133.

Quillet E., Panelay P. J., 1968. Triploidy induction by thermal shocks in the Japanese oyster, *Crassostrea gigas*. *Aquaculture*, **57**: 271–279.

Rhodas D. C., Lutz R. A., 1980. *Skeletal growth of aquatic organisms*, Plenum Press, New York.

Rodhouse P. G., Roden C. M., Burnell G. M., Hensey M. P., McMahon T., Ottway B., Ryan T. H., 1984. Food resource, gametogenesis and growth of *Mytilus edulis* on the shore and in suspended cultures: Killary Harbour, Ireland. *J. Mar. Biol. Ass. U.K.*, **64**: 513–529.

Stanley J. G., Allen S. K., Jr, Hidu H., 1981. Polyploidy induced in the American oyster, *Crassostrea virginica* with cytochalasin B. *Aquaculture*, **23**: 1–10.

Stanley J. G., Hidu H., Allen S. K., Jr, 1984. Growth of American oysters increased by polyploidy induced by blocking meiosis. I. *Aquaculture*, **37**: 147–155.

Tabarini C. L., 1984. Induced triploidy in the bay scallop, *Argopecten irradians*, and its effect on growth and gametogenesis. *Aquaculture*, **42**: 151–160.

Lubet P., Besnard J. Y., Faveris R., Robbins I., 1987. Physiologie de la reproduction de la coquille St-Jacques *Pecten maximus* L. *Océanis*, **13** (3): 265–290.

Medhioub M., Lubet P., 1988. Recherches cytologiques sur l'environnement cellulaire (tissue de réserve) des gonades de la palourde (*Ruditapes philippinarum*), Moll. Bivalve. *Ann. Sc. Nat. Zoologie, Paris*, **13** (9): 87–102.

O'Dor R. K., Wells J. M., 1978. Reproduction versus somatic growth: Hormonal control in *Octopus vulgaris*. *J. Exp. Biol.*, **77**: 15–31.

Peek K., Gabbott P. A., Runham N. W., 1989. Adipogranular cells form the mantle tissue of *Mytilus edulis* L.: seasonal changes in the distribution of dispersed cells in preformed Percoll density gradient. *J. Exp. Mar. Biol. Ecol.*, **126**: 217–230.

Pipe R. K., 1987. Ultrastructural and cytochemical study on interactions between nutrient storage cells and gametogenesis in the mussel *Mytilus edulis*. *Mar. Biol.*, **96**: 519–528.

Walne P. B., 1970. The seasonal variations of meat and glycogen content of seven populations of oysters (*Ostrea edulis*) and a review of the literature. *Fish. Invert. Minist. Agricul. Fish. Food, London*, II-**26**: 1–35.

Widdows J., Bayne B. L., 1971. Temperature acclimatation of *Mytilus edulis* with reference to its energy budget. *J. Mar. Biol. Ass. U.K.*, **51**: 827–843.

3

Reproduction in molluscs

1. BIVALVES

1.1 Gonad and genital ducts

1.1.1 Gonad

Bivalves possess a single gonad even though it is derived from two groups of mesoderm cells situated in the dorso-pericardial region. The gonad most frequently takes the form of a mass, surrounding the intestines in the visceral region (Ostreids, Venerids). In the mytilid mussels it also invades the mantle. In scallops the gonad is a discrete organ, most of which projects into the visceral cavity. The gonad has an acino-tubular structure; the genital tract is formed by a ciliated epithelium which is an extension of the tubules of the gonad. These latter come together to form two canals in the distal region; genital products are ejected either through two distinct symmetrical pores in a sagittal plane (mytilids) or through the terminal region of the ureters (scallops). The gonad volume may vary considerably from one species to another and within the same species in relation to the activity of the gonad.

1.1.2 Cytology (Fig. 28)

The primordial germ cells (10–12 μm) multiply actively to give spermatogonia and oogonia, small cells (7–10 μm) which are poor in cytoplasm with a reticulate nucleus and two nucleoli. These cells are always found in the terminal zone of the gonad tubules; their multiplication allows the growth of the gonad and its restoration.

Spermatogenesis takes place in the tubules in a centripetal manner. It is very rapid, taking around 10 days in mussels. The tiny spermatozoa (4–5 μm) have a long flagellum (15–20 μm) and a well differentiated acrosome. Their structure is similar in all bivalves. There are accessory cells situated in the walls of the tubules between the groups of spermatogonia: their role remains to be clearly defined (Pipe 1985).

Oogenesis: oogonia enter meiosis (oocyte 1) which is blocked in the prophase stage of the first mitosis (pachytene–diplotene): the lampbrush chromosomes are clearly visible. The nucleolus reappears and provides evidence of intense activity; there is significant RNA synthesis while the cytoplasm increases in volume and numerous ribosomes appear (previtellogenic stage) while the ergastoplasm becomes organized.

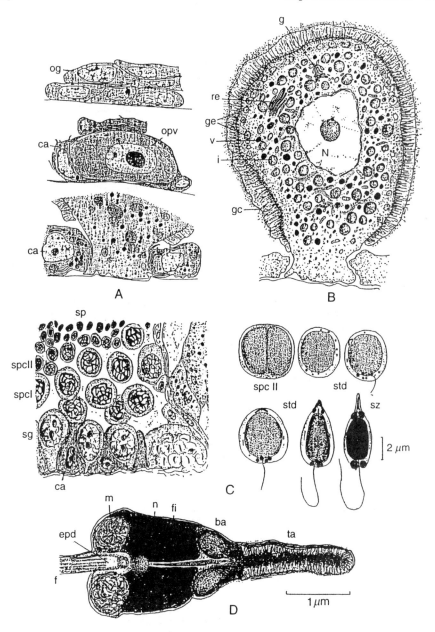

Fig. 28. *Mytilus edulis*. A: oogenesis (after Pipe 1985). B: ripe oocyte (after Albertini 1986). C: spermatogenesis (after Lubet 1959 modified). D: spermatozoid (after Bourcart *et al.* 1985). a = acrosome (of the), ca = accessory cell, cpd = proximal and distal centriole, fi = internal flagella, f = flagella, g = Golgi, gx = glycocalix, h = cortical grains, L = lipid globules, m = mitochondria, n = nucleus, og = oogonia, opv = previtellogenic oocyte, ov = oocyte, sg = spermatogonia, spcl-II = spermatocyte, I–II, std = spermatid, sz = spermazoid, ta = acrosome tube.

Vitellogenesis is characterized by an increase in cell volume (diameter 20–70 μm in mussels). The nucleus or germinal disc can reach 18–20 μm in diameter; chromatin becomes blurred while the nucleolus increases to a significant size (7–8 μm). The nuclear envelope becomes folded, especially towards the end of vitellogenesis.

The cytoplasm contains many mitochondria (with smooth cristae), an abundant ergastoplasm with Golgi apparatus, emitting vesicles at the extremities. The cortical region of the cytoplasm contains vesicles and cortical granules which are very important at the moment of fertilization.

The oocyte membrane folds to form microvilli which have microfilaments at their distal extremity. An abundant glycocalix forms a protective mucopolysaccharide layer.

The accumulation of reserves (yolk) is first marked by the appearance of lipid globules (triacylglycerides) which is related to mitochondrial activity. Almost simultaneously, with increase in volume of the ergastoplasm and dictyosomes, yolk grains of a glycoprotein or glycolipoprotein nature appear. These are stored in the Golgi vesicles. Yolk grains and lipid globules accumulate in the cytoplasm. Carotenes are most frequently linked to the lipids. Small quantities of glycogen appear at the end of vitellogenesis (Albertini 1986; Dorange 1989).

It is important to note that the amount of reserves varies in relation to the type of larval development. In bivalves, teleolithic eggs, such as those in gastropods or cephalopods, are never found. Eggs of oviparous species (50–80 μm diameter) have less yolk than those of viviparous species (80–150 μm) such as *Ostrea edulis*.

Biochemical studies have not revealed any essential differences between the male and female gonad in regard to the level of proteins. There is almost no glycogen at all in the male; and the level is very low in the female (Lubet *et al.* 1986c). There are, however, significant quantitative differences for lipids: the level of triacylglycerols was twice as high in oocytes as in spermatozoa while the level of phospholipids and cholesterol remains similar. Finally, as in the gonads of other marine invertebrates, there is a notable richness of polyunsaturated fatty acids, particularly 20 : 5 w 3 and 22 : 6 w 3 (Besnard 1988).

1.1.3 Atresia and lysis of oocytes

During the reproductive cycle, atresia and lysis of oocytes can be extremely important. These phenomena are particularly marked at the start of gonad activity (lysis of the earliest oocytes formed), after spawning and at the end of the breeding season. In certain species (*Ostrea edulis*, *Pecten maximus*) the lysis of oocytes can affect the whole gonad during certain parts of the year (Lubet *et al.* 1987). Lysozomal enzymes develop in the cortical zone: the oocyte membrane reduces, progressively losing the glycocalyx then the microfilaments. The rupture of the microvilli entails the extrusion of the cytoplasmic contents which invade the lumen of the tubules. The oocytes are expelled through the genital ducts. The product of lysis is homogeneous because of enzyme activity which continues in the lumen of the gonoducts. These ducts possess regions which are adapted for the resorption of lysates; their epithelium forms villi whose cells have microvilli and are the site of pinocytosis. The cells are rich in glycogen (*Pecten maximus*).

Elsewhere, throughout the sexual cycle, haemocytes are able to phagocytose lysed oocyte debris or spermatozoa, effecting a cleansing of the gonad.

1.2 Sexuality and reproductive strategy

1.2.1 Sex

Bivalves have the potential to develop as either males or females. As with crustaceans, there is a tendency for autodifferentiation of oocytes when primary germ cells are cultured *in vitro* (demonstrated in scallops of the pecten group). Differentiation of males depends on internal factors, the origin and nature of which remains to be determined (Allarakh & Lubet 1980).

The final development can be directed either towards the female sex or towards the male sex by the disappearance of the opposite cell line (*Ostrea*, *Venerids*).

Most species are gonochoric (mussels, clams, cockles, solenids) but various forms of hermaphroditism have been shown.

Simultaneous or synchronous hermaphroditism characterizes the situation where the gonad simultaneously produces the two types of germ cells, either in the same tubules or in two distinct zones in the gonad (*Pecten maximus, Chlamys opercularis, C. glaber*). However, the emission of gametes takes place most frequently in an asynchronous fashion which prevents self-fertilization.

Asynchronous hermaphroditism occurs when the male and female germ cells develop at different times. In larviparous oysters (*Ostrea*) several asynchronous cycles can take place in a single annual reproductive cycle: a first spermiation succeeds a spawning or vice versa (rhythmic consecutive hermaphroditism). In oviparous oysters (*Crassostrea* or *Sacostrea*) and scallops (*Chlamys varia*) the animals are generally protandrous at the start of their sexual activity and finally differentiate into males or females. This apparent gonochorism masks the existence of an alternative hermaphroditism (Le Dantec 1968). Changes of sex may occur eventually at the end of the annual reproductive period but only affecting a small fraction of the population (10–15%) for *Crassostrea angulata*, while in *Chlamys varia* the number of females in the population increases with the age of the animals (Lubet 1959).

For most species, there is no sexual dimorphism. The presence of dwarf males has been shown for several species, particularly the Argentine flat oyster *Ostrea puelchana*. Juvenile males attach themselves in the zone around the ligament of the shells of females which exercise a suppressing action on the growth of males.

The physiological mechanisms which control the determination of the change of sex are still unknown, as are the external factors which unlock them. Morton (1927) attributes the growth in the number of female broodstock in a population of flat oysters to good nutrition. These studies should be repeated systematically in order to conserve selected stocks of brood females in hatcheries. It often appears that brood flat and cupped oysters kept for too long show not only a drop in fertility but also the masculinization of certain individuals.

1.2.2 Strategies

A knowledge of the reproductive strategy of a species consists of determining the stages in the annual sexual cycle in various parts of the geographical distribution. Differences can be correlated with variations in the natural environment (reviewed by Lubet 1986a).

A. Methods

Methods consist of establishing the macroscopic development of gonads and, when this is possible, calculating the gonad/somatic index (GI):

GI = Weight of the gonad × 100/Weight of the soft tissues of the animal.

This latter is easy to establish for scallops but histological, stereoscopic methods must be used for mussels and flat oysters.

The histological and cytological study of the annual reproductive cycle allows the establishment of a series of stages in the development of the gonads and the prediction of the spawning season.

When the concentration of individuals in a population is important (oysters, scallops in Japan), investigation of larval production each year allows prediction of recruitment to the fishery.

B. Results

Bivalves have a great flexibility in their reproductive strategy, in relation to variations in the surrounding environment. It is possible to distinguish animals living in the relatively stable environments of the littoral zone and those from more coastal waters, lagoons and estuaries which are affected by sudden, major changes in the surrounding medium.

a. Relatively stable environments. The king scallop (*Pecten maximus*) is harvested from Norway to the south of Spain, from the infra-littoral to depths of over 100 m. Individuals can grow to a large size, thanks to fast growth extending over several years (Mason 1958) and can reproduce, without any reduction in fertility, for 5–6 years.

— A study comparing the gonad index has shown that animals coming from the shallowest waters (10–30 m, Seine Bay) show the greatest variations (maxima and minima) in the sexual cycle in gonad weight. There is a reduction in fecundity with depth (80–100 m, Cornwall). The same observations have been made for *Pecten magellanicus* (Barber *et al*. 1988).

—The annual sexual cycle is practically continuous throughout the distribution with the exception of the population in the Bay of St Brieuc (Fig. 29) where there is a long period of sexual dormancy in autumn and winter (Paulet *et al*. 1988).

— Sexually mature animals are present throughout the year in other populations. Partial spawning has been observed from winter to the following autumn (Norway to the British Isles) with the greatest concentration in early autumn (Mason 1958). This strategy increases the overall chance of successful recruitment by spreading it out over the year.

— In the Seine Bay and St Brieuc, several spawnings may also occur but the reproductive period is limited to summer and early autumn. There is a continuous cytolysis of oocytes from November to June (Seine Bay) and from April to June (St Brieuc). There is a turnover of oocytes within the gonad; several cohorts are destroyed. The products of lysis are resorbed *in situ* or in the gonadal tubes; they can also be released as a 'sterile' spawning in winter or spring (Bay of St Brieuc).

(a)

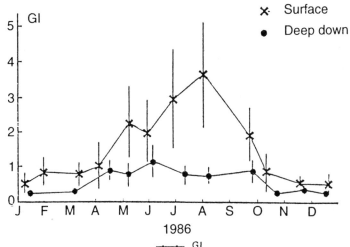

(b)

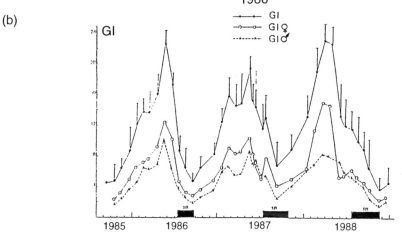

(c)

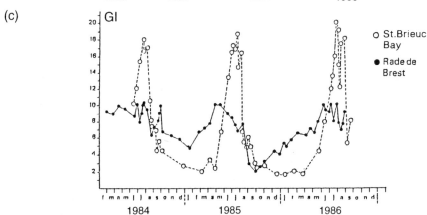

Fig. 29. Annual variations in the gonadic index (GI) in scallops. (a) *Placopecten magellanicus* (after Barber *et al.* 1988). (b) *Pecten maximus*—Seine Bay (after Lubet *et al.* 1987). (c) *Pecten maximus*—St Brieuc Bay (after Paulet *et al.* 1988).

Lysis decreases at the end of spring and mature oocytes are released. Depending on the year, spawning may be massive and synchronized (July) followed by a restoration of the gonad and a smaller spawning (August) or extended over several partial and asynchronous spawnings from July to October. This strategy is much more risky than that in which emissions are spread out through the year. The fecundity of these scallops (Seine Bay) is very high and has been estimated at between 7 and 15 million oocytes per spawning.

Sexually mature animals are present throughout the year in Norway, the British Isles, the Channel and the Atlantic. Spawning takes place at temperatures which vary through the scallop's distribution between 6–7 and 18–19°C. The larvae are planktonic, leading a pelagic existence for anything from 15 days to 3 weeks, depending on the temperature. The pedivelligers attach by the foot and then by the byssus to a substrate (calcareous algae, gorgonians) where they metamorphose. The post-larvae remain attached for around a month or longer, until they reach a diameter of around 1 cm. They detach through the loss of the byssus and disperse by swimming actively and eventually settle and bury into the bed of the sea, often a long way from the nursery (Mason 1958).

b. 'Unstable environments'. These are affected by sudden changes to environmental factors and are often characterized by a high primary productivity and a higher quantity of organic matter than is found in the littoral zone. However, on the debit side, massive mortalities can be brought about by the eutrophication of estuaries and lagoons. Two types of behaviour are demonstrated by farmed molluscs.

Mussel type. Mytilus edulis has a wide geographical distribution. Two subspecies can be distinguished (Lubet *et al.* 1986b).

— stocks of *M. edulis edulis* (ME) extend from the east coast of the USA through Canada, the south of Greenland and Iceland to the northern temperate Atlantic coasts of Europe; the Arcachon Basin in France constitutes the southern limit of distribution.

— *Mytilus edulis galloprovincialis* (MG) is found on the Atlantic coasts of Southern Ireland and Cornwall, on the Atlantic coasts of Europe and North Africa (Morocco), around the coasts of the Mediterranean (with the exception of the south east) and around the coasts of the Adriatic and the Black Sea.

Results (Fig. 30) from several authors show that ME has a spawning season limited to the summer in the most northern parts; this extends progressively to spring in the most southern regions (Channel, Atlantic). In the south, the end of spring and the summer are marked by a period of sexual dormancy and the accumulation of glycogen and lipids in reserve tissue.

MG has the reverse behaviour to ME, with an extended reproductive season in the northern part of its distribution in the Channel, then comparable to that of ME on the Atlantic, Mediterranean and Adriatic coasts of Europe. However, in the southernmost part of the range (Tunisia), mussels are sexually mature in winter and there is only one spawning event; the period of sexual dormancy and accumulation of reserves is extended.

Both subspecies have a high fecundity: $10–15 \times 10^6$ small oocytes (70–80 μm) per emission. Mussels which grow in a relatively stable environment where primary

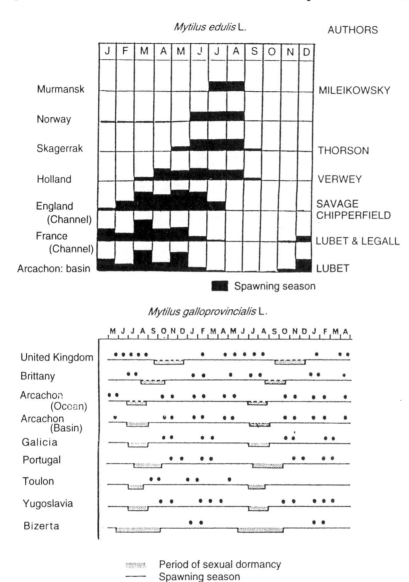

Fig. 30. Annual reproductive cycle in *Mytilus edulis* and *Mytilus galloprovincialis* throughout their distribution.

productivity is high, e.g. the rias of Galicia, have an almost continuous reproductive cycle and the reserve tissues are maintained without showing the fluctuations of those in mussels which are subjected to the effects of tides and drying out.

ME and MG have a great degree of flexibility in their 'opportunist' reproductive strategies. The emission of gametes in these animals is not limited by a specific minimum temperature and the mussels are able to reproduce naturally at temperatures between 2 and 16–17°C (ME) and 7–8 and 19–20°C (MG).

The larvae are planktonic and the pelagic stage lasts from around 15 days to 3 weeks. However, when conditions are unfavourable in winter, metamorphosis may be delayed. The pediveligers attach to substrates (algae), post-larvae and juveniles are able to undertake significant migrations, transported in water masses and settle in huge numbers on supports which are a long way from the nursery (Le Gall 1969).

Crassostrea (cupped oyster) type. This strategy is typical of many cultured species (*Crassostrea gigas, Ostrea edulis, Ruditapes philippinarum, R. decussatus, Mercenaria mercenaria*) and several which are exploited from the wild (*Chlamys varia, Cardium edule, C. glaucum, Mactrids, Venerids, Lucinids*).

Spawning is only possible above a minimum temperature.

The flat oyster, *Ostrea edulis*, has a wide geographical range (from Norway to the Mediterranean-Adriatic). The minimum temperature for the release of larvae can vary in relation to a latitudinal gradient (Korringa 1955): 14–15°C in the North Sea, 17–18°C in the Atlantic, 20–21°C in the Mediterranean. The differences may be associated with geographically separated races.

This species has a period of sexual dormancy in winter, a return to gametogenesis in April/May (France). Emissions take place from the end of June to the beginning of July: further emissions, due to the resumption of gonad activity can take place at the end of August or into September. In the most northerly part of the range, there is only one reproductive season but the length of the season is increased in the southern part (Galicia, the Mediterranean) where several emissions have been observed between May and October (Fig. 31).

The eggs, which are larger than those of most other bivalves (100–120 μm) are fertilized in the pallial cavity where the lecithotrophic larvae develop over a minimum period of around 12 days, up to the stage of the eyed veliger. Once released (spawned) they are pelagic, remaining in the plankton for 8–14 days, depending on temperature. They become attached to a substrate at the pediveliger stage (300–350 μm) by a 'cement' secreted by the byssal gland. The larvae are attracted by the mucus secreted by adult oysters: they have a tendency to become attached to the outer surface of shells. This behaviour pattern goes some way to explain the vulnerability of stocks which are exploited by dredging.

The cupped oyster *(Crassostrea gigas)* (Fig. 31) originates in the Pacific, and is also known as the Japanese oyster. It was introduced to Europe from 1970 to replace the Portuguese oyster, *Crassostrea angulata*, which had fallen victim to a major epidemic. Both of these oysters have the same mode of reproduction (Fig. 31): they are not distinct species and the 'hybrids' are fertile. They are now considered to be two geographical races, only differing in certain physiological adaptations.

The volume and activity of the gonad is extremely reduced from late autumn to the end of winter. There is no true period of sexual dormancy as gametogenesis continues its progress (mitosis of germ cells, previtellogenic development) very slowly. This allows the hatchery manager to alter temperatures in order to change the timing of spawning. Germ cells develop actively from the end of winter; the rate of development accelerates

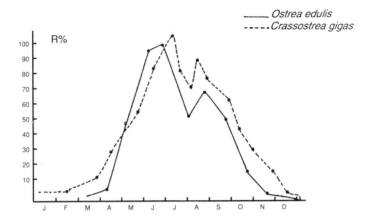

Fig. 31. Reproductive cycle of *Ostrea edulis* (L.) (France, Normandy) and *Crassostrea gigas* (France, Arcachon Bay).

in spring, leading to sexual maturity in July. Gametes are only released when the temperature exceeds a minimum of 20–22°C: this is higher than that required for the Portuguese oyster (17–18°C).

His (1976) studied the behaviour of cupped oysters *in situ* in the Arcachon Basin and showed that each animal can undergo several successive spawnings during the reproductive season (July–September). When these spawnings are large and synchronized throughout the population massive numbers of larvae appear, leading to a good recruitment on the spat collectors placed to catch them.

Fecundity is high and the animals can produce 7–10 million eggs (60 μm in diameter) in a single spawning; this increases with age (Héral 1990). The weight of gametes represents 7% of the flesh for a one-year-old oyster, 60% for a two-year-old and 80% for a three-year-old. The animals are oviparous and the planktonic larva leads a pelagic existence for 2–3 weeks, depending on the temperature. It becomes attached at the pediveliger stage (300 μm) and then follows a similar course to the flat oyster up to the stage of metamorphosis. It does not show the same attraction for the mucus produced by adult oysters but appears to be attracted to copper ions; this favours the use of certain types of collector such as tiles covered with a coating containing copper sulphate. This coating also has the advantage of reducing the settlement of development fouling organisms (barnacles, seaweeds).

The European clam (*Ruditapes decussatus*) and the Japanese clam (*R. philippinarum*) (introduced to Europe almost 20 years ago) have an annual reproductive cycle similar to that of the cupped oyster (Medhioub & Lubet 1988) (Fig. 32). The gonad, which is very much reduced in winter, is composed only of tubules and germ cells. These multiply at the end of winter and form strings which invade adjacent muscle tissue. Gametogenesis takes place from March to the end of June. The gametes are expelled during a first release at the end of July; this is followed by a partial restoration of the gonad which can result in a further spawning in September. In autumn, the volume and number of gonad tubules is reduced.

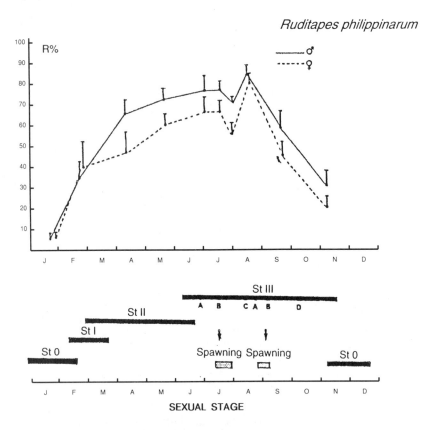

Fig. 32. Reproductive cycle of *Ruditapes philippinarum* in France (Normandy) (after Medhioub & Lubet 1988). St 0 = sexual dormancy, St I = initiation of gametogenesis, St II = gametogenesis, St III$_A$ = maturity, St III$_B$ = egg or sperm release, St III$_C$ = restoration of the gonad, St III$_D$ = end of the annual reproductive cycle.

Eggs are small (30–35 μm); in a major spawning event 2–3 million ova are released. This represents around 8–10% of the weight of the flesh of a three-year-old individual.

The reproductive cycle is very similar to that of the hard clam, *Mercenaria mercenaria* along the northern part of the east coast of the USA.

However, all these species have an extended spawning season in the more southern (warmer) part of their range. *Ruditapes decussatus* in North Africa and *M. mercenaria* in Florida (Hesselman *et al.* 1989) reproduce over a larger part of the year: autumn and the end of winter/beginning of spring. Gonad activity does not cease in winter but is interrupted and stopped during the summer months when temperatures approach the limit LT$_{50}$. This also occurs in the cockle *Cerastoderma glaucum* (Zaouali 1974).

1.3 Factors controlling reproduction
Environmental factors are here distinguished from internal factors although this separation is artificial as obviously one has an effect on the other.

1.3.1 Environment

A. *Temperature*

This appears to play a major role at the level of the strategy of the reproduction of species which have a vast geographical range in relation to latitude. Different authors have studied temperature variations and their effects on gametogenesis, the release of gametes or larvae and the development of the larvae of various species.

This information is essential for the management of broodstock in the hatchery.

a. Most populations of the scallops *Pecten maximus* and *P. opercularis* show practically permanent gametogenesis throughout the year. This is only altered and stopped if temperatures approach the lethal limit (LT_{50}). Release of gametes takes place at all temperatures between the limits. However, within this 'window' it appears that an increase in temperature speeds up gametogenesis. Fecundity depends on the balance in energy between the needs of the somatic tissues, particularly growth, which is temperature dependent and the requirements to cover the reproductive efforts (Lubet *et al.* 1987).

b. Most species which are of interest in aquaculture grow around the coasts and are characterized by an annual slowing down and then cessation of gametogenesis. This phase of sexual dormancy corresponds to a major regression of the gonad. It may also coincide with an increase in energy reserves to sustain future reproductive effort. The resumption of activity within the gonad is independent of the temperature of the external environment at the time: this happens through a complex neuroendocrine programme established at the end of the preceding sexual cycle and depends on temperature, nutrition and reserves (Gimazane 1971; Lubet 1980; Héral & Deslous-Paoli 1990).

Two types of behaviour have been demonstrated through ecophysiological studies.

b1. Mytilid (mussel) type (Fig. 33a). The interruption and then cessation of gametogenesis coincides with the lower LT_{50}, i.e. the lower lethal temperature causing 50% mortality in the population. At higher temperatures, anomalies in gonad development (lysis of oocytes, pycnosis of spermatozoa) and then cessation occurs 8–10°C before the upper LT_{50} is reached. Normal gametogenesis occurs over an optimum temperature range of 10–12°C. An increase in temperature within this range has no effect on the kinetics of gametogenesis, except at temperatures close to the lower LT_{50}. Gametes are released at all temperatures within the reproductive temperature range.

Cultured stocks of *Mytilus edulis* and *M. galloprovincialis* followed throughout the year at different temperatures which were maintained constant between 4 and 19°C have been used to demonstrate experimentally the annual reproductive cycle of natural populations from different latitudes, showing the great flexibility of these animals in relation to temperature (Lubet 1980; Lubet & Aloui 1987; Lubet *et al.* 1987).

In the natural environment, when the temperature reaches and then exceeds the level when gametogenesis is affected (spring), the activity of the gonad is reduced and then ceases. At this point the animals have exhausted the reserves which sustain the reproductive effort and growth uses up a significant quantity of energy. The gonad regresses, reserve tissue is restored and then accumulates the reserves which sustain the next

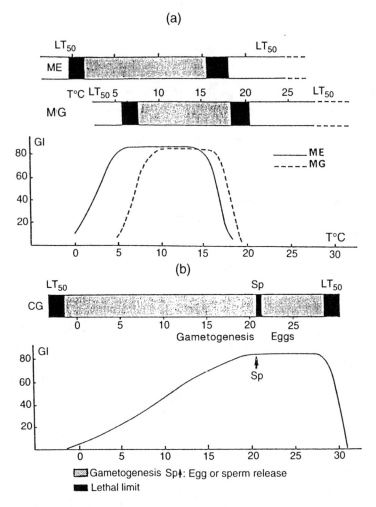

Fig. 33. Lethal temperature limits and temperature range for reproduction. (a) *Mytilus edulis* and *M. galloprovincialis*. (b) *Crassostrea gigas*. In black: alteration and then cessation of gametogenesis. In grey: gametogenesis, sp = minimum temperatures for gamete release.

reproductive cycle. The period of sexual dormancy corresponds to the warmest part of the year.

Animals growing naturally or in culture, under annual temperature variations which are inside the temperature range for reproduction, can have a continuous sexual cycle but this always shows a periodicity in the sequence of gametogenesis corresponding to the expression of an internal rhythm. When these bivalves are growing in zones where primary productivity is high (e.g. Galicia), they accumulate reserves throughout the year; the vesicular cells and adipo-granulose tissue coexist with germinal tissue, instead of alternating as in populations with annual sexual dormancy.

While the sexual cycle can be continuous, subject to the temperature conditions of the environment, it may also be discontinuous. Newell *et al.* (1982) have demonstrated this for mussels on the east coast of the USA. Nutrition always plays a fundamental role in reproductive strategy, and explains spatio temporal variations in relation to the need for a supply of energy to cover the requirements for reproduction.

b2. The flat oyster (Ostrea edulis) *type.* The temperatures at which gametogenesis deteriorates and then ceases occur well away from the LT_{50} values (i.e. significantly below the lower and above the upper LT_{50}). There are two sectors within the 'reproductive window':

— The widest extends from the lowest temperature permitting emission of gametes and larvae. In this zone, increases in temperature have a significant effect on the rate of gametogenesis. This is made use of in hatcheries to shorten the time taken for broodstock to mature. Bayne *et al.* (1975) quantified development time in terms of degree days (DD). Each species has its own requirement in order to arrive at sexual maturity and a certain quantity of energy, calculated in a cumulative fashion. Mean daily temperatures are used in making the calculations.
— Above the minimum temperature for spawning up to a zone where gametogenesis is impeded, the temperature interval for reproduction is restricted, the gonad can go through one or more restorations following the release of gametes. The cessation of reproduction may coincide with the exhaustion of the reserves which sustain the reproductive effort or with the action of high temperatures inhibiting the sexual cycle and often, in the southern part of the distribution, with the simultaneous effects of both of these parameters. The period of sexual dormancy extends from autumn to winter; energy reserves are built up in the autumn. As in mussels, the restoration of gonad activity depends on the programme established at the end of the previous reproductive cycle.

For *Ostrea edulis* in France, gonad activity ceases from the end of October to March. It takes around 1800 DD for emission of larvae starting from a temperature of 10°C which allows the reinstatement of gonad activity. The time taken is much shorter when temperature is high and gametogenesis is further advanced (Fig. 34). The temperature for release of larvae varies according to the stock: 14–15°C in the North Sea; 16–19°C in the Channel and Atlantic; 20–21°C in the Mediterranean—which implies the existence of distinct geographical races. Genetic studies on populations appear to bear this out; three populations (North Sea, Channel/Atlantic and Mediterranean) have been distinguished (Wilkins & Mathers 1974; Blanc *et al.* 1985). However, because oysters from different stocks have been introduced widely into culture basins, intermediate type behaviour patterns are often found.

The Japanese oyster (*Crassostrea gigas*) (Fig. 33b) also requires around 1900 DD to reach sexual maturity from the resumption of gametogenesis. However, if the temperature drops below 15°C during the period of sexual activity, gametogenesis slows down markedly and does not continue until the temperature increases again (Mann 1979a). The time for maturation is thus much shorter when temperature is higher, as long as it does not exceed 25°C (Loosanoff & Davis 1950). The time needed in the hatchery to obtain

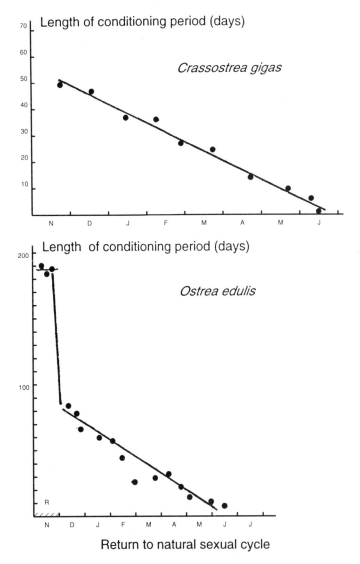

Fig. 34. Length of conditioning period in the hatchery (22°C) for *Ostrea edulis* and *Crassostrea gigas* (after Desvergee in Lubet 1980).

individuals ready to release gametes at an optimum temperature of 22°C is much shorter when the animals brought in are close to sexual maturity (Lubet 1980) (Fig. 34).

The minimum temperatures at which ova or sperm are released are around 20–21°C: these temperatures are higher than those for the Portuguese oyster, *Crassostrea angulata*, where release of gametes has been observed from 17–18°C (Le Dantec 1968). In oyster culture beds where summer temperatures are too low for spawning, gametes are resorbed *in situ* in autumn through cytolysis and phagocytosis by haemocytes. The metabolites thus recovered are used to sustain growth and are also stored in the form of glycogen.

When climatic conditions are favourable for the release of gametes it has been possible to establish a predictable relationship between the date of spawning (y) and the cumulative degree days (DD) from monthly temperatures from the autumn to the end of the following spring.

In the oyster culture beds of the Marennes-Oléron, Héral (1990) suggested the following relationship:

$$y = 228 - 2.87 \ (T \ \text{September–February}) + 1.078 \ (T \ \text{March–June})$$

It has been shown that autumn temperatures slow down the resumption of gonad activity and then spawning (negative correlation) while spring temperatures accelerate maturation and advance spawning (positive correlation). Winter temperatures have no significant action on these phenomena in the cupped oyster and other species (Loosanoff & Davis (1950), for the hard clam *Mercenaria mercenaria*; Sastry (1963), for the scallop *Pecten irradians*; Gimazane (1971) for the cockle *Cardium glaucum*; and Lubet (1980) for the flat oyster *Ostrea edulis*).

The duration of the larval phase and metamorphosis are temperature dependent; low temperatures extend these stages and slow down growth, which increases the chances of the destruction of these very vulnerable stages and therefore has an adverse effect on recruitment. According to Marteil (1960), the duration of the planktonic phases of the flat oyster is 7–8 days at 22°C and 14 days at 15–16°C. For the Japanese oyster, larval life lasts 15 days at 24–25°C and 28 days at 20–22°C. For both species the rate of survival of larvae can reach 10% under favourable conditions.

B. Salinity

The effect of salinity on reproduction has been frequently studied in natural populations although results have not always been consistent. Cultured species of bivalve live in coastal zones and adapt to changes in salinity between specific lethal limits.

Experiments carried out by Gimazane (1971) on cockles, and Lubet (1959) and Lubet *et al.* (1986b) on mussels have shown that within these limits reproduction continues normally as long as temperature and nutrition are optimal. The Japanese oyster is also able to reproduce over a wide range of salinities. Héral (1990) showed a correlation between temperature and salinity in the prediction of optimum spawning; the optimum salinity being around 25‰ and temperatures above 20–21°C.

For the Portuguese oyster, the years of good recruitment in the Arcachon Basin correspond to warm periods during which salinity dropped to around 20‰ (Le Dantec 1968). Salinity of around 35–37‰ does not allow any recruitment of the Portuguese oyster which does not reproduce in the Mediterranean or the Adriatic, while the Japanese oyster, introduced to these places, is occasionally able to reproduce (Etang de Thau, Bizerte), and even, as in Istrie, form good naturally occurring settlements.

Experiments on the American oyster (*Crassostrea virginica*) have shown that at the optimum temperature of around 19–20°C a drop in salinity increases the growth rate and decreases the time taken to reach the stage of metamorphosis.

C. Nutrition

Nutrition has a considerable effect on the reproduction of bivalves, affecting both fecundity and recruitment.

Various authors (reviewed in Lubet 1980) have established a correlation between the quality and quantity of phytoplankton and the resumption of gametogenesis. The results are not always convincing as other factors such as dissolved or suspended organic matter or bacteria have a part to play. Taking into account all the organic matter, Héral (1989) demonstrated the influence of food on the reproduction of the Japanese oyster. A drop in organic matter entails a reduction in the level of carbohydrates and lipids in the animals, resulting in either zero or low recruitment.

Bioenergetic studies have enabled an assessment of reproductive effort to be made (Héral 1990) which is higher for oviparous species with an extended reproductive cycle (mussels, scallops) than for oviparous species with a shorter cycle (*Ostrea edulis*).

D. Release of gametes or larvae

Spawning, sperm release and release of larvae depend on the synergy of various external factors. For many coastal species, the mechanical stimuli exerted by currents, waves, hydrostatic pressure differences and temperature stimuli are the most effective, particularly at the time of spring tides. Experiments (Lubet 1959) have shown that all sudden changes of external factors within certain limits are effective in initiating the release of gametes or larvae when bivalves are sexually mature. There is no 'privileged' status and the action of each of them alone is less effective than their simultaneous action.

1.3.2 Internal factors

External factors act through various routes at the level of the nervous and endocrine systems responsible for the regulation of sexual activity. The endocrine control of the reproduction of bivalves is poorly understood because of the difficulty of carrying out experimental studies. Recent progress has been possible through the use of bioassays and target cells.

A. Endocrine organs

The only endocrine organs that have been described are the neurosecretory cells (NSC) described for the first time by Gabe (1955) and subsequently by various authors (reviewed in Lubet *et al.* 1987).

a. The histology and cytology of the NSC has been studied in detail for *Mytilus edulis*. The use of thionine paraldehyde allowed Illanes (1979) to demonstrate three categories of basophilic neurosecretory cells: a_1 neurones (small size, 6–15 μm) which are elongated, piriform and unipolar; a_2 neurones (20–30 μm), spherical and multipolar; and a_3 neurones, which are irregular in form (20–25 μm) and unipolar.

Electron microscope studies (Fahrmann 1961; Illanes 1979; Damerval 1985) have given further information on the ultrastructure of the cells: granules, Golgi and Golgi vesicles. Neurosecretory products accumulate in the pericaryons and migrate via axons. At present the neuro-haemal organs have not been described.

b. The neurosecretory cells are situated in the cortical anterodorsal region of the ganglions (Fig. 35). The cerebroid ganglia make up 75% of the overall number or 1500–4000
cells which take up stain. Through the year the relative proportions of the different cells
are 73.5% for a_1, 1% for a_2 and 0.5% for a_3; these proportions remain almost constant.

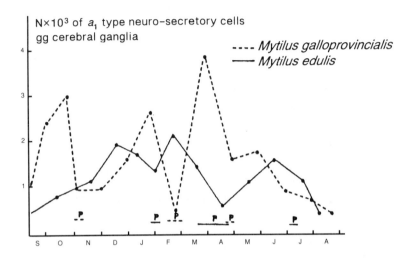

Fig. 35. Annual variations in the number of neurosecretory cells (a_1) active in the cerebral
ganglia of *Mytilus edulis* and *M. galloprovincialis* (after Lubet & Aloui 1987). P = spawning or
sperm release.

Illanes and Lubet (1980) have shown that a significant number of the a_1 cells release
their products before the spawning of eggs or sperm release. This research, together with
that of Lubet & Aloui (1987) has established (for mussels) a significant correlation
between the reproductive cycle and the number of active neurosecretory cells which
increases at the time of resumption of activity in the gonad, throughout gametogenesis up
to maturation and then decreases at spawning or sperm release to increase again on
resumption of gonad activity at the start of the new sexual cycle. These variations can
only be seen at the level of the cerebral ganglia.

This cyclical change in the a_1 neurosecretory cells is the same in animals which are
reared experimentally at constant temperature or those taken from the natural environment. The cycle of the a_1 is also the same when the gametes are not released. The gonad
then progresses to a new phase of gametogenesis; the unspawned, ripe oocytes are
cytolized and the volume of sperm increases in the male gonad. These observations
suggest that there is an internal rhythm which controls the activity of the a_1 neurosecretory cells.

B. Neurosecretory products
There are both aminergic and peptidergic secretions (Fig. 36).

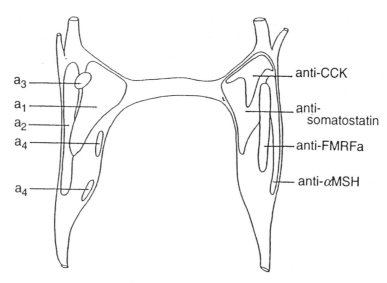

Fig. 36. Cerebral ganglia of *Mytilus edulis*: location of the main types of neurosecretory cells
(a_1–a_4) and immuno-reactive zones (after Mathieu & Van Minnen 1989).

a. The existence of *aminergic secretions* (Lubet *et al.* 1987) has been demonstrated in
the nerve ganglia of bivalves; serotonin and dopamine have been detected by
histochemical and biochemical techniques. These substances have been localized using
histofluorescence and immunocytochemistry. Aminergic cells have been identified in
different ganglia, particularly the pedal ganglia as well as the nerve endings at the junctions between nerve and muscle. More recently, the presence of opiates (encephalins) has
been shown in some neurones in *Mytilus edulis*.

b. The demonstration of *Peptidergic neurons* has been made fairly recently in molluscs
using immunocytochemical techniques. In bivalves, results to date are rather fragmentary. Fritsch *et al.* (1978) pinpointed cells of the same type in the digestive tract of *Unio
pictorum* and the swan mussel, *Anodonta cygnea*. However, other workers have failed to
identify the substance reacting to anti-cholecystokinin in the mussel digestive tract. A
calcitonin-like substance has been shown in juvenile king scallops (Le Roux *et al.* 1986).
Finally, a study carried out on the cerebral ganglia of *Mytilus edulis* (Mathieu & Van
Minnen 1989) has detected neurones which react positively to antibodies for different
hormones: FMRFa-cholecystokinin, aMSH-somatostatin (Fig. 36).

C. Endocrine control
Culture of isolated explants or tissue cultured in association with autologous or
heterologous nervous ganglia have shown that isolated explants degenerate (Houtteville
& Lubet 1975). Neuroendocrine factors from the cerebral ganglia and, to a lesser degree,
from the visceral ganglia, once liberated, are essential for the development of gonad
mitosis, meiosis (spermatocytes) and vitellogenesis (Lubet & Mathieu 1978; Mathieu
1987). These factors do not appear to be sex-specific as cerebral ganglia taken from

females sustain the development of the male line and vice versa, nor species specific as the continuation of the development of the gonad of the Japanese oyster, the flat oyster and the clam have all been maintained *in vitro* in association with the cerebral ganglion of a mussel. However, this technique cannot be studied using bioassays because of the heterogeneity of gonad tissue which does not allow for reliable comparison of results.

a. Endocrine control of the male line (Mathieu 1987). A bioassay has been developed making use of the activity of an enzyme-key for the synthesis of pyrimidine bases (aspartate *trans*-carbamylase: ATC) whose activity is highly significantly correlated with the activity of the male gonad (Fig. 37).

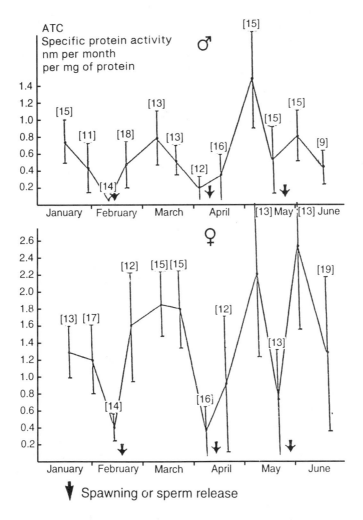

Fig. 37. Activity of ATCase (aspartate trans-carbamylase) during the annual reproductive cycle of *Mytilus edulis* (Normandy) (after Mathieu 1987).

Another bioassay consists of incorporating 3H thymidine. An autoradiographic study has shown that this substance is only incorporated into germ tissue.

The biological material serving as a target is made up of cells detached from the mantle of a mussel whose fractions show an excellent homogeneity of response which is not true of explants of cultures from organs which are extremely heterogeneous.

An activating factor in gonad mitosis has been demonstrated from cerebral ganglia taken from mussels; the effect is a typical dose–response. Partial purification has shown that the substance is a protein with a molecular weight of around 5 kDa. A substance with the same biological properties has been detected in the haemolymph.

b. Endocrine control of the female line. The culture of the female gonad of several species of bivalve (mussel, oyster, clam) has shown the existence of a factor of cerebral origin which sustains the development of oogonia (Lubet & Mathieu 1978). There is also at least one factor of cerebral origin, distinct from the previous one, which is essential for vitellogenesis. In its absence, oocytes lyse. This factor significantly increases the synthesis of RNA and phospholipids (Lubet *et al.* 1986a). However, the factor is neither responsible for the reinitiation of meiosis and the maturation of oocytes nor for spawning.

c. Oocyte maturation and spawning. Sato *et al.* (1985) have demonstrated the existence of a factor in the gonad of *Spisula* (= *Mactra*) which has not been identified but unlocks the final oocyte maturation process (rupture of the germinal vesicle, reinitiation of meiosis). Through the injection of serotonin (5-hydroxytryptamin, 5HT) at a concentration of 2 μM into the gonad of *Spisula* or the addition of this substance to a suspension of oocytes, Hirai *et al.* (1988) obtained the same effects on maturation. Kadam & Koide (1989), using nerve ganglia and haemolymph from *Spisula*, demonstrated that 5HT was the main factor responsible for the rupture of the germinal vesicle, after purification with HPLC. There are specific receptors on the surface of the oocyte. We have seen that serotonin is present in nerve ganglia; Stefano & Catapane (1979) have shown that the maximum concentration of 5HT is found at the time of genital maturity.

Other amines may also be involved in the physiological processes associated with spawning. Osada *et al.* (1987) showed a significant increase in the level of dopamine in several species of bivalve (*Mytilus edulis, Patinopecten yesoensis, Crassostrea gigas*).

Sudden temperature, mechanical and ionic changes in the external environment stimulate spawning by inducing the discharge of monoamine neurotransmitters. Lubet (1959) also demonstrated the importance of nerve ganglia in unblocking the release of gametes.

The induction of the release of gametes in the hatchery using sea water treated with UV is a widely used procedure; spawning can also be induced with hydrogen peroxide (Morse *et al.* 1977). This is through action on the peroxidation enzymes, particularly endoperoxidase prostaglandins which transfer polyunsaturated fatty acids (linolenic acid, arachidonic acid and eicosopentaenoic acid) respectively to prostaglandins E_1, E_2 and E_3. The presence of these constituents (PGF2a) has been shown in marine invertebrates by various authors, particularly by Nomura & Ogata (1976) for *Crassostrea gigas* and Ono *et al.* (1982) for *Patinopecten yesoensis*.

Their concentration increases significantly in the gonads of both sexes at the time of sexual maturity. Results relating to this effect are still confusing. While spawning has been initiated by injection of PGE and PQF at a concentration of 3×10^{-12} for *Mytilus californianus*, Matsutani & Nomura (1982) were unable to show any effect on the induction of gamete release in *Patinopecten yesoensis*, neither have they been able to show any effect in *Pecten maximus*.

Since the work of Galtsoff (1940) on *Crassostrea virginica* all authors have agreed that introducing sperm into a tank containing mature female bivalves facilitates spawning. This procedure is also used to stimulate females together with a temperature shock (rise of 5–6°C). Few studies have been carried out to determine the reason for this effect since those of Nelson & Allison (1940) who extracted a substance (diantline) from sperm. This appeared to act as a pheromone, the nature of which is as yet unknown. This has the effect of inducing generalized spawning in natural stocks after the first release of sperm: synchronized spawning leads to improved recruitment.

2. GASTROPOD REPRODUCTION (ABALONE)

2.1 Morphology, morphogenesis and cytology of gonads

2.1.1 Morphology
Abalone are gonochoric although there are rare cases of hermaphroditism in juveniles or simultaneous hermaphroditism in adults (Girard 1972). The sex ratio evens out at around 50% from the fourth year. The genital gland, uneven in form and turned back on the right side, is greenish in the female and greyish white in the male; it is visible from the exterior without damaging the animal after moving aside the tissues which mask it (Fig. 36).

2.1.2 Morphogenesis
Haliotis tuberculata, in French waters, does not show precocious sexuality although the time taken to reach maturity may be shortened in the hatchery (Cochard 1980). Girard (1972) showed that in Brittany, the first maturation took place in males at 2 years old (20–40 mm shell length) and in 3-year-old females (40–50 mm). Similar results have been found for abalone in the Pacific. The juvenile gonad forms inside two leaves of a cavity similar to a coelom (Bolognari 1953: *H. lamellosa*) from a mass of tissue or a conical appendage.

Each of the 'leaves' has a mass of cells, including primordial germ cells together with haemocytes (granulocytes, monocytes). When the gonad develops in the female form (Girard 1972) the cavity becomes broken up into follicles by the development of connective tissue bars. The primary germ cells and haemocytes migrate along these walls. The multiplication of the germ cells covers these connective tissue bars with a continuous layer of cells.

Oocytes develop in a centripetal pattern, pushing against the connective tissue walls. They form progressively along the connective tissue bars (trabeculae) which are roughly parallel. Germ cells, oogonia oocytes and then accessory cells become embedded along the connective tissue.

The morphogenesis of the male gonad follows a very similar pattern. However, the walls of the coelomic cavity only develop elongated 'sacs' which are organized into a network of tubules whose walls are covered with primary germ cells and spermatogonia.

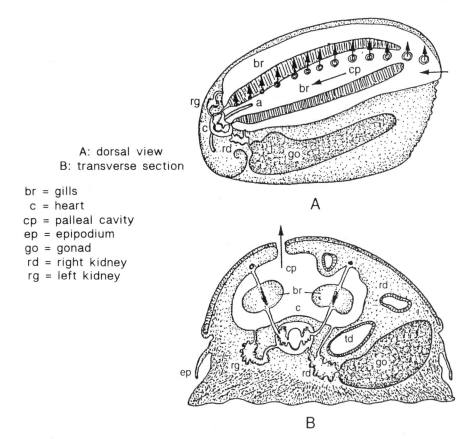

A: dorsal view
B: transverse section

br = gills
 c = heart
cp = palleal cavity
ep = epipodium
go = gonad
rd = right kidney
rg = left kidney

Fig. 38. Reproductive system of *Haliotis tuberculata* (after Croft 1929 modified).

2.1.3 Cytology

The development of the germ cells resembles that described for bivalves. However, primary germ cells survive into the adult stage; they multiply and part of them differentiate into spermatogonia and oogonia.

Spermatogonia (8–10 μm) and then primary spermatocytes (6–8 μm) and secondary spermatocytes (4–5 μm) accumulate in the lumen of the tubules. Spermatozoa measure 6 μm and have an acrosome 2.5 μm in length (Lewis *et al.* 1980).

Primary germ cells (4–6 μm) can be distinguished from oogonia by their oval form. Oogenesis does not differ from that of bivalves although there is no major accumulation of yolk (triacylglycerides, glycoproteins) in the cytoplasm. At the end of vitellogenesis, oocytes measure around 200 μm in diameter and are only attached to the wall by a fine stalk. The accessory cells are alongside the oocytes; their function remains to be determined precisely.

Germinal products are evacuated through the oocyte duct or, in the case of sperm, the renal cavity (Fig. 38) and then the urinary pore in the pallial cavity where the exhalant current disperses it into the environment through the holes in the shell. As with all archegastropods there is no copulatory organ and fertilization is external. Fertilized eggs are protected until the trochophore larva stage by a gelatinous coat which allows them to float.

2.2 The sexual cycle and fecundity

2.2.1 The sexual cycle
The annual gonad cycle has been followed histologically and through the gonadic index (Fig. 39).

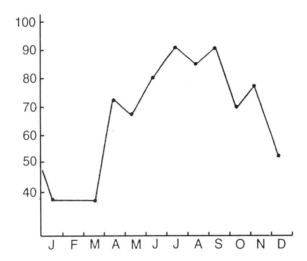

Fig. 39. Annual variations in the gonad volume index of *Haliotis tuberculata* (after Girard 1972).

With the exception of *H. rufescens* (Boolotian *et al.* 1962), which has a continuous sexual cycle, almost all other species are seasonal with a period of sexual dormancy in winter and spawning in summer and autumn (Pena 1986).

Girard (1972) has shown that *H. tuberculata* in Brittany is sexually dormant in winter (November to the end of March). The resumption of gonad activity is marked by a massive multiplication of cells. Oogenesis and spermatogenesis take place from April to the end of June. From July to November, the female gonad is emptied progressively, without any restoration between spawnings, while after the first sperm release in July, the male gonad is able to undergo further spermatogenesis in August, emptying again up to November.

Shepherd & Laws (1974), for the same species, showed that the timing of spawning and maturation varied in relation to latitude and occurs earlier in the warmer parts of the distribution.

2.2.2 Fecundity
The fecundity of female abalone depends on age; a good relationship has been established between the number of eggs released and the shell length or the weight of the

animal (Fig. 40; Girard 1972): 5×10^5 eggs for a size of 7.5 cm, 5×10^6 for 10.5 cm, and over 5×10^6 for 11.5 cm (*H. tuberculata*).

Fertility depends on the concentration of spermatozoa. In the hatchery the studies carried out by Kikuchi & Uki (1974) on *H. discus* showed that a concentration of 1 to 2×10^5 spermatozoa ml^{-1} gave the best results.

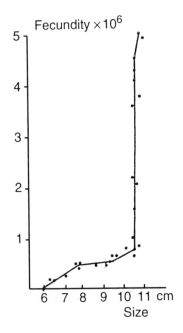

Fig. 40. Relationship between fecundity and animal size for *Haliotis tuberculata* (after Girard 1972).

2.3 Factors controlling reproduction

Research on the conditioning of broodstock has given information on the effect of temperature and photoperiod. As with bivalves, salinity appears not to have any particular effects on the maturation of females within the limits tolerated by the species.

2.3.1 Environment

A. Temperature

Girard (1972) studied the change in temperature during the sexual cycle of *H. tuberculata*. A minimum temperature of 9–10°C coincides with the period of sexual dormancy; the maximum (17–18°C) coincides with that of spawning. The fact that this species is not found in the British Isles suggests that there is a minimum temperature for the release of gametes which limits the distribution to the north. When this temperature is not reached, oocyte lysis occurs.

Kikuchi & Uki (1974) established a linear relationship between temperature and gonad weight for two Japanese species (*H. discus* and *H. discus-hannai*). The latter species takes 1500 degree days from 7°C to reach sexual maturity. Cochard (1980) found that 1800 DD were required for *H. tuberculata* from 10°C.

However, some species such as *H. cracherodi* and *H. gigantea* (Kikuchi & Uki 1974) show no response to a change in temperature.

B. Photoperiod
This constitutes a limiting factor. Webber & Giese (1969) studied *H. cracherodii* and Kikuchi & Uki (1974) *H. discus-hannai* and showed that a photophase of 12 h (12L/12D) stimulated gametogenesis. Experiments carried out by Cochard (1980) on *H. tuberculata* showed that when animals are reared at 18°C, maturation was prevented for 7 to 8 months by total darkness (0L/24D) while maturity was reached after 4 months with a photophase of 12 h (12L/12D). The optimum appears to be between 14 and 16 h illumination. However, a photophase of 24 h in *H. discus* resulted in the blocking of gametogenesis and gonad regression (Pena 1986).

The optimum conditions of illumination in a hatchery are around 150 lux, comparable to daylight at a depth of 6–8 m under water.

C. Nutrition
This has a major effect on maturation (Uki & Kikuchi 1982). In the hatchery, a daily consumption of algae (**macroalgae**) equivalent to 10% body weight appears to be appropriate for broodstock kept under optimum environmental conditions. A mixture of several species of algae is better than a single species given alone

2.3.2 Internal factors
Very little is known concerning the endocrine factors which control the sexual activity of the neurosecretory cells in the cerebral mass of *H. lamellosa*: experiments on the culture of isolated gonad extracts (Potrel & Lubet, unpublished) have shown that during the period of sexual dormancy the isolated explant, in a medium without hormones, maintains itself but does not develop. The addition of one or several ganglia removed in winter does not cause any change. In order to induce gametogenesis in the explants, it is necessary to add ganglia removed from animals with developing gonads. This brings about mitosis of germ cells which suggests that there is a mitogenic cerebral factor, as has been shown in other prosobranchs such as the slipper limpet, *Crepidula fornicata*, and in bivalves.

Filling tanks containing mature broodstock with UV-treated water precipitates spawning or sperm release (Kikuchi & Uki 1974; Kan-No 1975; Morse *et al.* 1977).

Oxygenating water to give a final concentration of 5 mM induces spawning; the action is facilitated by the addition of tris-(hydroxymethyl-methylalanine) at 6 mM. The animals are kept in the solution for 2–2.5 h and then rinsed and returned to sterile sea water; spawning takes place about 1 h later. This method is widely used to obtain synchronized spawning of large numbers of eggs in a sterile environment.

Prostaglandins may be involved in this process: spawning has been obtained in *H. rufescens* by injection at a concentration of 3×10^{-12} (Tanaka 1978).

3. CEPHALOPOD REPRODUCTION

The cuttlefish, *Sepia officinalis*, has a wide geographical range from the south of Norway, the North Sea, the Channel and the Atlantic coasts of the British Isles, Europe and North Africa as far as the south coast of Morocco.

The cuttlefish is gonochoric, as are all cephalopods. Sexual dimorphism is visible at the level of the hectocotyle arms (or copulatory organ) (Fig. 41); there are colour differences between the two sexes. The sepions have grooves which distinguish between immature and mature individuals (Fig. 42).

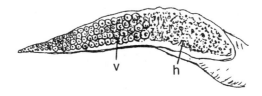

Fig. 41. *Sepia officinalis*, hectocotyle arms. v = suckers, h = hectocotylus.

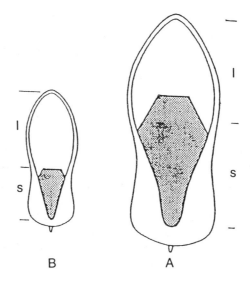

Fig. 42. *Sepia officinalis* beaks. Forms A and B (after Mangold 1966, modified). A = mature animal, B = immature.

3.1 Morphogenesis, cytology, functional organization

3.1.1 Morphogenesis
The gonad remains undifferentiated up to the end of larval development but the rudiments of the nidimental glands are already visible before hatching, allowing the sexes to be distinguished (Lemaire & Richard 1970). The general outline of the gonad is made up

of germ cells forming a mass inside the epithelium of the coelomic wall and an adjacent mass of connective tissue or stroma projecting into the coelom.

The stroma develops in the coelomic cavity forming a ribbon of tissue which gives rise to the ovary. The germ cells multiply and invade the stroma; they differentiate into oogonia. Connective tissue cells surround the young oocytes forming the follicles. The ovary is fed by a dense network of capillaries which run between the follicles.

The testis is a globular organ within the coelomic cavity, divided by connective tissue rays which are covered with spermatogonia.

The genital glands represent around 1% of the body weight of juveniles. At the time of maturity the ovary may reach up to 150 g (16% of body weight).

3.1.2 Cytology
Six stages of gametogenesis were defined by Richard (1971) and described at the ultrastructural level by Dhainaut & Richard (1976).

A. Oogenesis (Fig. 43)
At the start of the development of the gonad, germ cells multiply and make up a stack of oogonia which ensures the production of oocytes in the gonad (stage I). Meiosis is

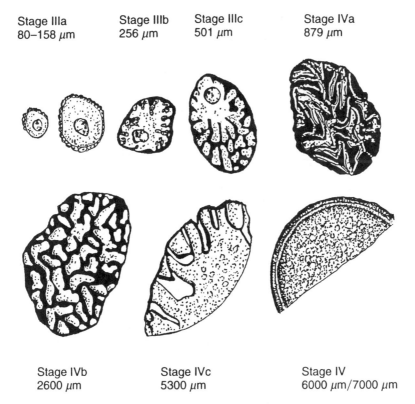

Stage IIIa 80–158 μm   Stage IIIb 256 μm   Stage IIIc 501 μm   Stage IVa 879 μm

Stage IVb 2600 μm   Stage IVc 5300 μm   Stage IV 6000 μm/7000 μm

Fig. 43. Development of the oocytes of *Sepia officinalis* (from Dhainaut & Richard 1976) (follicle cells shown in black).

blocked at the prophase of the first division (stage II). Oocytes thus accumulate large quantities of RNA in the cytoplasm (previtellogenesis—stage III) while the follicle cells arrange themselves around the periphery of the oocyte (stage IIIa). These cells multiply actively (stage IIIb) and form strings which penetrate the interior of the cytoplasm of the oocyte. Yolk synthesis begins (stage IV) and the oocyte increases in size, as does the weight of the ovary. At the end of vitellogenesis follicular cells regress and disappear; a chorion forms around the oocyte (stage V). Yolk, made up of lipoproteins, is very abundant. Finally, during the spawning period, follicles (oocytes in previtellogenesis or vitellogenesis) become atretic.

B. Spermatogenesis
Connective tissue is invaded by germ cells which multiply and give rise to spermatogonia (stage I). These cover the connective tissue rays and form seminiferous tubules. Meiosis takes place in a synchronous fashion (spermatocyte I, stage II and spermatocyte II, stage III) followed by spermatogenesis (stages IV, V and VI) at the same time as the development of the hectocotylus (copulatory arm).

C. Nidamental gland
The principal anterior and posterior nidamental glands have the same structure as the oviduct gland. They are made up of stacked lamellae formed from a space in the glandular wall. During the second year of the cuttlefish's life, from September to April, there is an intense multiplication of cells in all the glands, into the epithelium of the mantle, formed from tubules buried in connective tissue. The activity is synchronized with that of other glands.

Colour is an excellent indication of ovarian activity: whitish-cream (immature), cream-orange (oogenesis) and orange/coral (mature); this last colour is due to the presence of symbiotic bacteria (Richard 1971). The changes in all these glands are synchronized with that of the ovary.

3.1.3 Functional organization

A. Mature oocytes
These pass into the coelomic cavity and are ejected by the coelom duct. In the oviduct gland, they are surrounded by a layer of albumen. In the pallial cavity, products secreted by the nidimental gland form an external envelope for the eggs. These protein structures are liberated by two pores, situated in front of the urinary orifices, the secretions of the anterior gland pour out into the evacuation canals for the posterior glands. These proteins are deposited in layers around the egg. They are coloured with black by the melanins from the ink sac, the product of which is liberated simultaneously. The size of the eggs released varies from 5–6mm in small females to 7–9 mm in larger ones (Mangold 1963).

B. The testicular orifice
This opens in the coelomic cavity. A thin, deferent canal links the gonad to the male genital tract; it winds its way to the anterior part of the animal and opens out into the genital vesicle. To this is attached a secondary seminal vesicle. The seminal cord ends as

a long tertiary vessel, in the accessory gland. Spermatozoa enter the deferent canal and then the vesicles, the walls of which secrete a viscous substance which forms the walls of the spermatophore. These accumulate in 'Needham's sac'; they vary in size from 5 to 16 μm as a function of the size of the males (Mangold 1963).

C. Fecundity

This has been estimated, depending on the size of the animal, to be between 500 and 1000 mature eggs. Copulation is preceded by an approach behaviour pattern; animals are characterized by typical colorations. They position themselves head to beak by attaching all the arms except for the male copulatory tentacle which introduces spermatophores into a special pouch, situated under the buccal mass of the female.

Spawning occurs shortly after coupling, the male remains in visual contact with the female. Eggs are ejected at regular intervals (2–3 min) over several hours. The female empties her entire ovary over several days.

The methods of fertilization are still unknown: spermatozoa enter the oocyte through a central canal through the surrounding envelope which ends up by the micropyle.

Eggs are evacuated by the siphon at the base of the ventral arms which attach them to a substrate. They form black clusters ('sea grapes').

3.2 Reproduction

3.2.1 The sexual cycle

There have been many studies on the reproductive cycle of the cuttlefish around the coasts of France (reviewed in Boletzky 1983). Apart from rare exceptions, these animals reproduce only once in their short lifetime (1–2 years) at different sizes depending on where they are found within their geographical distribution, sometimes at different times in the same area. The problem is complicated by the fact that cuttlefish are migratory animals, arriving in coastal waters in the spring and returning to oceanic waters in autumn.

3.2.2 Development of reproductive organs

These have been studied using various methods. The most frequently used are those of the gonado-somatic index (GSI: weight of gonads × 100/body weight), the nidamento-somatic index (NSI = weight of the nidamental gland × 100/body weight), the ovo-ovarian index (OOI = weight of smooth eggs × 100/weight of the ovary). The relationship between the smooth part and the striated part of the sepion is also an excellent index of maturation. Cuenot (1933) recognized two forms: A, corresponding to mature individuals, and B to immature (Mangold 1966) (Fig. 44).

In the southernmost part of the distribution (Mediterranean: France, Tunisia, Senegal coast), large adults, ready to spawn, arrive at the coast at the start of February, followed at intervals of greater or lesser duration by medium-sized cuttlefish with developing gonads. The youngest do not arrive until the summer and remain up to November and December. The reproductive season is extended over several months (Mangold 1966; Bakhayokho 1980; Najai 1983).

(a)   (b)

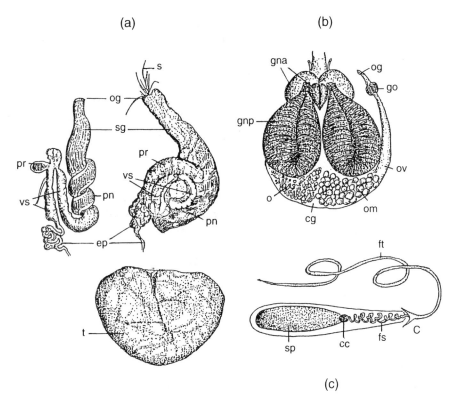

(c)

Fig. 44. Reproductive apparatus of *Sepia officinalis* (after Richard 1971, modified). (a) Male. a = accessory to prostate gland, cp = deferent proximal canal, og = genital orifice, pn = Needham's sac, pr = prostate, s = spermatophores, sg = genital sac, t = testicle, vs I, II, III = vesicles. (b) Female. cg = genital coelom, gna = accessory nidamental gland, gnp = principal nidamental gland, go = oviduct gland, o = oocytes, og = genital orifice, om = mature oocycts, ov = oviduct. (c) Spermatophores. c = couvercle, cc = cement body, fs = spiral filament, ft = filament terminal, sp = spermatozoids.

In the Bay of Biscay, Lafont (1868), Cuenot (1933) and Jeon (1982) distinguished three categories of animals: a spring generation from large mature animals which arrive at the coast in April, a summer generation from these same animals which continue to spawn, and a precocious fraction which is made up of the summer generation of the next cycle.

In the North Sea and the Channel (Richard 1971; Medhioub 1986) the cycle is similar to that in the Bay of Biscay.

Two-year-old individuals reproduce; males arrive at the coast before the females. Spawning takes place in spring, egg incubation lasts from two-and-a-half to three months with hatching between 15 July and 15 August. Juveniles leave coastal waters in the autumn and carry out migrations of a greater distance than the cuttlefish from the Mediterranean.

In the following spring, young, immature animals approach the coasts, arriving three months after the two-year-old adults. They grow very rapidly because of the summer temperatures and the excellent feeding. Migration is food related. The males develop their gonads and become mature. In females, the development of the oocytes remains blocked in previtallogenesis during the summer. These animals return to deeper waters in the autumn. Oogenesis begins again in December when the female is 15–16 months old; this winter phase is a period of intense sexual maturation in females and sexual development in males. The animals then perform another spring migration; spawning takes place in coastal waters from April to June (Fig. 45).

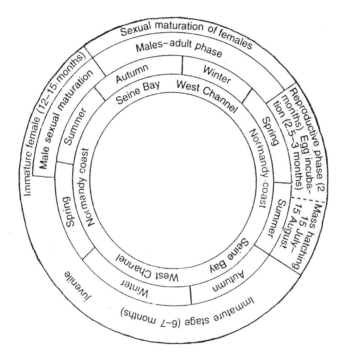

Fig. 45. Life cycle of the *Sepia officinalis* L. in the Channel (after Medhioub 1986).

3.3 Factors controlling reproduction

The works of Wells & Wells (1959) and Young (1971) on octopus have established that gonads develop under the influence of a hormone secreted by the optic glands. These, together with the 'brain', only liberate this factor under certain external conditions (short days, low illumination).

Richard (1971, 1975) analysed the effect of the combined action of light and temperature. The first growth of the gonad, as with somatic growth, is temperature dependent. Nutrition favours the growth of the body. Light, particularly short wavelength (blue–green), has a controlling effect on maturation. This has been demonstrated by *in vitro* studies of the optic gland in association with juvenile gonads (Richard 1975). Long photoperiods stop, via the brain, the development of the female gonad, although in mature animals they favour the release of gametes and the deposition of eggs on substrates.

These researches have led to a better understanding of the reproductive cycle of cuttlefish. Strong summer light blocks the maturation of juvenile females while high temperatures and good feeding favour rapid growth in the coastal zone.

In autumn, in shallow zones, growth is stopped but very low winter light intensities allow the maturation of the ovary and the testis. The migration which follows in spring or in summer allows the release of eggs in the coastal zone.

REFERENCES

Bivalves

Albertini L., 1986. Etude cytologique, cytochimique et expérimentale de la vitellogenèse chez la moule, *Mytilus edulis* L. Th. Doct. Sp., Univ. de Cae: 1–125.

Allarakh C., Lubet P., 1980. Analyse expérimentale de la différenciation du sexe chez quelques *Pectinidés*. *Arch. Anat. Microscop. et Morph. Exp.*, **70** (1): 47–58.

Barber B. J., Getchell R., Shumway S., Schick D., 1988. Reduced fecundity in a deep water population of the geant scallop, *Placopecten magellanicus*, in the Gulf of Maine, U.S.A. *Mar. Ecol. Prov. Ser.*, **42**: 207–212.

Bayne B. L., Gabbott P. A., Widdows J., 1975. Some aspects of the effect of stress in adult of the eggs and larvae of *Mytilus edulis* L. In: *Effects of temperature on ectodermic animals*, Wiesser (ed.). Berlin, Springer-Verlag.

Besnard J. Y., 1988. Etude des constituants lipidiques dans la gonade femelle et les larves de *Pecten maximus* L. Th. Doct., Univ. Caen: 1–154.

Blanc F., Attard J., Pichot P., 1985. Genetic variability in the European oyster *Ostrea edulis*: geographic variation between local french stocks. *Aquaculture*.

Bourcart C., Lavallard R., Lubet P., 1965. Ultrastructure du spermatozoïde de la moule *Mytilus perna* (Van Ihering). *C.R. Acad. Sci.*, **260** (12): 5096–5099.

Coe W. E., 1943. Sexual differentiation in molluscs I. Pelecypods. *Quart. Rev. Biol.*, **18** (2): 154–164.

Damerval M., 1985. Identification et rôle physiologique des 'inclusions pigmentees' dans le système nerveux de la moule et de la crépidule. Th. Doct. Sp., Univ. Caen: 1–156.

Dorange G., 1989. Les gamètes de *Pecten maximus* L. Th. Doct., Univ. Brest: 1–133.

Fahrmann W., 1961. Licht und Electronenmikroskopische untersuchungen des Nerversystem von *Unio timidus* (Ph) unter besonder Berüch-sichtigung des Neurosekretion. *Zeits. für Zellfors.*, **54**: 689–716.

Fritsch H., Van Noorden S., Pearse G., 1978. Localisation of somatostatin and gastrin like immunoreactivity in gastrointestinal tract of *Ciona intestinalis*. *Cell. Tiss. Res.*, **186**: 181–185.

Galstoff P. S., 1940. Physiology of reproduction of *Ostrea virginica*. *Bio. Bull.*, **78** (1): 117–135.

Gimazane J. P., 1971. Introduction à l'étude expérimentale du cycle sexuel de *Cardium edule*. Th. Doct. Sp., Univ. Caen: 1–125.

Héral M., 1990. L'ostréiculture traditionnelle. In: *Aquaculture*, G. Barnabé (ed.). Lavoisier (Tec. Doc.) Ed., Paris: 348–397.

Hesselman D. M., Barber B. J., Blake N. J., 1989. The reproductive cycle of the adult hard clam *Mercenaria* spp. in the Indian River lagoon Florida. *J. Shellfish. Res.*, **8** (1): 43–50.

Hirai S., Kishimoto T., Koide S., Kanatani H., 1988. Induction of spawning and oocyte maturation by 5-hydroxytryptamine in the surf clam. *J. Exp. Zool.*, **245**: 318–329.

His, E., 1976. Contribution à l'étude biologique de l'huître dans le bassin d'Arcachon, *Crassostrea angulata* et *Crassostrea gigas*. Th. Doct. Sp., Univ. Bordeaux: 1–60.

Houtteville P., 1974. Contribution à l'étude cytologique et expérimentale du cycle annuel du tissu de réserve de la moule *Mytilus edulis* L. Th. Doct. Sp., Univ. Caen: 1–135.

Houtteville P., Lubet P., 1975. The sexuality of Pelecypods molluscs. In: *Intersexuality in the animal kingdom*, REINBOTH (ed.). Springer-Verlag: 179–187.

Illanes J., 1979. Recherches cytologiques et expérimentales sur la neurosécrétion chez la moule *Mytilus edulis* L. Th. Doct. Sp., Univ., Caen: 1–135.

Illanes J., Lubet P., 1980. Etude de l'activité neurosécrétrice au cours du cycle sexuel annuel de la moule *Mytilus edulis* L. *Bull. Soc. Zool.*, *France*, **105** (1): 141–145.

Kadam A. L., Koide S. S., 1989. Characterization of a factor with oocyte maturation inducing activity in *Spisula. Biol. Bull.*, **176**: 8–13.

Korringa P., 1955. Water temperature and breeding throughout the geographical range of *Ostrea edulis* L. *Ann. Biol., Paris*, **33**: 1–17.

Le Dantec J., 1968. Ecologie et reproduction de l'huître portugaise, *Crassostrea angulata* Lmk., dans le Bassin d'Arcachon et sur la rive gauche de la Gironde. *Rev. Trav. I.S.T.P.M., Paris*, **32**: 1–126.

Le Gall P., 1969. Etude des moulières normandes. Th. Doct. Sp., Univ. de Caen: 1–71.

Le Roux S., Bellon-Humbert C., Lucas A., 1986. Mise en évidence d'une substance apparentée à la somatostatine dans le système nerveux, le rein et la glande digestive de juvéniles de *Pecten maximus* (Moll. Bivalve). *C.R. Acad. Sci., Paris*, **302** (3): 191–196.

Loosanoff V. L., Davis H. C., 1950. Conditioning *Venus mercenaria* for spawning and breeding larvae in the laboratory. *Biol. Bull.*, **98**: 60–65.

Lubet P., 1959. Recherches sur le cycle sexuel et l'émission des gamètes chez les *Mytilidés* et les *Pectinidés* (Moll. Bivalves). *Rev. Trav. I.S.T.P.M., Paris*, **23** (3): 387–548.

Lubet P., 1980. Influence des facteurs externes sur la reproduction des Lamellibranches. *Océanis*, **6** (5): 469–489.

Lubet P., 1986. Strategies of reproduction in bivalve molluscs. *Advances in Invertebrate reproduction*, M. Porchet, J. C. Andries & A. Dhainaut (eds). Elsevier: 401–408.

Lubet P., Albertini L., Robbins I., 1986a. Recherches expérimentales au cours de cycles annuels, sur l'activité gonadotrope exercée par les ganglions cérébroïdes sur la gamétogenèse femelle chez la moule *Mytilus edulis* L. *C.R. Acad. Sci. Paris.*, **303** (13): 575–580.

Lubet P., Aloui, N., Karnaukhova N., 1986b. Recherches expérimentales sur l'action de la température sur le cycle de reproduction de *Mytilus galloprovincialis*, comparaisons avec *M. edulis. C.R. Acad. Sci., Paris*, **303** (12): 507–512.

Lubet P., Brichon G., Besnard J. Y., Zwingelstein G., 1986c. Sexual differences in the composition and metabolism of lipids in the mantle of the mussel, *Mytilus galloprovincialis* Lmk. *Comp. Bioch. Physiol.*, **48** (B) (3): 279–288.

Lubet P., Besnard J. Y., Faveris R., 1987. Physiologie de la reproduction de la coquille St-Jacques. *Océanis*, **13** (3): 265–290.

Lubet P., Aloui N., 1987. Limites létales thermiques et action de la température sur les gamétogenèses et l'activité neurosécrétice de la moule *Mytilus edulis* et *M. galloprovincialis. Haliotis*, **16**: 309–316.

Lubet P., Mathieu M., 1978. Experimental studies on the control of annual reproductive cycle in Pelecypod molluscs. *Gen. Comp. Endocrinol.*, **34**: 109.

Mann R., 1979a. Some biochemical and physiological aspeces of growth and gametogenesis in *Crassostrea gigas* (Thg) and *Ostrea edulis* L., grown at sustained elevated temperatures. *J. Mar. Biol. Ass., U.K.*, **59**: 95–110.

Mann R., 1979b. The effect of the temperature on growth, physiology and gametogenesis in the Manila clam *Tapes philippinarum* (A. and R.). *J. Exp. Mar. Biol. Ecol.*, **38**: 121–133.

Marteil L., 1960. Ecologie des huîtres du Morbihan: *Ostrea edulis* et *Gryphaea angulata. Rev. Trav. I.S.T.P.M., Paris*, **24** (3): 329–446.

Mason J., 1958. The breeding of the scallop *Pecten maximus* L., in the Manx waters. *J. Mar. Biol. Ass. U.K.*, **37**: 653–671.

Mason J., 1983. Scallop and Queen Fisheries in the British Isles. *Buckland Foundation Books, Fish.* New Books: 1–142.

Mathieu M., 1987. Etude expérimentale des contrôles exercés par les ganglions nerveux sur l'évolution des gamétogenèses et les processus métaboliques associés chez la moule *Mytilus edulis* L. Th. Doct. Etat, Univ. Caen: 1–218.

Mathieu M., Van Minnen J., 1989. Mise en évidence par immunochimie des cellules neurosécrétrices peptidergiques dans les ganglions cérébroïdes de la Moule (*Mytilus edulis*). *C.R. Acad. Sci., Paris*, **308** (3): 489–494.

Matsutani T., Nomura T., 1982. Induction of spawning by serotonin in the scallop *Patinopecten yesoensis* (Jay). *Mar. Biol. Letters*, **33**: 353–358.

Matsutani T., Nomura T., 1984. Localization of monoamines in the central nervous system and gonad of the scallop, *Patinopecten yesoensis. Bull. Jap. Soc. Sci. Fish.*, **50**: 425–430.

Medhioub N., Lubet P., 1988. Recherches cytologiques sur l'environnement cellulaire (tissus de réserve) des gonades de la Palourde, *Ruditapes philippinarum. Ann. Sc. Nat. Zoologie, Paris*, **162**: 299–310.

Morse D. E., Hooker N., Duncan H., Jensen L., 1977. Hydrogen peroxyde induce spawning in molluscs, with activation of prostaglandin endoperoxydase. *Science*, **196**: 293–300.

Morton J. H., 1927. Observations and experiments on the sex change in the European oysters. *J. Mar. Biol. Ass. U.K.*, **14** (2): 101–117.

Nelson T. C., Allison J. B., 1940. On the nature and the action of Diantlin, a new hormone-like substance carried by the spermatozoa of the oyster. *J. Exp. Zool.*, **85** (2): 299–338.

Newell R., Hilbish T., Koehn R., Newell C., 1982. Temporal variations in the reproductive cycle of *Mytilus edulis* from localities on the east coast of the United States. *Biol. Bull.*, **162**: 299–310.

Nomura T., Ogata H., 1976. Distribution of prostaglandins in the animal kingdom. *Bioch. Biophys. Acta*, **431**: 127–131.

Ono K., Osada M., Matsutani T., Mori K., Nomura T., 1982. Gonadal prostaglandin E_2a profile during sexual maturation in the oyster *Crassostrea gigas* Tnbg. *Mar. Biol. Letters*, **3**: 223–230.

Osada M., Matsutani T., Nomura T., 1987. Implication of catecholamines during spawning in bivalve molluscs. *Int. J. Invert. Reprod. Dev.*, **12**: 241–252.

Paulet Y. M., Lucas A., Gerard A., 1988. Reproduction and larval development in two *Pecten maximus* populations from Britanny. *J. Exp. Mar. Biol. Ecol.*, **119**: 145–156.

Pipe R. K., 1985. Seasonal cycles and effects of starvation on egg development in *Mytilus edulis*. *Mar. Ecol. Prov. Ser.*, **24**: 121–128.

Plisetskaia E., Kazakov V., Soltitskaya L., Leibson R., 1978. Insulin producing cells in the gut of freshwater bivalve molluscs *anodonta cygnea* and *Unio pictorum*, and the role of insulin in the regulation of their carbohydrate metabolism. *Gen. Comp. Endocrinol.*, **35** (2): 133–145.

Sastry A. N., 1963. Reproduction of the Bay Scallop *Aequipecten irradians* Lmk. Influence of the temperature on maturation and spawning. *Biol. Bull.*, **125** (1): 146–153.

Sato E., Wood D., Sahni M., Koide S., 1985. Meiotic arrest in oocytes regulated by a *Spisula* factor. *Biol. Bull.*, **169**: 334–341.

Stefano G. B., Aiallo E., 1975. Histofluorescent localisation of 5 hydroxytryptamine and dopamine in the nervous system and the gills of *Mytilus edulis*. *Biol. Bull.*, **148** (1): 141–156.

Stefano G. B., Catapane E. J., 1979. Enkephalins increase dopamine levels in the central nervus system of marine molluscs. *Life Science*, **24**: 1617–1622.

Wilkins N., Mathers N. F., 1974. Enzyme polymorphism in the European oyster *Ostrea edulis*. *An. B. Groups Bioch. Genetics*, **4**: 41–47.

Zaouali J., 1974. Les peuplements malacologiques dans les biocœnoses lagunaires de Tunisie—Etude de la biologie de l'espèce pionnière *Ceratodesma glaucum*. Th. Doct. Etat, Univ. Caen: 1–130.

Gastropods

Bolognari A., 1953. Richerche sulla sessualità di *Haliotis lamellosa* (Moll. Gast. pros). *Arch. Zool. Ital.*, **38**: 361–402.

Boolotian R. A., Farmanfarmaian A., Giese A. C., 1962. On the reproductive cycle and breeding habits of two western species of *Haliotis*. *Biol. Bull.*, **122** (2): 183–193.

Cochard J. C., 1980. Recherches sur les facteurs déterminant la sexualité et la reproduction chez *Haliotis tuberculata*. Th. Doct. Sp., Univ. Brest: 1–167.

Girard A., 1972. La reproduction de l'ormeau (*Haliotis tuberculata* L.). *Rev. Trav. I.S.T. Pêches Marit.*, **36** (2): 163–184.

Kan-No H., 1975. Recent advances in abalone culture in Japan. *Proc. First Int. Conf. Aquacult. Nutr.*: 195–211.

Kikuchi S., Uki N., 1974. Technical study on artificial spawning of abalone (genus *Haliotis*). I—Relation between water temperature and advancing sexual maturity of *Haliotis discus-hannai*. *Bull. Tohoku Reg. Fish. Res. Lab.*, **33**: 69–78.

Lee J. Y., 1974. Gametogenesis and reproductive cycle of abalones. *Publ. Mar. Lab. Busan Fish. Coll.*, **7**: 21–50.

Lewis C. A., Leighton D. L., Vacquier V. D., 1980. Morphology of abalone spermatozoa before and after acrosome reaction. *J. Ultrastructure Res.*, **72**: 39–46.

Morse D. E., Hooker N., Duncan H., Jensen L., 1977. Hydrogen peroxyde induced spawning in molluscs, with activation of prostaglandin endoperoxydase. *Science*, **196**: 293–300.

Pena J. B., 1986. La gonada de *Haliotis discus* Reeve (*Gast. Prosobr*), y los factores que influen en su maturacion. *Iberus*, **6**: 2.

Shepherd G. A., Laws H. M., 1974. Studies on Southern Australian abalone (genus *Haliotis*). *Austr. J. Mar. Freshwat. Res.*, **25**: 49–62.

Tanaka Y., 1978. Spawning induction in the abalone, *Nordotis gigantea*, by addition of hydrogen peroxyde. *Proc. 7th Jap. Sov. Joint. Symp. Aquacult.*: 207–211.

Uki N., Kikuchi S., 1982. Technical study on artificial spawning of abalone—genus *Haliotis*. *Bull. Tohoku Reg. Fish. Res. Lab.*, **44**: 83–90.

Webber H. H., Giese A. C., 1969. Reproductive cycle and gametogenesis in the black abalone *Haliotis cracherodii* (*Gastropoda prosobranchiata*). *Mar. Biol.*, **4**: 152–159.

Yahata T., Takano K., 1970. On the maturation of the gonad of the abalone, *Haliotis discus-hannai*, I. A comparison of the gonadal maturation of abalones from Matsumal and Rebun in Hokkaido. *Bull. Fac. Fish., Hokkaido Univ.*, **21** (3): 198–199.

Cephalopods

Bakhayokho M., 1980. Pêches et biologie des céphalopodes exploités sur les côtes du Sénégal. Th. Doct. Sp., Univ. Brest: 1–119.

Boismery J., 1988. Structure et développement des glandes annexes de l'appareil génital de la seiche femelle *Sepia officinalis*. L. *Bull. Soc. Zool. France, Paris*, **113** (3): 321.

Boletzky S. Von, 1983. *Sepia officinalis*. In: *Cephalopods life cycles*, Boyle (ed.), Vol. 1. Academic Press, London: 31–52.

Cuenot L., 1933. La seiche commune de la Méditerranée, Etude de la naissance d'une espèce. *Arch. Zool. Exp. Gèn.*, **75**: 319–330.

Dhainaut A., Richard A., 1976. Vitellogenèse chez les Céphalopodes décapodes; évolution de l'ovocyte et des cellules folliculaires au cours de la maturation génitale. *Arch. Anat. Micros.*, **65** (3): 183–208.

Jeon I., 1982. Etude des populations de seiches *Sepia officinalis* L. (1758), due Golfe de Gascogne. Th. Doct. Sp., Univ. Nantes: 1–20.

Lafont A., 1868. Journal d'observations faites sur les animaux marins du Bassin d'Arcachon pendant les années 1866–1868. *Act. Acad. Sci. Bordeaux*, **30**: 630.

Lemaire J., Richard A., 1970. Evolution embryonnaire de l'appareil génital: différenciation du sexe chez *Sepia officinalis* L. *Bull. Soc. Zool. France*, **95**: 475–478.

Mangold K., 1963 Biologie des Céphalopodes benthiques et nectomique de la Mer Catalane. *Vie et Milieu*, suppl. **13**: 1–285.

Mangold K., 1966. *Sepia officinalis* de la Mer Catalane. *Vie et Milieu*, **A17**: 961–1012.

Medhioub A., 1986. Etude de la croissance et du cycle sexuel de la seiche (*Sepia officinalis* L.) des côtes normandes. Th. Doct. Sp., Univ. Caen: 1–117.

Najai S., 1983. Contribution à l'étude de la biologie des pêches de Céphalopodes en Tunisie, Application à l'espéce *Sepia officinalis* L. Th. Doct. Sp., Univ. Tunis: 1–229.

Richard A., 1971. Contribution à l'étude expérimentale de la croissance et de la maturation sexuelle de *Sepia officinalis* L. (Moll. Céphalopode). Th. Doct. Etat., Univ. Lille: 1–304.

Richard A., 1975. L'élevage de la seiche *Sepia officinalis* L. (Moll. Céphalopode). *10th Eup. Symp. Mar. Biol., Ostende*, **1**: 359–380.

Van Den Branden C., Gillis M., Richard A., 1980. Carotenoid producing bacteria in the accessory nidamental glands of *Sepia officinalis*. *Comp. Biochem. Physiol.*, **66(B)**: 331–334.

Wells M. J., Wells J., 1959. Hormonal control of sexual maturity in *Octopus*. *J. Exp. Biol.* **36**: 1–33.

Young J. Z., 1971. The anatomy of the nervous system of *Octopus vulgaris*. Oxford Univ. Press (Clarendon), Oxford.

Part III
Crustacean farming: the biological basis

A. Van-Wormhoudt and
C. Bellon-Humbert

Introduction to crustacean farming

The physiology and biology of crustaceans are dominated by three major functions which are closely interlinked: these are moulting, reproduction and development. There is a clear relationship between the three as they are dependent on a number of common factors.

One of the characteristics of these functions is their rhythmic pattern. They are all under hormonal control which regulates them in a cyclic manner and are modulated by external factors, among which are the daily lunar cycle (12 h), a day/night cycle (nycthemeral) (24 h), seasonal and annual cycles. These factors are essentially connected with temperature and light or photoperiod.

These functions (moulting, reproduction and development) depend on digestion, laying down reserves and the quality of the food available; these will be discussed in Chapter 4 of this part.

Two types of crustacean are of particular interest:

1. Those which are an important component of commercial fisheries. These are either large individuals which command a high price, or crustaceans which can be harvested easily in large numbers (see Table 1).
2. Crustaceans which are important for aquaculture. This is for one of two main reasons:
    (a) they may (at least in some stages) be cultured as prey for other crustaceans or farmed fish;
    (b) they may be cultured as food for man.

Most of the crustaceans in the first category are very small and have a high rate of reproduction. They include cladocerans, artemia, daphnids, amphipods such as *Gammarus* species and also several shrimps such as *Palaemon varians* for sea trout. *Macrobrachium nipponense* is used in Russia as a supplementary feed in the rearing of certain species of catfish. Species in the second category include *Reptantia*, reared in water courses in temperate regions. Penaeid shrimps of several species are reared in subtropical and tropical parts of South East Asia, Latin America, the USA and Pacific islands. In fresh water, *Macrobrachium* (comparable to Penaeids in sea water) grows to a large size and can be cultured together with rice or various species of fish.

Table 1. Crustacean catches, 1985

	tonnes $\times 10^3$
Fresh water	
Crayfish	65
Carids	143
Sea water	
Crabs	887
Lobsters	55.5
Crayfish	83.8
Langoustins	61.4
Galathea (squat lobster)	10.3
Penaeids	1556
Deep water penaeids	62
Pandalids	226
Crangonids	55.8
Stomatopods	5.3
Euphausids	191
Others	69.9
Total	3472.2

Several other species of crustaceans have been used in rearing trials. These include lobsters, where juveniles are reared for restocking. The demand for lobsters and crayfish is largely met by catches from commercial fisheries in the Atlantic (Mauritania), off southern Africa, Cuba and Canada, although these fisheries are subject to quotas and the price therefore remains high. However, there are problems of cannibalism during the nursery and ongrowing stages with lobsters and also disease problems. At present, immediate prospects for commercial lobster farming are poor. Some experimental results are not widely known because they have only appeared in Russian literature. These concern the culture of small crustaceans which serve as food for other farmed species and the experimental production of shrimps for chitin, a product with around 400 known industrial uses.

In 1985, 3.4 million tonnes of crustaceans were captured; 55% of these were shrimps; crabs and spider crabs made up a further 26% (Table 1).

On top of these totals, 116 000 tonnes of shrimps were cultured in the marine environment and a further 3000 tonnes of the prawn, *Macrobrachium*, in fresh water. Most of the species cultures are the tropical shrimps *Penaeus monodon*, *P. vannemi* and *P. stylorostris*.

1

Crustacean development

The development of crustaceans takes place in two stages: one inside the egg—this is embryonic development; and the other following hatching—this is the larval stage which is separated from the adult stage by metamorphosis. However, there are many variations on this basic development plan between species.

1. EMBRYONIC DEVELOPMENT

Although there is some variation at the different stages, the embryonic development of crustaceans shows a remarkable fundamental uniformity in the formation of the blastula and the presence of a nauplius stage. For some groups this stage marks the end of embryonic development, at which point hatching takes place. In general, this is true for species which have small eggs and therefore poor yolk reserves. For species with heavy eggs, rich in yolk, development generally continues within the egg and the stage which hatches is the zoaea, mysis or even a juvenile which is very similar in appearance to the adult.

The crustacean egg generally shows centrolecithal cleavage around a central yolk. Segmentation is spiral, leading to the formation of a blastula where, in all crustaceans, cells take up clearly defined positions on the surface of the egg. There is a remarkable consistency in the pattern of cell division and migration between taxonomic groups.

Embryonic development is best studied in species which incubate their eggs in special pouches (ephippium) or in hatcheries. Species studied include cladocerans, copepods, cirripedes and some peracarids, as well as decapods. The sequence of stages appears to be the same in all species where the eggs develop directly. It is at its fastest at the beginning, in the egg. This pattern of development takes place in other crustaceans; copepods, ostracods, euphausiids and penaeids of the decapod group.

1.1 Embryonic development of the shrimp, *Palaemon serratus*

The first division of the egg, which measures around 600 μm in diameter, quickly establishes a polarity, with a vitelline pole and a protoplasmic pole, through migration of cells under the egg membrane (Fig. 1a, b).

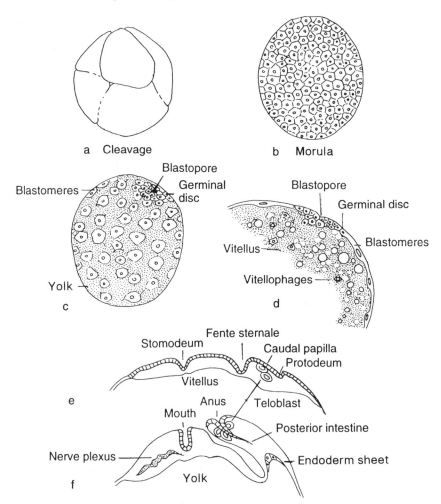

Fig. 1. Development of the egg of the shrimp *Palaemon serratus* (C. Bellon *et al.* in prep.). a:
Cleavage. b: Morula. c: disaggregation and migration of blastomeres. d: formation of the
blastopore and the germinal disc. e and f: median part of a more advanced embryo, showing the
formation of the caudal papilla and the start of differentiation of the nervous system and the
digestive tube.

Segmentation follows the formation of an animal pole made up of micromeres and
macromeres. It is not complete, leaving the central mass of the egg, the yolk, undivided.
On top of this mass, the blastoderm completes the embryo (Fig. 1a, b; Fig. 2a, b). This
forms an elongated disc comprising the areas which form the mesoderm, ectoderm and
endoderm which have already been determined. A small depression appears in the disc;
this is the blastopore and underneath it a series of yolk-digesting (vitellophagic) cells
develop and deepen. These cells begin to digest the yolk; this entails a movement of the

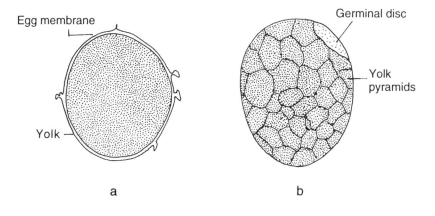

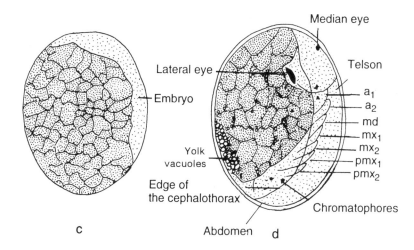

Fig. 2. Macroscopic observation of *Palaemon serratus* egg at the same stages of development (a, b and c) and at a more advanced stage showing the appearance of eye pigment, chromatophores and the differentiation of appendages (d).

blastoderm whose elements converge to form two cellular bands on the surface of the egg. These form a rough V-shape around the blastopore (Fig. 1c, d; Fig. 2c). This is the first indication of the embryo. Within this outline the endoderm is represented by the vitellophages, the mesoderm by a cellular mass coming from the proliferation of cells under the blastopore, and the majority of cells on the surface form the ectoderm of the embryo. This stage, which is remarkably similar in crustaceans, represents gastrulation. It can be defined as the movement of the presumptive areas in the blastoderm and their settling in the places where the organs will form. These movements are coordinated in time and space.

Very rapidly, in the posterior region of the blastopore, a band of cells develops and proliferates covering the blastopore and the mesodermal components. This caudal strip is the origin of the telson (Fig. 1e, f). The protodeum appears as a small hole in the pretelsonic ectoderm (in other crustaceans this is the remainder of the ectoderm) while at the front of the blastoderm a further small depression appears; this is the stomodeum.

During this phase the vitellophages arrange themselves in a regular layer around the yolk and are already forming the beginnings of the hepatopancreas, while the mesoderm cells proliferate forwards and towards the ventral part of the embryo (Fig. 2d). At this stage the first signs of the appearance of the external form of the crustacean can be seen; this is the nauplius. The two embryonic bands come together, each divides into three lobes, the first of which is the origin of the eyes and antennules, the second the antennae, while the third is post-oral. The labrum also differentiates. The three pairs of appendages, which are bi-ramic develop very rapidly and are covered with hairs.

In the small embryo the digestive tube consists of the following parts, developing along a gradient from the tail towards the head. The endoderm cells (vitellophages) come into contact in the caudal strip with the anus, and arrange themselves to form part of the intestinal tube which joins up with the wall formed by the vitellophages which envelop the yolk (Fig. 3a). At the anterior end the stomodeum has budded off cells which form the oesophagus. These cells, of ectodermic origin, become associated with mesoderm cells which form the musculature of the stomach. The junction is in the mid part of the intestine (Fig. 3b).

The nervous system develops from the proliferation of small cells from the anterior ectoderm which form a plexus, then a ventral chain of paired ganglia. Protocerebral ganglia can be distinguished from antennular (deutocerebrum) and antennal (tritocerebrum) and a large median postoral ganglion, the mandibular ganglion. The differentiation of the nervous system takes place from the head backwards along a cephalocaudal gradient (Fig. 3a, b) and follows a totally opposite path from that of the digestive system.

In this nauplius, the mesoderm forms a mass under the ectoderm which progressively invades the anterior region to form the three first somites. Behind this the dividing cells arrange themselves symmetrically in the subterminal region. These cells ensure the growth of the posterior region of the embryo (Fig. 1e, f).

Some crustaceans hatch as nauplii; ostracods, copepods, cirripedes, cephalocarids, branchiopods and penaeids. Their caudal region is developed to a varying degree. Other crustaceans such as Palaemon serratus continue to develop within the egg. At this point the embryo is essentially composed of a head and a telson turned forward with the flexure which is characteristic of decapod embryos and with a ventral neural surface and a dorsal yolk sac. Other appendages begin to appear: maxillae 1 and 2, then three pairs of maxillipeds and the buds of the first three pairs of pereiopods in front of the swimming appendages. In the region where the caudal strip or telson curves towards the cephalothorax, the ecto- and mesoteloblasts are arranged in a ring and are constant in number for a given species, forming a region where the cephalothoracic and abdominal somites proliferate. The abdominal musculature, the musculature of the appendages and the endoskeleton differentiate from the cellular masses. Haemocoeliac cavities, notably the heart, appear between the digestive tube and the ectoderm, in front of the

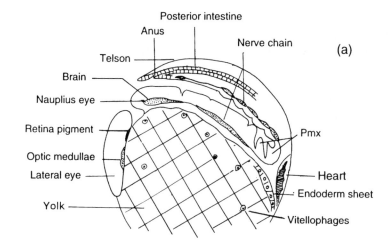

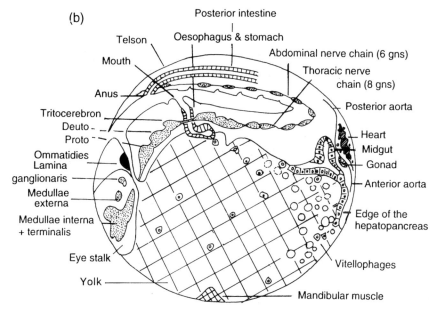

Fig. 3. Cross-section embryo formation of *Palaemon serratus*. (a) Differentiation of the lateral
eye (eye stalk), nervous system, heart, posterior intestine; epithelialization of the vitellophages in
the endoderm sheet. (b) Formation of the stomach, mid gut and central nervous system.

hepatopancreas. The anterior and posterior aortas differentiate rapidly (Fig. 3e, f). Other
paired cavities form in the antennal segment; these are the excretory antennal glands.

The primary germ cells differentiate under the pericardium and constitute the rudiments of the gonads.

At the same time, the nervous system has become highly differentiated. While the
somites are forming, the suboesophageal ganglia and the abdominal nerve chain develop.

The eye differentiates with all its visual components. The Y organ is present and cuticulogenesis is active with the development of hairs and spines. A paired dorsal organ exists for only a short time but may represent the point of dorsal attachment for the mandibular muscles. In other species there is a form of embryonic gill. The appearance of this in shrimps coincides with the formation of the four lobes of the liver and the dorsal epithelium which then covers the remaining yolk (Fig. 3b, 4a). The yolk, which is extra-embryonic at the start of development, thus becomes intra-embryonic (Fig. 4b, c).

At this stage (the pre-zoaea), the embryo still lacks gills.

Hatching takes place following a moult inside the egg. The zoaea larva can be recognized by the flattened blade at the end of the telson, the smooth rostrum and the sessile eyes (Fig. 5a).

For the offspring produced by a female during a single spawning, hatching takes place simultaneously, even though eggs may have been released on two successive nights. The larvae are dispersed by movements of the mother's pleiopods and swim actively with their maxillipeds.

1.2 Other crustaceans

Other crustaceans may hatch as more advanced zoaea (without pleiopods) or at the mysis stage (with pleiopods). This is true for eucarids. Some species hatch at the juvenile stage (Leptostraca, Syncarids, Peracarids, Astacura). Some crustaceans have a very characteristic larval form at the moment of hatching, such as the phyllosoma of spiny lobsters and the megalopa of crabs.

Embryonic development is of variable duration and relates not only to the amount of yolk present but also to temperature and to season.

In summary, the general pattern of embryogenesis is the same for all crustaceans with several variations; segmentation is spiral, the blastula stage is characterized by the formation of presumptive areas and by cellular migration. The development of the nauplius may be completed in the egg or outside through growth of the post-naupliar region. The formation of the organs in each presumptive area of the blastula is unique to arthropods and differs from the pattern in annelids and molluscs where segmentation is also spiral.

2. LARVAL DEVELOPMENT

Direct development is the general rule in cladocerans, phyllocarids and peracarids, i.e. hatching gives rise to young which resemble their parents although they are not sexually mature. Indirect development, which requires a series of transformations through a succession of larval stages, is normal for decapods with the exception of the Astacura (crayfish and lobsters) and fresh-water crabs (Potamon). Most crustaceans follow this indirect pattern of development and there is a wide range of larval stages within the phylum.

2.1 Larval development of the shrimp, *Palaemon serratus*

On hatching from the egg, the tiny larva is transparent with a few chromatophores on the ventral surface. The larva swims on its back, head first. It exhibits positive phototropism and does not feed until the yolk is absorbed. The larva has two large sessile eyes,

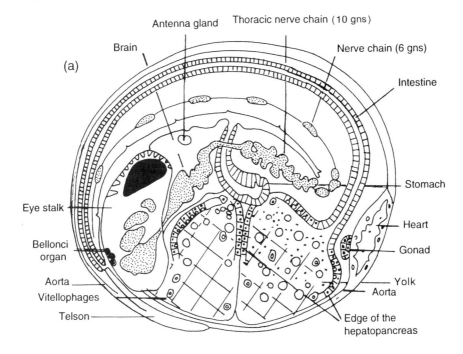

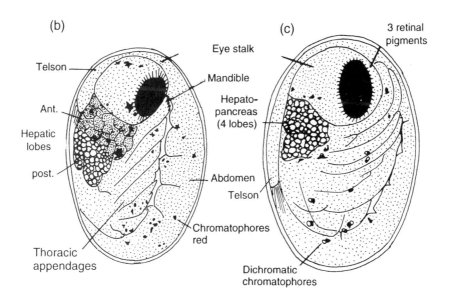

Fig. 4. Continuation of development of *Palaemon serratus*. (a) Section of an advanced embryo showing the formation of the hepatopancreas, differentiation of gonad, antennae and components of the eye. (b) and (c) Macroscopic observation of the egg before hatching, showing the embryo, thoracic appendages, the size of the eye stalk and the resorbic of vitellus (yolk) becomes intraembryonic (hepatopancreas).

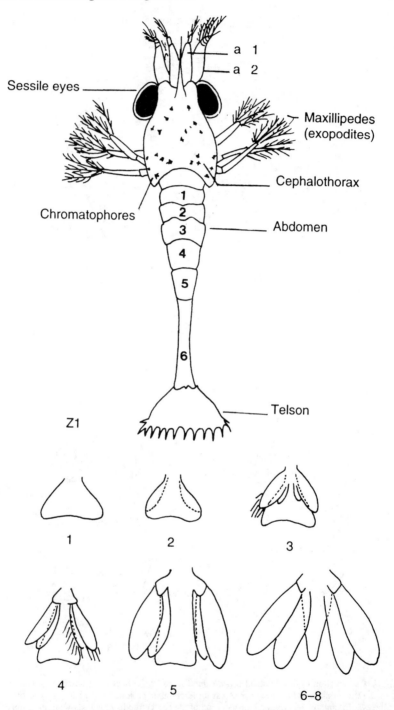

Fig. 5. Zoaea 1 of the shrimp *Palaemon serratus* after hatching and the development of the telson during the larvae stages (1 to 6–8).

antennae, a smooth rostrum and a buccal apparatus which is already similar to that of the adult. The larva moves by beating its exopodites and the three pairs of maxillipeds. It does not possess peropods, pleiopods or gills and respires through its very thin skin. The telson is a triangular, flattened blade. This larval form is termed a zoaea 1 and is succeeded, following a moult, around three days later (at 18°C) by a more developed form with two pairs of periopods, eyes on peduncles and the beginnings of the uropods inside the telson. Successive moults bring about changes in the other periopods and the release and development of the uropods (Fig. 5b). The larva thus becomes a mysis, after the appearance of the pleopods. The larvae are pelagic; a further metamorphosis gives rise to juveniles.

Metamorphosis changes the way in which the shrimps move about. The exopodites of the maxillipeds and periopods are resorbed. The animal then walks with thoracic legs and swims with pleopods. The carapace acquires the characteristics of the adult (spines, chromatophore system). The branchial system develops. The gills are typically those of adults. The changes also affect the form of the internal organs: lobes of the liver form, as do sensory pores, statocysts and the neurosecretory system. The gonoducts develop in their final place. The juvenile which emerges from this metamorphosis is sexually differentiated, is an omnivore and perfectly capable of responding to variations in its environment. Little by little, the internal organization arrives at that of the adult animal; this is achieved by the time of first sexual maturation when the shrimp can truly be described as an adult (Fig. 6).

2.2 Crustacean larval forms
Other than in the crustacean species which have direct development, all species pass through a succession of larval stages from hatching to development; these are separated by moults. Even at the point of hatching all are not at the same stage of development and different forms are recognized which correspond to a stage which is advanced to a greater or lesser degree.

The most frequently encountered forms are the: nauplius, zoaea, megalopa, and juvenile. Each of these forms can be characterized by its method of propulsion which corresponds to the degree of development of its appendages.

The nauplius larva (euphausiids, penaeids, sergestids, copepods, cirripedes) swims with cephalic appendages (three pairs of leaf-like appendages, see 'Artemia nauplius', page 72). When the antennae or the rudimentary thoracic appendages are used the larva is designated a *metanauplius*. At this stage, the nauplial eye, a small, median organ in front of the brain, is always present.

The zoaea larva is a characteristic stage of crabs and carids. Many shrimps hatch at this stage. Penaeids pass through this stage after that of the nauplius. Zoaea move using thoracic appendages, i.e. long exopodites fringed with hairs on the three pairs of maxillipeds and periopods at the end of the stage. The uropods are not free and appear in stage 3 (but are not present in brachyurids). In crabs, metazoaea are a final stage when the thoracic appendages appear in front of the maxillipeds.

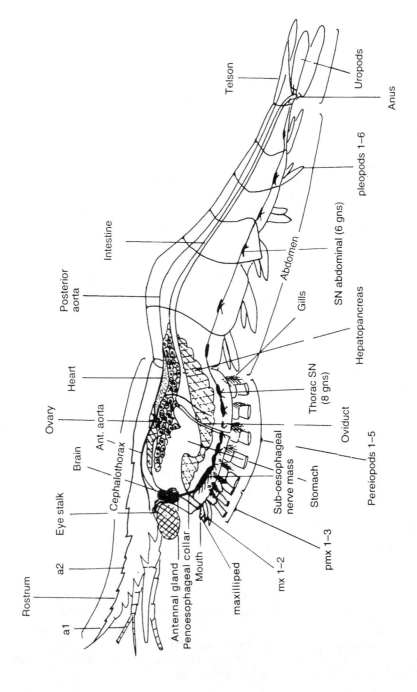

Fig. 6. Internal structure of the shrimp *Palaemon serratus* (adult).

The mysis larva is a larva whose pleopods are differentiated in the Natantia. It is really a zoaea in its final larval stages. The true pleopods are only found in the malacostracans. In penaeids and segerstids, a metamorphosis separates the two forms between stages 3 and 4. This metamorphosis is characterized by a marked change in the locomotory appendages; at the start these are the antennae and first maxillipeds, these are succeeded by the thoracopods.

The complexity of the forms at hatching and the series passed through has led to the use of a range of terms (see Table 2). The number of larval stages is constant in some groups of crustaceans, for others it varies according to the surrounding environment.

The megalops stage can be attained in one or several moults. It corresponds to the development of sexual identity and is separated from previous forms by metamorphosis.

There are extensive variations in larval forms which affect the appearance and form of the appendages and body which may be very different from that of the adult (spiny lobster phyllosome larva, crab zoaea). Although many of these stages are separated by metamorphosis, the question as to whether there is a true final metamorphosis to the adult stage must be asked.

3. TRANSITION TO THE ADULT FORM (METAMORPHOSIS)

Metamorphosis is a drastic change during the life of an animal which radically alters its form and biology.

It is possible that originally the development of crustaceans was a continuous process punctuated by moults, adding on somites and appendages to arrive at the adult form. This occurs in the anostraca where the biology of the animal does not change during its development.

However, other groups such as copepods, ostracods, cirripedes and malacostracans have a free-living larva which, after several moults, assumes the adult form which is very different from the first stage.

In copepods, the simplest example, the small copepodite larva is a nauplius which already resembles the adult and moults five times before sexual maturation. Apart from the final moult there is therefore no clear-cut metamorphosis. Decapods are different in that metamorphosis is very marked. There is a tendency for a slowing down in the development of the larvae when the eggs have been incubated. Only in the penaeids is larval development still very primitive.

In the nauplius stage, cephalic appendages help the larva to move through the water; in zoaea, movement is aided by the thoracic appendages and in the mysis stage by the abdominal appendages. This comes about through a structural reorganization with the development of some appendages and the regression of those which have become redundant (exopodites on the periopods). For carids, there is a morpho-functional reshaping which could be termed a metamorphosis. In penaeids there is no real distinction between larvae and post-larvae and the adult form is reached with no further well-defined change. Brachyurous types are very different: metamorphosis involves an enlargement of the cephalothorax and specialization of the thoracic appendages at the same time as the reduction in size of the abdomen.

Among the crustaceans there are some attached forms such as barnacles (cirripedes) and parasitic forms (isopods, Rhizocephalia) which exhibit the most spectacular metamorphosis. This entails a modification of appendages and other parts of the body because of the completely different mode of life of the adult animal. In contrast to this, the astacura (crayfish, lobsters) show, at hatching, appendages arranged on the same general plan as adults. The only missing organs are the uropods which are acquired after three moults.

Because the larvae are not highly specialized, crustaceans undergo progressive changes between the larval and the adult stages. Their metamorphoses differ from that of insects, both in duration and in extent. They are limited to the first part of the animal's life and have phylogenetic associations. They are both an adaptation and a method of dispersal through the habitat as well as a method of feeding. However, as with insects, the changes are directed towards the adult form and are associated with physiological and ecological modifications.

Table 2. Larval stages

Groups	Nauplius	Protozoea	Zoea	Mysis	Megalops
Copepoda	6	5 (Copepo-dites)		0	0
Cirripedia	6		1 (Cypris)		0
Euphausiacea	3–6	3 (Calyptopis)		2–5 (Furcilia)	n (Cyrtopia)
Penaeidea	5–8	3		2–5	1 (Parva)
Segerstids	2	3 (Elapho-ceres)		2 (Acantho-some)	n (Masti-gopus)
Cridea	0		5–9		
Palinuras	0		6–15 (Phyl-losomes)		1 (Puerullus)
Nephrops	0		3		1
Anomura	0		4–7		1 (Glaucothoe)
Maia	0		0–2		1
Brachyura	0		3–6		1–2
Squilla	0		0		6–12 (Alima or Erichtus)
Astacura	0		0		Juveniles

2

Moulting

Moulting is the result of a series of metabolic and morphological processes which, with the casting off of the exoskeleton (cuticle, carapace, together referred to as exuviae) allows the crustacean to grow. Growth takes place over a short period of time, immediately after moulting. It is a periodic phenomenon which occurs with varying frequency throughout the life of crustaceans. Moulting also allows vital processes other than somatic growth (the increase in size and weight of an individual) such as cleansing of the filtering apparatus which is used in feeding, casting off of parasites, regeneration of appendages, reduction in the density of the organism as an aid to dispersal (notably for planktonic crustaceans) and the reduction in weight of crustaceans which have a heavy carapace (lobsters). It is also necessary at the time of fertilization and the release of eggs.

1. GROWTH AND MOULTING IN ADULTS

1.1 Moulting frequency

In all crustaceans growth is discontinuous and takes place over a relatively short period following the casting off of the rigid carapace which surrounds the animal. The interval between two moults or the intermoult cycle is of variable length and is a function of the size of the animal and also of the time of year. When an animal is smaller and younger it moults more frequently than an older animal.

However, moulting rhythm is influenced by other external and internal factors. Constant conditions of light or shade, temperature drops, starvation (fasting) all slow down the moulting frequency. On the other hand, a rise in temperature, and extension of the photoperiod to an optimum level accelerate the frequency of moulting. The ectoparasite *Bopyrus* does not affect the frequency with which the host moults. Internal factors also have an effect; intensive regeneration of an appendage also induces moulting. In females, the moult preceding spawning is generally drawn out and incubation entails a major slowing down of the cycle equivalent to the blocking of a moult in gravid females. As a general rule, females have a longer moulting cycle than males. Finally, in some crustaceans, moulting ceases after sexual maturity has been reached (e.g. in majid crabs). This is termed anecdysis.

1.2 Growth

After each moult there is an increase in size which is around 10% for young animals. However, the growth index (the relationship between the increase and the initial size) decreases as the animal grows. Growth in weight follows the pattern of growth in length in a linear fashion. Some conditions are unfavourable for growth; animals which are starving or have a dietary deficiency do not grow. However, in other conditions which retard moulting, growth in length may be increased. While external parasitism has no influence on the index of growth, internal parasites (*Nectonema* in shrimps) have a bad effect, reducing the size of the hepatopancreas and the ovaries. During the period of reproduction, growth is often found to be reduced or nil. Finally, it should be pointed out that conditions in captivity may be unfavourable for growth.

1.3 Calculation of longevity

A knowledge of the mean growth index of the animal during each moult and the annual moulting rhythm allows an estimation of the lifespan of the animal to be made (Nouvel 1960). This is of major importance in culture and is also of assistance in growth studies.

In the shrimp *Palaemon serratus*, in the natural environment, the maximum size for males caught in the fishery is 71 mm and the lifespan is two years. Females reach a length of 80–106 mm and live for at least three years.

2. THE MORPHOLOGY OF MOULTING

Moulting is the process whereby the animal casts off its tegument; it is the term for a regular, identical operation and involves a coordination of metabolic and morphological processes.

These processes affect the tegument and are accompanied by changes which are easy to observe visually as well as changes to the hepatopancreas and blood, which play a major part in the metabolism of the individual.

2.1 Composition of the tegument

On a general plan, from the interior to the exterior, the tegument of crustaceans is made up of the following layers:

— The epidermis. Large cells on top of the hypodermis which contains the chromatophores, which are coloured cells whose pigment is capable of moving within a system of ramifying branches. The epidermis lies on top of a basement membrane.
— The endocuticle. This is made up of a bed formed by a chitino-protein complex which is not calcified or tanned, but is overlain by a calcified layer and finally a pigmented layer containing granules and tanned proteins.
— A thin epicuticle, without chitin, made up of lipoprotein. This forms the pocket surrounding the hairs (Fig. 7).

What happens to the tegument during the intermoult cycle? Microscopic observation of the fine appendages such as the pleopods, uropods and scaphognathites has helped to construct a picture of the development of the new exoskeleton underneath the old one and the destruction of the latter (Fig. 8). Drach (1944), for *Palaemon serratus*, defined five

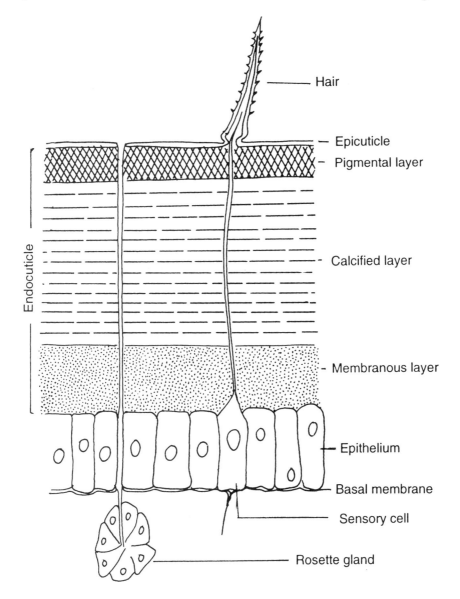

Fig. 7. Diagram of the structure of the tegument (after Dennell 1960).

major stages in the moulting process. These have been found in all crustaceans (Drach & Tchernigovtzeff 1967).

2.2 Metabolic processes during the intermoult cycle
The softening and decalcification of the old cuticle and the construction of the new one require a series of well-defined metabolic processes.

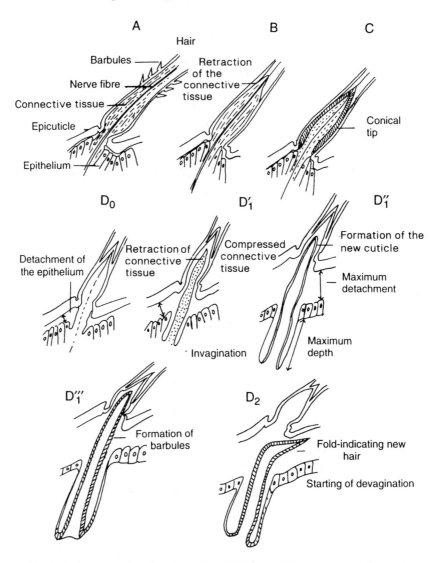

Fig. 8. Moulting stages, based on observation of the hairs of *Palaemon serratus* (after Drach 1944).

2.2.1 Post moult

In stage A (see Table 3) the animal which has just moulted begins to absorb water gradually. The water passes into the tissues and the hepatopancreas. During this stage a new endocuticle is secreted, up to stage A_2; this calcified layer is only built up during the post-moult period.

In stage B the calcified part of the endocuticle becomes mineralized with salts coming from the hepatopancreas and the pigmented layer becomes calcified in its turn. A major component of the mineral salts is drawn from the external environment. The tegument

Table 3. Stages during the intermoult cycle (in *Palaemon serratus*)

Stage A	2.5% of the period: Postmoult.
A_1	Skeleton very soft, protoplasmic matrix full of hairs.
A_2	Branchiostegites are still flexible under pressure from the finger. The matrix shrinks to the middle of the hair.
Stage B	16.5% of the period.
B_1	The branchiostegites are semi-rigid. The exoskeleton has a parchment like texture.
B_2	A conical case begins to form at the base of the hairs.
Stage C	21% of the period: Intermoult *sensu stricto*. The exoskeleton is completely formed, and has hardened. The conical casing of the hair has been completed.
Stage D	60% of the period: Premoult.
D_0	Short stage. The epithelium draws away from the cuticle.
D_1'	21% of the period. Start of the formation of the hair: Withdrawal of the nerve fibre from the cavity of the old spine.
	Formation of a circular fold around the fibrillar bundle.
D_1''	14% of the period. The circular fold becomes deeply invaginated and reaches its maximum depth. Start of the secretion of a new layer of cuticle.
D_1'''	6.5% of the period. Formation of the fine structures of the hair (hooks, barbs) and coloration of the new cuticle.
D_2	17% of the period. Formation of the pre-exuvial layer of the new exoskeleton.
D_3	Resorption of the old skeleton.
D_4	Opening of the 'moulting crack'.
Stage E	0.5% of the period: Moulting or exuviation.
	Throwing off the tegument, the animal extracts itself from the old exoskeleton which it then abandons.

hardens and the permeability of the epicuticle decreases. Tissues gradually lose their water, a phenomenon which continues throughout stage C. This water is gradually replaced by proteins. This laying down of living matter which leads to a growth in weight is only achieved at stage D_0.

2.2.2 Intermoult
In stage C reserves of organic matter are built up again and accumulate in the hepatopancreas at the same time as the new tegument is formed and tissues increase in

size. The reserves are chiefly in the form of lipids, fatty acids and glycerol (glycogen, glucose, glucosamine and acetylglucosamine which are polymerized to form chitin). Glycogen serves both as the basis for the formation of the new tegument (chitin) and as a source of energy. Proteins are used for somatic growth at this stage and are thus stored in the hepatopancreas to a lesser extent than lipids. The levels of calcium phosphate and magnesium phosphate in this organ also increase, up to stage D_1'. Minerals are largely extracted from the surrounding environment.

2.2.3 Premoult

In stage D_0 the epithelium separates from the membrane layer, and its cells enlarge and show signs of secretory activity. At the same time the level of glycogen in the hepatopancreas begins to decrease while oxygen consumption increases. The resorption of mineral and organic substances from the endocuticle has begun. The epithelium begins to secrete a liquid containing chitinase and a phosphatase. These processes continue through stage D_1.

In stage D_2, the endocuticle becomes detached and the contents pass into the blood. At the same time, decalcification continues and the new layers of the future tegument form: the epicuticle, beginning at stage D_1, and the pigmented and calcified layers. The level of proteins and lipids in the hepatopancreas shows a huge decrease. Part of these reserves is used to meet the energy requirements; the rest is used in building up the new exoskeleton. At this stage, the animal reduces its food intake.

In stage D_3, lines of resorption develop at the base of the claws and appendages, muscular attachment to the skeleton breaks, as does the articulating membrane joining the cephalothorax to the abdomen. The tegument has become thin and fragile.

2.2.4 Moult

There are two distinct phases in stage E: one is passive and corresponds to water absorption. The level of cholesterol increases sharply, and the level of glucose, protein and lipids in the blood increases. This entails an increase in osmotic pressure and the absorbtion of a huge quantity of water. The water is largely ingested via the mouth, then filtered and passed into the haemolymph. The volume of the blood increases, the animal 'inflates', pushing back the carapace. An active phase thus succeeds the passive one: the tegument opens and the animal emerges from the carapace. This requires energy which is supplied by oxidation of blood pigments, the level of which increases in the haemolymph. At the time of emergence, the respiratory surface is not functioning. The animal must remain still because of the drop in metabolism and the increase of Ca^{++} and Mg^{++} ions in the blood.

3. HORMONAL CONTROL OF MOULTING

The progress of the metabolic stages during the moulting cycle is thought to be regulated by the interplay of two hormones.

(a) One is a neuropeptide produced in the eyestalk or the protocerebrum in crustaceans with sessile eyes. This is the Moulting Inhibition Hormone (MIH). It is formed in the neurosecretory cells of the medulla which are grouped into X organs (Hanström 1989) and transported along axons to a gland, the sinus gland, where most of the neuropeptides synthesized in the optic ganglia accumulate before being liberated into the haemolymph. This hormone appears to act in a complex manner at the level of the cyclic AMP and also at the level of crustecdysone. In the absence of eyestalks, moulting is accelerated.

(b) However, other neuropeptides can interfere in the control of moulting: they include GIH (Gonad Inhibiting Hormone) which, more accurately, inhibits vitellogenesis. In the absence of this hormone, moulting is delayed in females during vitellogenesis. There is therefore an antagonistic effect between MIH and GIH.

(c) Other peptide hormones which are involved in metabolism are: hyperglycaemia hormone (CHH); hormones controlling the synthesis of proteins; a hormone similar to calcitonin which probably plays a role in the calcification of the tegument; a hormone similar to vasopressin which participates in water metabolism and the intake of Cl^-, Na^+ and Mg^{++} which is present at the time when the tegument is cast off; and a hormone similar to gastrin which has many effects on the digestive system. These hormones, which have a molecular structure similar to the corresponding hormone in vertebrates, have been investigated using immunocytochemistry and are isolated and purified from several species of crustaceans, particularly decapods where they are found in the eyestalk, an organ which is often compared with the vertebrate hypophysis because of its hormonal complexity. Study of these substances has shown that synthesis varies significantly throughout the intermoult cycle but their role is still poorly understood.

(d) Moulting hormone itself is a steroid. Several steroids are well known: 2O—OH ecdysone comes from the ecdysone synthesized from cholesterol in the Y-organ; ponasterone comes from 25-desoxyecdysone. The Y-organ is an endocrine gland in the maxillary segment, close to the gill chamber (Gabe 1953; Echalier 1954, 1955, 1956). This hormone has the function of triggering the premoulting process, acting on the epithelium and on mitotic division. In its absence the premoult stage is blocked and the crustacean remains in the intermoult; for crabs this is stage C or C_4.

As has been shown, some crustaceans cease moulting once maximum size has been reached. In *Maja squinado*, for example, this anecdysis is accompanied by an atrophy of the Y-organ, linked to a blockage brought about by the eyestalk (Fig. 9). In practice it is possible to unblock the premoult by injecting ecdysterone and accelerating intermoult by cutting the eyestalks, as long as external conditions, especially temperature, are optimal.

4. QUANTITATIVE ESTIMATES

It goes without saying that in farmed conditions it is essential to control the sequence of events leading up to moulting to provide the optimum yield or conversion. For this it is necessary to calculate the production of organic matter and energy for the formation of the eggs and the teguments which can be regarded as losses of organic matter to the environment. Changing conditions can lead to an increase in the weight of soft tissue (somatic growth), cuticular matter which is periodically cast off during moulting (growth of exuviae) and sexual products released during spawning (germinal growth).

Quantitative analysis of these factors (Khmeleva & Goloubev 1986) hinges on measuring the flow of material and energy in individual animals and in populations. Measurement of exuvial and germinal growth allow a calculation to be made of the overall waste

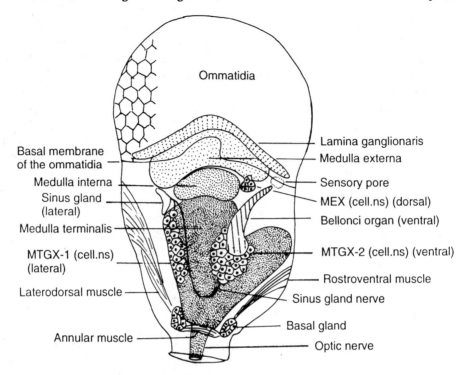

Fig. 9. Diagram of the structure in the eye stalk (*Palaemon serratus*).

production by crustaceans in the ambient environment and from this a deduction of total somatic growth. It has been estimated that the organic part of the exoskeleton of crustaceans falling to the sea bed each year weighs several billion tonnes (moulted exoskeletons and whole dead animals). The losses of material and energy are therefore highly significant.

4.1 Relative weight of the exuviae

The weight of the exoskeleton increases in proportion to the weight of the animal which is moulting. It is therefore possible to make a comparison between the weight of the exuviae (W_{ex}) and the weight of the animal before it has moulted (W). Exuvial growth or the relative growth of the exuviae is defined by the relationship W_{ex}/W. This weight is linked to body weight: 2.5% for planktonic branchiopods, 55–61% for benthic crabs and lobsters. The relationship varies during ontogenesis. Thus, for the crayfish *Astacus*, the index is 10% for larvae, 37–44% for adults and at its highest in females. The index is extremely high in the Reptantia (decapods) because the tegument is highly mineralized and very low in planktonic crustaceans. It is also linked to feeding: starvation increases the index by a factor of 1.3 in the shrimp, *Palaemon serratus*. It is also temperature-related, varying outside the optimum temperature limits. The index cannot therefore be regarded as constant.

4.2 Calorific value of the exuviae

The mean energy content of the exuviae is around 2.01 kcal g^{-1} dry matter and varies inversely with the degree of mineralization of the tegument. The most energy-rich exuviae are those of the cladocerans, anostracans and the planktonic stages of decapods, i.e. the crustaceans which have delicate teguments which are rich in organic matter. Energy content varies from 1.11 to 4.14 kcal g^{-1}. An increase in temperature leads to a decrease in the energy content. It is therefore possible to define the relationship between W_{ex}/W and the reserves in calories in terms of the growth of the tegument and the degree of mineralization of the carapace.

4.3 Duration of the intermoult stage (D_{ex})

This varies depending on the physiological state of the animal and factors such as temperature. Body weight is an important factor in the determination of this period.

$D_{ex} = PW^q$, where p, q are empirically derived factors which are calculated from experiments. As a function of temperature, it is possible to write:

$D_{ex} = D_{ex_0} \cdot q10^{-1}/10$ ($q10$ is the Van't Hoff factor = 2.60 for all classes of crustaceans. D_{ex_0} is the length of the intermoult at 0°C).

4.4 Growth rate of the exuviae (I_{ex})

This represents the quantity of matter or energy equivalent expended to increase the weight of the exoskeleton over a given period of time.

For 270 species of crustaceans of weight varying between 0.006 and 3380 mg, Khmeleva & Goloubev (1986) found the relationship (expressed in calories) to be:

$$I_{ex} = 0.51 W^{0.729 \pm 0.15}$$

W = dry weight.

4.5 Growth rate (dW/dt)

$dW/dt = A - T$, where A is the energy partitioned to assimilation and T is the energy expenditure in metabolism. T is a function of body weight, thus $T = a_2 W^{b_2}$. A is linked to the dietary ration R by a coefficient of assimilation, U^{-1} (constant). Thus $A = R.U^{-1}$ but R is also related to body weight: $R = aW^{b_1}$, from which $A = aW^{b_1}.U^{-1}$, where aU^{-1} is a constant = a_1:

$$A = a_1 W^{b_1}$$

Finally, the net growth rate can thus be expressed as:

$$dW/dt = a_1 W^{b_1} - a_2 W^{b_2} - I_{ex} - P_{ov}$$

I_{ex} corresponds to the exuvial rejects, expressed in calories and P_{ov} represents the energy equivalent production of gonad tissue during the reproductive period. These facts can be integrated to calculate the exuvial growth of populations at a given moment.

5. PRACTICAL APPLICATION OF INFORMATION RELATING TO MOULTING

The evaluation of the production of exuviae and the duration of intermoult stages are essential for several reasons:

1. *In the culture of crustaceans for the table* (a major industry), it is important to make the maximum possible reduction in the energy which goes into the wastes, favouring the use of energy for somatic growth. However, in shrimps, a decrease in the rate of assimilation results in a high production of exuviae. It is therefore necessary to increase the food supply, providing a source of energy, often in the form of sugar, to increase somatic production and obtain lighter, thinner exuviae. Shortening the intermoult period by extending the photoperiod, increasing temperature and cutting the eyestalk reduces chitin formation. Diet enriched with protein has a similar effect: for lobsters the frequency of moulting can be increased by 40–60% and the relative weight of the exuviae is decreased. It is a good idea to synchronize moulting; this prevents cannibalism on the animals which are weakened in the final stages of moulting. In Russia there is a trend towards the production of crustaceans as prey for various species of fish. It is essential to produce prey which are rich in organic matter and which have a thin cuticle.

2. *In trawling for shrimps* there are enormous losses at some time of the year because of mass moulting in warm waters. Low temperatures block the process of casting off the exoskeleton. It is therefore sensible to go fishing before or some time after moulting. Animals which are moulting do not survive transport well. Blocking metabolism by cooling down the water or using CO_2 reduces the respiration rate and stops the casting off of the exoskeleton.

3. Finally, *significant quantities of chitin* are produced in tanks. Chitin which falls into the water is destroyed by both aerobic and anaerobic chitinophagic bacteria of the genera *Flavobacterium* or *Actinobacterium*. Chitin is broken up by two enzymes, firstly a chitinase and then a chitinobiase which completes the destruction of the N-acetyl glucosamine compound, then to glucosamine which forms humic amino acids in silt (melamines): through this device, chitin plays a role linking living organisms. Minerals (calcium carbonate, magnesium carbonate, etc.) are returned to suspension in the water. Organic matter and minerals are consumed by other invertebrates and the exuviae can become part of the diet of detritivores (see Part I, Chapter 2). Some species of crustaceans and fish consume the exuviae to resorb the mineral salts (*Macrobrachium, Oronectes, Daphnia*).

Apart from the natural role of chitin in ecosystems, industry has a high demand for the substance. From 670 000 tonnes of crustaceans captured from the wild annually, 28 000 tonnes of chitin are extracted to produce a compound called chitosan.

Most of this chitin is extracted from krill (euphausiids). Chitosan, a natural organic compound, has various technical and medical applications. It is used in the purification of polluted waters and in the manufacture of filters for the recovery of particles of uranium, cadmium, lead and zinc in waters in colloid form. It increases the active life of pesticides, the hardness and viscosity of hydrosoluble colours and improves the quality of cellulose fibres in paper. In the food, medical and cosmetic industries it is used as a protective coating. It is also used in the manufacture of cosmetics and in medicines which are used for wound healing.

The world demand for chitosan shows no sign of decreasing. Most of this demand is met from sea fisheries but it is possible that exuviae from crustacean farms and large lakes could also be used to meet the demand.

REFERENCES

Dennell R., 1960. Integument and exoskeleton. In: *The physiology of crustacea. I. Metabolism and growth*, T. Watterman (ed.). Academic Press, New York: 449–472.

Drach P., 1944. Etude Préliminaire du cycle d'intermue et son conditionnement hormonal chez *Leander serratus* (Pennant). *Bull. Biol. France Belg.*, **78**: 40–62.

Echalier G., 1956. Effets de l'ablation et de la greffe de l'organe Y sur la mue de *Carcinides maenas* L. *Ann. Sci. Nat. Zool. Biol. Anim.*, **18**: 153–154.

Gabe M., 1953. Sur l'existence chez quelques crustacés Malacostracés d'un organe comparable à la glande de mue des insectes. *C.R. Acad. Sci. Paris*, **237**: 1111–1113.

Hanström B., 1989. *Hormones in invertebrates*. Oxford, Univ. Press. London and New York: 198 pp.

Khmeleva N. N., Goloubev A. P., 1986. La production chez les crustacés. Rôle dans les écosystèmes et utilisations (Transl. from Russian): C. Bellon. IFREMER: 198 pp.

Nouvel L., 1960. Moulting and its control. In: *The physiology of crustacea. I. Metabolism and growth*, T. Waterman (ed.). Academic Press, New York: 473–536.

3

Reproduction

Reproduction is the function which ensures the survival of the species and the multiplication of individuals.

Crustaceans, with the exception of some protandrous hermaphrodites such as the Tanaidacea, the cirripedes (barnacles), some parasitic isopods and, in the decapods, some shrimps belonging to the Pandalidae and to the Hippolytides are all dioecious, i.e. individuals are either male or female. However, parthenogenesis has been observed in cladocerans and ostracods. One species of isopod has been shown to be ovoviviparous. Strategies for reproduction and survival vary between species, and there is little uniformity in the length of the period of development and the stage when the young are released. This is largely dependent on the amount of yolk contained in the egg, the number of eggs spawned and the number of embryonic larval stages which are included or missed out.

Crustacean reproduction is a series of periodic events under hormone control which can also be influenced by external factors: photoperiod, temperature, salinity and diet.

Some crustaceans breed only once in their lifetime. These are called monocyclic (copepods). Others breed many times: these are termed polycyclic.

1. THE RELATIONSHIP BETWEEN MOULTING AND REPRODUCTION

In females the growth of the ovary and vitellogenesis determine the capacity for reproduction; in males the equivalent process is spermatogenesis. Reproduction and somatic growth in both sexes require the intake of building material and energy. The programme is highly organized in crustaceans. The hepatopancreas provides the lipid and glycogen reserves which are mobilized during moulting and vitellogenesis; the level of organic compounds show significant variations during these events (crabs, shrimps, crayfish).

In several of the Majids (*Maia* and *Libinia*) the moulting which occurs at the onset of sexual maturation (puberty) is the last moult which the animals undergo; from then on reserves are only required for reproduction as there is no further somatic growth. In other crustaceans (*Carcinus*, *Pachygrapsus*, *Palaemon*), the moulting process may cease three or four months after the moult which accompanies the first sexual maturation, but most crustaceans continue to moult and to spawn at the same time. There are seasons for reproduction during which, in some species, the moulting process is slowed down, or in others when moulting starts again after spawning. This happens in crabs in Tasmania, in *Pachygrapsus crassipes* and for *Carcinus maenas* in Brittany which spawns in February and moults between May and August.

The nature of the antagonism between moulting and reproduction is variable. For example, in crabs the period of somatic growth is entirely separate from that of ovarian growth. The Reptantia have a long intermoult (C = 66%) and a short premoult (D = 25%). This long intermoult period favours ovarian growth and the crabs are thus able to spawn one, two or three times during a single intermoult period. During the premoult phase they are in a period of sexual dormancy. The reproductive strategy of Natantia and amphipods is entirely different: intermoult only represents 25% of the duration of the moulting cycle for the first group and 40% for the second. The females are thus unable to achieve any oocyte growth during this short period. They must therefore complete ovarian maturation at the same time as somatic growth, i.e. during the building up of a new exoskeleton (D = 70–60% of the total cycle). The reserves in the hepatopancreas must therefore be sufficient to respond to both needs simultaneously.

In summary, antagonism between reproduction and moulting is extremely pronounced in the Reptantia where the long intermoult period allows at least one reproductive cycle. It is much less evident in the Natantia where energy reserves serve both the requirements of moulting (long premoult) and ovarian growth.

In the male, spermatogenesis is seasonal in many crustaceans, but in prawns (*Macrobrachium*) and shrimps (*Penaeus*) there is a coincidence between the production of spermatozoa and seminal fluid and moulting. Their production also requires a supply of energy and it is possible that there is antagonism between moulting and the production of sperm, at least in some species.

2. THE GENITAL APPARATUS

Typically, there are two genital glands which evacuate their products to the exterior of the animal via two collecting canals. These are the primary sex characters. There may be sexual dimorphism at the level of the secondary sexual characteristics which may be either permanent or transitory.

2.1 Female genital apparatus

There are two ovaries which are generally fused in the median sagittal plane. They are lodged in the dorsal part of the cephalothorax and are attached to two oviducts which open to the exterior; in the malacostracans the opening of each oviduct is situated at the level of the coxa of the third pair of periopods. Ovarian morphology can vary from one species to another, from a single cord-like body in *Palaemon serratus* to the multilobed body found, for example, in *Penaeus japonicus* (Fig. 10, 1 & 2).

In the anostracans (including Artemia) the ovaries are long, swollen vesicles and the ovaries join to form a large uterus which protrudes to the exterior, in an evagination of segments, forming an egg pouch.

Female copepods have a single ovary served by two oviducts which open at their head into a single orifice, the spermatheca, to which the male attaches the spermatophore. In the hermaphrodite cirripedes the ovaries are contained in a peduncle in clusters and the oviducts open between the first pair of thoracic limbs. There is also great variability in the secondary sexual characteristics. Permanent characteristics include the length of the pleopods, ciliation of the internal appendix which appears with maturation in female malacostracans, the presence of a seminal receptacle, the form of the cuticular plate of the thelycum in penaeids.

In species which incubate their eggs there is an incubation chamber which varies in size and shape between types of crustaceans. In the Rhyzocephales the eggs develop in the pallial cavity while in cladocerans (daphnia) the incubation chamber is in the cephalothorax and the oviducts open into the space between the body and the carapace.

The malacostracans which incubate their eggs do this in a type of abdominal incubation chamber which is formed by an elongation of the dorsal plate and the gap in the pleiopods. Female Stomatopoda (e.g. *Squilla*) carry their eggs stuck in a disc between their modified maxillipeds. Other characteristics are merely transitory and disappear after incubation has been completed, i.e. after the casting off of the tegument which follows the hatching of the larvae.

2.2 Physiology of the ovary

Within the ovary a series of biochemical and cellular processes takes the immature oocytes to a state of maturity.

Meiotic division is completed during oogenesis and the yolk is laid down. In several crustaceans the ovary differentiates in two regions, a germarium where the first stages of oogenesis take place (mitotic proliferation of primary germ cells, meiosis) and a vitellarium where the yolk is formed and transferred to the egg (cladocerans, copepods, *Orchestia*, *Clibanarius*, *Pachygrapsus*), but in most species this demarcation is less clear or does not exist at all. In the shrimp *Palaemon serratus*, the oocytes produced at the germinal crest increase in size gradually and change colour as they fill with yolk. This also happens in the shrimp *Penaeus japonicus* where the division of the oogonia takes place throughout the ovary followed by the accumulation of lipid and protein reserves in the oocyte (vitellogenesis). This period, termed ovarian maturation, occurs for the first time after a moult when the individual has reached a certain size or age. In many crustaceans, ovarian maturation and spawning take place seasonally.

During vitellogenesis yolk accumulates in the oocytes by a variety of processes and the composition varies in different crustacean groups.

Lipovitellin is a lipoprotein with an attached group (carbohydrate or carotenoid), but it may also contain proteins or carbohydrates. It accumulates in the form of platelets surrounded by a membrane. Both glycoprotein granules and lipoglycocarotenoprotein granules can be found in the same egg. Two phases in the formation of yolk can be distinguished. The first is slow; this is the phase of oocyte growth, the second, or secondary vitellogenesis, is characterized by the accumulation of yolk in the true sense in the egg.

Where does the yolk come from? At the start of vitellogenesis nuclei show signs of intense activity (fragmentation of the nucleoli) and the first granules are synthesized by the ergastoplasm of the perinuclear cytoplasm (*Oronectes, Homarus*). In *Libnia*, granules have been observed to form in the nucleus where they accumulate before passing to the cytoplasm. In *Orchestia gammarella*, the separation of the two phases is very clear: the glycoprotein component of the oocyte is synthesized by the ergastoplasm at the beginning of vitellogenesis; this then takes part in the incorporation of a lipoprotein compound from the haemolymph which penetrates the oocyte by pinocytosis. The biochemical and immunological similarity between lipoprotein and vitellogenin, the protein which appears in the blood of females which are becoming sexually mature and which is not found in the blood of males, is striking. Vitellogenins are proteins which may or may not have a carbohydrate group attached. These are carotenolipoproteins. Different studies which have been conducted on amphipods, isopods and Natantia show that these form in the fat bodies of crustaceans (cephalothoracic or abdominal). However, it is possible that the hepatopancreas, which empties during the second part of vitellogenesis, plays an active part in the formation of yolk.

2.3 The male genital apparatus

This is often more complex than the female equivalent in crustaceans. Typically, it comprises two testes served by two deferent canals which are joined together in the Malacostraca to the coxae of the fifth pairs of periopods. Attached to this, the deferent canal has, in its terminal portion, the androgen gland which is responsible for the differentiation of the secondary sexual characteristics. There may also be accessory glands.

In shrimps, crabs and crayfish, the male genital apparatus is very simple: The testes are two elongated twisted glands which are extended by a sperm duct which has the androgen gland on its terminal part; this is small and pyramid shaped (*Palaemon*).

In penaeids the testes fuse to form a lobed body from which come the deferent canals which terminate in a 'bulb' (Fig. 10: 1A, B). The canal is divided into two parts by a horizontal septum in the preterminal and terminal region. The greatest complexity in the male genital apparatus is found in the lower crustaceans. Sperm ducts in the anostracans have projecting gonopods, while in the ostracods, four seminal tubes join together in a long, crooked tube, closed at one tip but with a diverticulum in the middle which has a seminal vesicle, and ending in a penis. Deferent canals in copepods have accessory glands and a seminal vesicle and open into a glandular pocket, the spermatophore sac, which communicates with the exterior. In *Lepas*, a hermaphrodite cirripede (barnacle), the testis surrounds the digestive tube and the sperm duct ramifies to give five pairs of sperm vesicles in the cirrae which are linked to a long penis.

The secondary sexual characteristics in the male are permanent. They affect the first two pairs of pleopods in malacostracans (forming a petasma by modification of the endopodites of the first pairs that fuse in a strip which is unique to penaeids, for example). The second pleopod has a club-shaped male appendage, alongside the internal appendage carried by the endopodite.

2.4 Spermatogenesis

Testes are made up of extremely twisted tubes inside a very thin connective tissue sheath. Spermatogenesis takes place within the tubes. The germinal zone is divided into islets in a mesodermal tissue which is formed from nursery cells. The germinal zone is joined to the basal strip of the gonad. The spermatogonia can be distinguished by their large nuclei. The secondary spermatogonia are adjacent to each other, not separated by interstitial cells. There are many primary and secondary spermatocytes in the lumen of the tubules at

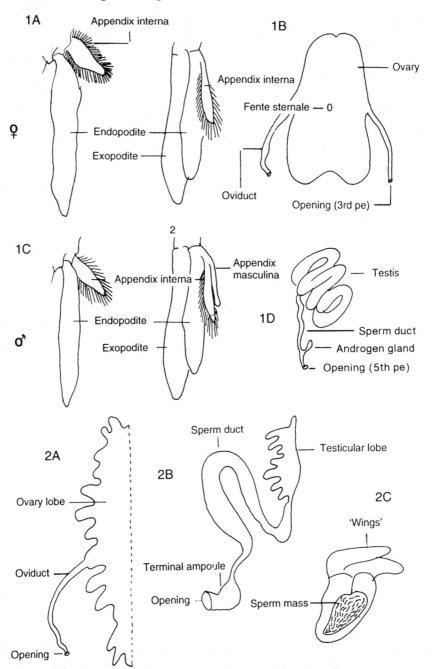

Fig. 10. Primary and secondary sexual characters. 1A: Pleopods 1 and 2 of the shrimp *Palaemon serratus* (female). 1B: Ovary. 1C: Pleiopods 1 and 2 of the shrimp, male, same species. 1D: Testis. 2A: Ovary of the shrimp *Penaeus japonicus*. 2B: Testis. 2C: Spermatophore of the same shrimp.

different stages of meiosis. Further down the tubes, spermatids have a smaller nucleus and indistinct chromatin. Spermatozoa, with the exception of those in the Mystacocarids, Cirripedes and Branchiura are not flagellated cells and are very different morphologically.

While the crustacean spermatozoa has a distinct head with an acrosome, a nucleus in the proximal part with protein inclusions, Golgi apparatus, mitochondria and a distal segment containing a chromatoid body of unknown function and an axial filament, there are many variants (Fig. 11: 1–4). In the decapods, spermatozoa may have a number of non-motile arms surrounding a central zone where the nuclear material is distributed; the middle part may be lacking and there may be no nuclear membrane. Where this occurs, chromatin is dispersed in the form of fibrils (*Penaeus setiferus*) or it may be the chromosomes which disperse in a spermioplasm after fusion of the nuclear membrane and the plasma membrane (crayfish). Crabs are even more complex.

As a general rule, spermatozoa show a great capacity for survival, even in anaerobic conditions. After meiosis, chromatin becomes condensed and reorganized with specific,

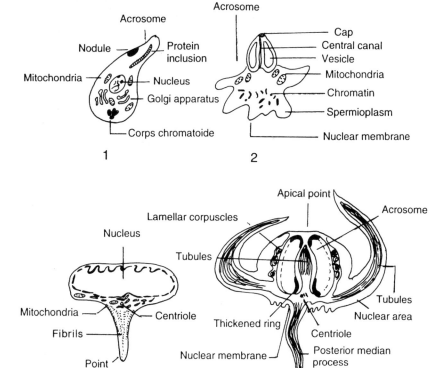

Fig. 11. Examples showing the diversity of structure of crustacean spermatozoa. 1: *Praunus* (Mysidaces). 2: *Astacus* (Reptantia). 3: *Palaemon serratus* (Natantia). 4: *Libinia* (Brachyura).

highly variable but characteristic proteins: these are protamines and histones. As intracellular reserves are low, the seminal fluid, containing phospholipids and glycogen ensures the survival of spermatozoa. It may also be rich in K^+, Na^+, glucose, phosphates, amino acids (taurine and lysine) and ascorbic acid as has been found in barnacles. The liquid is produced either by the epithelial secretions of the sperm ducts (crabs, crayfish) or in specific annex glands (cirripedes).

2.5 Spermatophores

In a large number of crustaceans, spermatozoa are not transferred to the female in a fluid but are protected by an envelope secreted by an accessory gland or by the sperm duct.

In the isopod *Porcellio*, the cells close to the orifice of the testicular follicule secrete a mucoprotein which envelops the spermatophore. In various decapods, this is made throughout the length of the sperm duct and the spermatozoal mass, during its progress through the canal, becomes coated with several layers of secretions (*Penaeus*). In decapods, spermatophores generally have protein walls which may or may not be tanned with quinones, tyrosines and phenols. These are deposited in the female's spermathecae, which may or may not be pierced, where conditions are favourable for survival. These probably allow a change in motility and the density of the membrane of the spermatozoa. Other crustaceans, such as carididae, have external fertilization which does not require such a means of transmission.

3. SPAWNING AND FERTILIZATION

Spawning is linked to a mechanism modulated by factors in the ambient environment in order that embryonic development should take place in the best possible conditions. Thus, heating up the water and extending photoperiod induce spawning: in crabs, the moment of spawning is determined by the tide. The stage during the intermoult cycle is also important. In the Natantia, the female must be emerging from moulting and at the start of the post-moult phase. However, in the Brachyura, spawning takes place during intermoult. In Natantia, there is often a second spawning immediately after the eggs hatch and the moult at the end of incubation. In this case there is an acceleration of the vitellogenesis of a batch of eggs lasting a single moult cycle, while the first batch to be spawned is being incubated.

The simplest process is that of external fertilization where spermatozoa, emitted directly by the male, fertilize eggs expelled to the exterior through the female's oviduct. The fertilized eggs are attached to hairs on the abdomen and pleopods by secretions from epithelial glands in these regions. The role of this cement is not only to ensure the attachment of the eggs but also to protect them. It may also have a bactericidal function (*Palaemon serratus*). In penaeids, copulation takes place after the female has moulted but the process is different: the male deposits two spermatophores into the thelycum and it should be noted that fertilization is independent of the state of maturation of the oocytes. If environmental conditions are favourable, oogenesis is accelerated, oocytes mature rapidly and are fertilized; they are then released at the same time as the spermatozoa from the spermatheca when spines pierce the wall of the spermatophore through the action of muscular contractions. Fertilization takes place externally and the eggs are thus dispersed in the water where development takes place. If, on the other hand, conditions are unfavourable, the ovary fails to mature in the intermoult following the attachment of the spermatophores. These are cast off with the exuviae and the cuticle covering the spermatheca at the following moult and the animal is ready to begin the fertilization cycle again.

The behaviour of the animals is very characteristic for each species during fertilization and spawning and includes a variety of postures and behaviour during aggression and courtship.

4. FACTORS INFLUENCING REPRODUCTION

4.1 External factors, genetics and nervous control
Apart from external factors such as temperature, salinity, photoperiod and diet, the following factors influence reproduction:

— Genetic factors: differentiation and development of the androgen gland responsible for male sex characteristics and behaviour are under genetic control. In crustaceans, genes responsible for the synthesis of vitellogenin are present in both sexes but are suppressed by the androgen gland in the male (Charniaux-Cotton 1960).
— Reproduction is also under nervous control. External factors, perceived by sense organs are transmitted to the central nervous system which thus modulates hormone action on the organs involved in reproduction: initiation or suppression of gametogenesis, production of vitellogenin and sexual receptivity. The androgen gland itself may be able to act on the brain affecting the perception of visual stimuli or pheromones.
— The most important factors acting on the control of reproduction are hormonal.

4.2 Hormonal factors
The eyestalk (Fig. 10) contains neuropeptides which inhibit both moulting (MIH) and the gonad (GIH). However, the process unblocked through eyestalk ablation varies greatly with species, age and season, in terms of somatic growth, ovarian growth and precocious spermatogenesis. The brain and thoracic ganglion appear to be linked to the synthesis of a hormone which stimulates the gonad (GSH) while the moulting hormone (MH), secreted by the Y organ, unlocks the moulting process.

During the breeding season, moulting occurs when MIH and GSH are at very low levels in the organism and GIH and MH are at their optimal. This is particularly true for the Natantia (Fig. 12: 1). It appears that a very small level of GIH is enough to block the processes of vitellogenesis and spermatogenesis. GIH acts indirectly through the action of the androgen gland on the testes. It blocks the maturation of the oocytes in the ovary. Ecdysteroids (ECD) responsible for the induction of the premoult stage are essential in oogenesis and the division of oogonia, in the synthesis of vitellogenin and the growth of oocytes.

In the Brachyura and Anomura, where somatic growth and reproduction are antagonistic, the Y organs are not essential for the synthesis of vitellogenin. However, high levels of ECD are found in the ovaries, confirming its role in the proliferation of oogonia which takes place during intermoult (*sensu stricto*). These ecdysteroids do not induce moulting

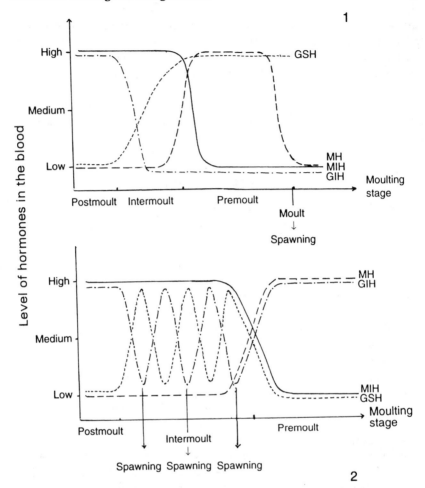

Fig. 12. Interaction between the moulting cycle and reproduction in Natantia (1) and Brachyura (2). Operation of hormones controlling the moulting cycle and reproduction. MIH = moult inhibiting hormone (———), MH = moult hormone (- -), GIH = gonad inhibiting hormone (—·—), GSH = gonad stimulating hormone (· · ·). After Adiyodi K. G. & R. G. 1970).

in the above-mentioned case but it is not known whether they come from the Y organs and are secondarily captured by ovarian cells.

In amphipods and isopods, somatic growth and ovarian growth go on at the same time and the Y organs are essential for vitellogenesis. Moulting occurs so frequently in the cirripedes that a single reproductive cycle is often interrupted by exuviation. Oocyte growth recommences at each premoult as a function of ECD. However, it should not be forgotten that ecdysteroids are primarily moulting hormones which only have a second-ary effect on reproduction.

It is probable that there are also ovarian hormones which control transitory sex charac-teristics (hairs, incubation chambers). Oestrogen activity has also been found in some crustacean ovaries, and oestrogen has recently been isolated and an enzyme system for the hydroxylation of progesterone demonstrated.

Finally, methylfarnesoate (juvenile hormone), synthesized by the mandibular organs (Laufer & Borst 1988) may also play a part in the control of reproduction and development.

The control of reproduction in crustaceans can therefore be seen to be extremely complex.

5. QUANTITATIVE ASPECTS OF REPRODUCTION

In the farming of crustaceans it is essential to be able to estimate the egg and sperm production in order to be able to calculate the cost to the individual and to the population. Several techniques are used to determine the relative and absolute fecundity in terms of weight or calorific value, the duration of the reproductive period, reproductive output in relation to the size of the animal and the temperature and the rate of gonad development. These parameters have recently been calculated (Khmeleva & Goloubev 1986).

REFERENCES

Adiody R. G., Adiyodi R. G., 1970. Endocrine control of reproduction in decapod crustacea. *Biol. Rev. Cambridge Philos. Soc.*, **45**: 121–165.

Charniaux-Cotton H., 1960. Sex determination. In: *The physiology of crustacea. I. Metabolism and growth*, T. Waterman (ed.). Academic Press, No. 11: 411–447.

Dennell R., 1960. Integument and exoskeleton. In: *the physiology of crustacea. I. Metabolism and growth*, T. Watterman (ed.). Academic Press, New York: 449–472.

Drach P., 1944. Etude préliminaire du cycle d'intermue et son conditionnement hormonal chez *Leander serratus* (Pennant). *Bull. Biol. France Belg.*, **78**: 40–62.

Drach P., Tchernigovtzeff, 1967. Sur la méthode de détermination des stades d'intermue et son application générale aux crustacés. *Bull. Lab. Arago. 'Vie et milieu'*, **28**, 3A.

Echalier G., 1954. Recherches expérimentales sur le rôle de l'organe Y dans la mue de *Carcinus maenas* L. (Crustacé decapode). *C.R. Acad. Sci. Paris*, **238**: 523–525.

Echalier G., 1955. Rôle de l'organe Y dans le déterminisme de la mue de *Carcinides* (*Carcinus*) *maenas* L. Exp. d'implantation. *C.R. Acad. Sci. Paris*, **240**: 1581–1583.

Echalier G., 1956. Effets de l'ablation et de la greffe de l'organe Y sur la mue de *Carcinides maenas* L. *Ann. Sci. Nat. Zool. Biol. Anim.*, **18**: 153–154.

Khmeleva N. N., Goloubev A. P., 1986. La production chez les crustacés. Rôle dans les écosystèmes et utilisations (Transl. from Russian): C. Bellon. IFREMER: 198 pp.

Laufer H., Borst D. W., 1988. Juvenile hormone in Crustacea. In: *Invertebrate endocrinology*, Vol. 2.

4

Digestion in crustaceans

1. THE DIGESTIVE TUBE

This is a simple tube made up of an oesophagus, a stomach, an intestine and a rectum. The branched digestive gland or hepatopancreas is situated in the midgut. Only this midgut comes from the embryonic endoderm and is not delimited by a chitin membrane.

1.1 The oesophagus and the stomach

In copepods this is a simple tube; the maximal complexity of this organ is found in the eucarids. In the langoustine (*Nephrops*), the oesophagus is short and has a tegumentary gland whose function is poorly understood.

The stomach is divided into two parts:

— The cardiac chamber, which, at some points, has a calcified cuticle forming ten ossicles making up a 'gastric mill'.
— The pyloric chamber, separated from the cardiac chamber by a valve taking the form of a dorsal caecum. This forms a filter which allows small particles to pass towards the hepatopancreas and contains the gastric juice.

In the Brachyura there are three clearly defined teeth: this is where the gastric mill is at its most complex. In the penaeid shrimps, which are very primitive, only the main ossicles are present.

1.2 The intestine

In calanoid copepods the intestine is partitioned into three sections, the walls of which have the same cell types as are present in the hepatopancreas of decapods. In decapods, the intestine is usually short. It may even be absent as in *Galathea* (the squat lobster) or slightly longer in *Homarus*, the true lobster.

The wall secretes a peritrophic membrane: two types of cells have been described. Its role in osmoregulation has been demonstrated in penaeids and *Carcinus,* as well as that in the absorption of lipids. The caecae are branched in this intestine. They number between one (penaeids) and three (brachyura).

In terrestrial amphipods two posterior caecae play a part in calcium regulation and excretion. Paracrine secretory cells have been described from here (Graf 1982).

1.3 The hepatopancreas

This takes the form of a tubular, glove-like structure which is diffuse in some crustaceans (crayfish) and compact in others (shrimps). Its main function is the secretion of digestive enzymes and the absorption of nutrients. No 'classic' endocrine cells have been described, except in *Carcinus* (Loizzi 1971). The hepatopancreas also secretes emulsifying agents. It is surrounded by muscles and nerves which effect the contraction of the tubules and the transport of enzymes and nutrients.

Four types of cell have been described: the embryonic E type, the fibrilla (F) type, and R and B types with many vacuoles (for reviews, see Gibson & Barker 1979 and Dall & Moriarty 1983). Classically, the embryonic cells give rise to R cells (absorption cells) which degenerate to form emulsifying agents (acylsarcosine, taurine or taurochemo-desoxycholic acid) and to F cells which play a part in the synthesis and secretion of digestive enzymes, themselves the origin of B cells.

B cells take part in intracellular digestion by means of pinocytosis (Al Mohanna & Nott 1987). Secretion may also be holocrine (in *Astacus*), merocrine (*Oronectes*), or apocrine. In penaeids, R cells are packed with lipids (Table 4).

Table 4. Development of cell types in the hepatopancreas

2. ENZYMES

2.1 pH of the digestive juice

In shrimps this is generally around 7 but it may be slightly acid (pH 5) in lobsters and crayfish.

2.2 Digestive enzymes

Approximately 80–95% of the digestive enzymes are found in the hepatopancreas. Some enzymes may be localized in the F cells (amylases and esterases). Each species has a characteristic enzyme spectrum which relates to its diet.

• **Carbohydrases**   Crustaceans have a wide range of such enzymes: around twenty have been identified although few have been the subject of detailed study.

Amylases: These represent 1–2% of the proteins in the digestive juice (lobsters, shrimps) although they are present in greater quantities in herbivores (*Crangon*,

Table 5. Characteristics of crustacean carbohydrates

Carbohydrates	Bond
Laminarine	$\beta1$–3
Cellulose	$\beta1$–4
Glycogen	$\alpha1$–4
Dextran	$\alpha1$–6
Agarose	
K carragenane	
Chitin (N-acetyl glucosamine)	$\beta1$–4
Pectin	
Disaccharides	
Laminaribiose	$\beta1$–3
Cellobiose	$\beta1$–4
Gentiobiose	$\beta1$–6
Maltose	$\alpha1$–4
Trehalose	$\alpha1$–1
Lactose	$\beta1$–4 gal–glc
Melibiose	$\alpha1$–6 gal–glc

palaemonids) than in carnivores (crayfish, lobsters). The number of isoenzymes is variable: from two in *Penaeus japonicus* to seven in *Palaemon elegans*. As in vertebrates, they are activated by NaCl and calcium ions.

Chitinases and cellulases: Chitinases and cellulases are produced by the hepatopancreas in many species (*Procambarus*) but it is possible that the bacteria in the digestive tract also play a part in the digestion of chitin and cellulose.

Laminarinases: In herbivorous and omnivorous crustaceans, laminarinases are important and take part in the depolymerization of many of the compounds from the phytoplankton; their action is completed by endo- and exo-1,3-glucanases.

• Proteolytic enzymes: endopeptidases

Crustacean trypsin is better understood than the endopeptidases: this has been recently sequenced and shows many similarities with vertebrate trypsin (43% similarity with bovine trypsin). In penaeids this enzyme represents 6% of the protein content of the digestive tube and in *Uca* it may represent over 25%. Its main characteristic (in contrast to vertebrates) is to be inactivated by acid pH and there is no sign of zymogen.

Chymotrypsin is present in many species, particularly the pink shrimp, *Palaemon serratus*.

Collagenase has been purified: it is a serine protease with a trypsic and chymotrypsic action but is also capable of hydrolysing collagen.

Protease, of low molecular weight is very important in crustaceans: it plays the same role as pepsin in vertebrates. It has a wide specificity. The sequence has recently been determined.

Besides these endopeptidases which are well known there are also cathepsines and rennet, the function of which is not yet well understood in crustaceans.

• Many **exopeptidases** have been described. The best known of these are carboxypeptidases A and B. Aminopeptidases also have an important role.

• **Lipases and esterases** hydrolyse triolein and triglycerides more slowly. Esterases act rapidly on ethyl butyrates. Fourteen different forms of esterase have been demonstrated in the pink shrimp, *Palaemon serratus*.

• **Nucleases** hydrolyse nucleic acids and have a significant activity in the digestive tube. Phosphatases and phosphodiesterases complete the action of the nucleases.

2.3 Variations in the activity of digestive enzymes
There are variations in the activity of digestive enzymes during growth, the intermoult cycle and circadian rhythms which allow a study to be made of variations in the diet of these animals and for the diet to be adapted to suit their needs: this is extremely important in aquaculture.

2.3.1 Variations during growth
In penaeids, enzyme activity is very low in nauplii but increases in the zoaea stage. Disharmonies in growth are evident during the zoaea–mysis transition, at the metamorphosis where the secondary sexual characteristics appear. One characteristic which has been found in some of the species that have been studied is an increase in the amylase/protease ratio as the animal increases in size.

2.3.2 Variations during intermoult
Intermoult in crustaceans is only comparable in animals under optimum growth conditions (temperature, nutrition). In winter, the slowing down of the moult is expressed by the absence of rhythms of variation in enzyme activity. In contrast, in summer maximum synthesis is measured during premoult. This synthesis is expressed by major enzyme activity or through the appearance of new isoenzymes.

2.3.3 Circadian rhythms
Endogenous circadian variations can be modulated by numerous environmental parameters. Several classic works and that of Van Weel (1960) have shown the importance of food intake on the rhythm of digestive enzymes. We have been able to confirm these results on shrimps and have also shown the importance of photoperiod in *Palaemon serratus* or darkness in penaeids.

In the wild, under natural conditions, there are certain endogenous rhythms: circadian biphasic rhythms (*Palaemon serratus*) or circatidal rhythms (*Palaemon elegans*). The interaction of these two factors, tides and circadian rhythms, can be difficult to predict.

Research has pinpointed the best time for feeding in farms, which is important when food is based on granules which have a limited life in water.

2.3.4 Role or environmental factors

A. Temperature
The regulation of the activity of amylase takes place through an increase in the affinity of an enzyme for its substrate (Km) measured *in vitro* at low temperature. The maximal affinity corresponds to the optimum temperature for the species. The adaptation temperature is the same as that at the level of synthesis of digestive enzymes: a direct compensation mechanism has been demonstrated for *Palaemon serratus*.

In other species, the nature of isoenzymes may change. Also, with a variation in the concentration of metabolites, different strategies are used in poikilotherms to bring about the thermal acclimation which permits survival. This is also valid for metabolic enzymes. In *Euphausia superba* for example, the optimal temperature for amylase is around 37°C.

B. Light
The quantity of light has an effect on digestive enzymes and on growth. The stimulatory role of green light and the inhibitory effect of red light have been demonstrated in *Palaemon serratus*. Light (through photoperiod) also has a synchronizing effect on circadian rhythm.

C. Nutrition
In crustaceans, the activity of amylases and proteases is modulated by the feeding regime. The induction of their synthesis in relation to diet has been demonstrated.

Optimum specific activities correspond to optimum quantities of nutrients in the granular diet. In *Palaemon serratus*, proteases have a maximum activity for 45% of proteins and amylases for 3% starch (Van Wormhoudt *et al.* 1980). The type of starch given is of prime importance; pregelatinized starch is not easily digested. In lobsters, high concentrations of starch have an inhibitory effect on amylase activity (Hoyle 1973). In the crayfish, *Procambrus*, the level of starch has no effect on amylases.

2.4 Hormonal regulation of the secretory activity of the hepatopancreas
The hormones which act on the digestive tube are also those which act during the intermoult cycle. Among these are the ecdysteroids, moulting inhibition hormone (MIH) and the hyperglycaemic hormone (CHH) (Van Wormhoudt *et al.* 1984).

Eyestalk ablation leads, particularly during premoult, to a drop in the activity of digestive enzymes which suggests that there is a stimulatory factor. 2O–OH ecdysone may be this factor (Van Wormhoudt *et al.* 1985). The existence of several factors which inhibit protein synthesis has also been demonstrated: one of these factors may be MIH which, *in vivo*, shows an inhibitory effect on the production of ecdysteroids by the moulting gland (Y organ). *In vivo*, 2O–OH ecdysone stimulates protein synthesis in the hepatopancreas in a linear fashion for concentrations between 1 and 100 ng g^{-1} PF.

Recently, it has been shown that purified hyperglycaemic hormone can act *in vitro* on isolated prawn hepatopancreas, stimulating the liberation of digestive enzymes. Future progress will depend on obtaining purified hormone which will allow the testing of the work of Skinner (1985) on the existence of an inhibiting hormone or that of Fingerman *et al.* (1967) and McWhinnie & Mohrherr (1970) of a stimulating hormone.

As well as these hormones which are specific to crustaceans, other hormones have been purified by antibodies against vertebrate hormones (Table 6).

Table 6. Hormones found in the crustacean digestive tube

	Biological activity	Authors
Calcitonin	Hypocalcaemia	Arlot *et al.* 1986
Gastrin/CCK	Secretagogue	Favrel *et al.* 1987
Insulin	Glycogenesis	Sanders 1983
Growth hormone	—	Toullec & Van Wormhoudt 1987

Peptides of the vertebrate type are ubiquitous in the nervous system (eyestalk, brain, stomogastric system, post-commissural organ, pericardial organ, nerve cord) and in the digestive tube. Of these, it is accepted that the gastrin/CCK plays a major part in the synthesis of digestive enzymes, their liberation and also in the contraction of the gastric mill.

The subtle interplay of these different hormones allows the control of the activity of the digestive tube.

3. ABSORPTION AND STORAGE OF METABOLITES

3.1 Absorption of metabolites
In decapods the midgut is too small to play an important role, but the hepatopancreas plays an essential part. The R cells absorb small metabolites (Ahearn *et al.* 1983) and take part in laying down reserves (lipids and glycogen) used in metabolism (Al-Mohanna & Nott 1987). Digestion of microparticles takes place in B cells in penaeids (Al-Mohanna & Nott 1986).

3.2 Laying down reserves
Many elements are stored in the hepatopancreas. The amount stored varies in relation to the stage of the moulting cycle. Thus for *Cancer*, proteins can go from 0.4 to 1.1 g, lipids from 0.3 to 2 g and glycogen from 0.016 to 0.3 g.

3.2.1 Lipids
There are major variations depending on the species: the hepatopancreas in pagurids may contain 74% by (dry) weight of lipids, while in penaeids the content is around 10%. These lipids constitute an important source of energy. Among the lipids, steroids and fatty acids can be distinguished.

Lipids can be synthesized from carbohydrates. Also, glucose can be transformed to glycerophosphate, a precursor of lipids.

Table 7. Lipid level in *Penaeus japonicus* (after Guary *et al.* 1974)

Phospholipids	7.7%
Mono and diglycerides	9.1%
Sterols	6.9%
Free alcohols	3%
Free fatty alcohols	2.2%
Triglycerides	53%
Esters	11%
Others	7.1%

The fatty acids are divided into:		
	saturated	28.4%
	monounsaturated	25.7%
	polyunsaturated	45.9%

The $\omega 3$ fatty acids are the most important (20–30%) followed by $\omega 9$ (15–20%) and by $\omega 7$ (10%).

Lipids can themselves take part in gluconeogensis. The control of this synthesis is under hormonal influence. The eyestalks contain an inhibiting factor (O'Connor & Gilbert 1968; Bollenbacher & O'Connor 1973).

3.2.2 Carbohydrates

The level of carbohydrates is generally low. The level of glycogen varies from 1.2 to 5% dry weight in the hepatopancreas of some species of penaeids. It is lower, around 1%, in the pink shrimp *Palaemon serratus*. The level is highest at the end of premoult. In *Carcinus* the glycogen in the hepatopancreas is at its highest in winter and its lowest in spring. In this same period the value of haemocyte glycogen is at its highest (circulating haemocytes) (Loret *et al.* 1989).

Free glucose is transformed to L-acetate, then pyruvate before giving the G^6P intermediary in the synthesis of glycogen (transformation to G^1P and uridine-phosphate glucose, UDPG).

Glycogen is mainly used in the synthesis of chitin. The hydrolysis of glycogen is carried out by glycogen phosphorylase or by pyrophosphatase (UDPG pyrophosphatase). During periods of fasting, reserves drop sharply. In *Astacus* which have starved for six weeks, 82% of glycogen reserves have been used compared with 18% of lipid reserves.

The only hormone concerned in carbohydrate metabolism which has been purified and sequenced is hyperglycaemic. This increases the level of circulating two hours after injection (5 pmol) *in vivo*; it acts through the intermediary of cyclic AMP and cyclic GMP in the hepatopancreas by decreasing the activity of glycogen synthetase and increasing the activity of phosphorylase.

Insulin-like compounds are likely to have a role in glycogenesis but they do not appear to have a role in the regulation of the level of glucose in the haemolymph (Sanders 1983).

3.2.3 Minerals and trace elements

Calcium is often deposited in the form of crystals of calcium phosphate in the cytoplasm of the R cells. In some species, it forms gastroliths in the stomach functioning with the gastric mill. This is resorbed in premoult to form the new exoskeleton in certain species. Only in some species of terrestrial or intertidal crustaceans (*Orchestia*) is the level high (8%) in organs which are specialized for the recovery of calcium (posterior caecum). In sea water the requirement is low. Copper and iron can also accumulate in the R and F cells respectively, either linked with the synthesis of digestive enzymes or with the synthesis of haemocyanins (pigments).

4. NUTRITIONAL REQUIREMENTS

The approach to these requirements has come from an understanding of artificial diet. Requirements have been most closely studied in Artemia. Kanazawa *et al.* (1970), Ceccaldi (1986) and Cahu (1986) were the first to examine the dietary regime of penaeids; Guillaume (1986 and 1988) has reviewed this work.

4.1 Proteins

The essential amino acids for crustaceans have been identified by the injection of radio-active acetate for the first time with *Crangon* in 1971. They are the same as for vertebrates: arginine, threonine, methionine, valine, isoleucine, leucine, histidine, lysine, phenylalanine and tryptophan.

The breakdown of protein results, essentially, in the production of ammonia. The digestibility of proteins in the diet is chiefly a function of the proportion of essential amino acids in the protein and, even more, of limiting amino acids. The proteolytic 'equipment', which varies from one crustacean to the next, may however be important for the kinetics of the liberation of these amino acids. The acids which are most frequently limiting are threonine, methionine and arginine for casein and histidine and tryptophan for protein from cod. However, it may be difficult to supplement granules with amino acids which are soluble in sea water: it is better to mix protein from different sources.

The quantitative protein requirements of shrimps varies with species and their natural dietary regime; from 20–30% for *Penaeus vannemi* up to 50–60% for *Penaeus japonicus* (New 1980). These requirements also vary with the size of the animals which become increasingly less carnivorous with age (Lee *et al.* 1984). Finally, the quality of the protein is also important; identical growth has been obtained with 30% crab protein and 40% casein protein in lobsters.

The mediocre quality of some fish meal where the amino acid composition is comparable to hydrolysates (CPSP) may be due to the presence of some inhibitory factors.

4.2 Lipids (see also Part IV, Chapter 2)

Decapod crustaceans synthesize saturated fatty acids and monounsaturated fatty acids from cellular acetate but are incapable of producing linoleic acid (18 : 3 ω6) and linolenic acid (18 : 3 ω3). However, they can desaturate these and lengthen them to form corresponding higher polyunsaturates. The speed at which these reactions take place is, however, too slow to cover the fatty acid requirements. The acids of the linolenic family have a high nutritional value: requirements for these compounds must be satisfied by the diet. Generally, supplements of lipids of marine origin such as cod liver oil, rich in long-chain polyunsaturated fatty acids (20 : 5 ω3 and 22 : 6 ω3) are added to the diet.

Crustaceans are not capable of synthesizing sterols but can rapidly convert dietary phytosterols into cholesterol: this compound is essential for growth. Decapods absorb sterols more effectively than mammals because of the presence of a specific emulsifying compound in their digestive tube.

Phospholipids also play an important role in lipid transport and are essential. The cholesterol requirement is generally 0.5–1% while that for lecithins (from soya or of animal origin) varies from 1 to 3%.

Cod liver oil or soya oil is often added at concentrations of 7% in order to be certain of covering the energy requirements.

4.3 Carbohydrates
Optimum levels of carbohydrates have been reported for certain species although these substances are by no means essential. These levels lead especially to a speed of hydrolysis or synthesis of carbohydrate constituents which vary from one species to another. Glucosamine has a beneficial role, due to the slow speed of the synthesis of the precursor of chitin in crustaceans. The addition of di- or triholosides appears to be more favourable than that of glucose. The addition of starch varies with species, rate of incorporation, the nature of the starch and its method of presentation. Its role as a binder may limit the leaching of certain nutrients and improve the nutritional value of the diet.

4.4 Minerals
Many studies have focused on the role of calcium and also of phosphorus. In fact the level of calcium in sea water is sufficient to satisfy the needs of marine crustaceans (Cheng & Guillaume 1984); however the requirement for phosphorus is very high: around 1.5%.

Minerals are given as a saline mixture with the pH generally adjusted to 7.

4.5 Vitamins
Little practical research has been carried out. The use of vitamin supplements is often enough to counteract any deficiencies. Carotene is the most frequently used source of Vitamin A. Choline, vitamin D and tocopherol are necessary. Deshimaru & Kukobi (1979) have shown that thiamin is essential as well as pyridoxine, B group vitamins and ascorbic acid (vitamin C).

These requirements may be different in adults in farms and in hatchery larvae where the presence of bacteria and symbionts may make up for the absence of vitamins in the diet.

4.6 Other factors
Some amino acids (taurine, betaine), peptides and nucleotides have an attractant effect, improving the appetence of a diet and, as a consequence, its effectiveness.

Other factors with a molecular weight of between 1000 and 5000 Da, such as extract of squid, are capable of stimulating growth at levels below 0.1% of the diet. These factors are not attractants and may even have a negative overall effect in aquaculture by stimulating the intake of excess food.

Finally, the possibility that certain hormones added to the diet may stimulate growth (as with growth hormone) cannot be ignored.

4.7 Diet manufacture

Purified artificial diets use a source of pure protein (soya, casein) to which are added purified products (cellulose, starch, cod liver oil) as well as a vitamin and mineral mix. Agar is generally used as a binder and helps to make a moist paste, although many other substances are also used (alginate, pectins, rubber, gelatine, chitin) and even synthetic binders (carboxymethyl cellulose, formol-urea).

Compound diets are obtained from many sources of proteins (fish, soya cake), carbohydrates (maize flour, wheat) and lipids (liver from various fish species).

Compound diets are manufactured under pressure: this produces granules which can be stored and handled more easily than fresh diets (see also Part IV, Chapter 2).

The different techniques used are extrusion, compression using moisture, cooking/extrusion which combines high temperature and pressure and, most recently, agglomeration which allows manufacture at high pressure but lower temperatures.

In larvae there is a move towards the use either of microparticles or microcapsules which are less polluting.

BIBLIOGRAPHY

Adiody R. G., 1985. Reproduction and its control. In: *The biology of crustacea, Vol. 9: Integument, pigments and hormonal processes*, D. E. Bliss (ed.). Academic Press, New York: 147–216.

Ahearn G. A., Monckton E. A., Henry A. E., Botfield M. C., 1983. Alanine transport by lobster hepatopancreatic cell suspensions. *Am. J. Physiol.*, 150–162.

Al-Mohanna S. Y., Nott J. A., 1986. B cells and digestion in the hepatopancreas of *Penaeus semisulcatus*. *J. Mar. Biol. Ass. UK*, **66**: 403–414.

Al-Mohanna S. Y., Nott J. A., 1987. R Cells and the digestive cycle in *Penaeus semisulcatus*. *Mar. Biol.*, **95**: 129–137.

Anderson D. T., 1982. Embryology. In: *The biology of crustacea, Vol. 2: Embryology, morphology and genetics*, D. E. Bliss (ed.). Academic Press, New York. 1–42.

Arlot Y., Van Wormhoudt A., Favrel P., Fouchereau-Péron I., 1986. Calcitonin-like peptides in *Palaemon serratus*. *Experientia*, **42**: 419–420.

Bellon-Humbert C., 1985. Développement embryonnaire de *Palaemon serratus*. Résultats préliminaires. Bases biologiques de l'aquaculture. Montpellier, 1983. IFREMER, actes de colloques No. 1: 181–194.

Bellon-Humbert C., Van-Herp F., Van Wormhoudt A., 1984. Localisation immunocytochimique de neuropeptides et d'amines biogenes dans le pédoncule oculaire de la crevette *Palaemon serratus* (Pennant). *Ann. Soc. Roy. Zool. Belg.*, **114**, suppl. 1: 164.

Bollenbacher W. E., O'Connor J. D., 1973. Production of an ecdysone by crustaceans Y-organ *in vitro*. *Amer. Zool.*, **13**: 1274–1276.

Ceccaldi H. J., 1986. Crustacés et nutrition: de la chaîne trophique à la consommation de l'homme. Colloque (ed.). CNERNA: 1–16.

Cahu C., 1986. Nutrition de larves de crustacés. Colloque (ed.). CNERNA: 37–56.

Charniaux-Cotton H., 1962. Androgenic gland of crustaceans. *Gen. and Comp. Endocr.*, Suppl. 1: 241–247.

Charniaux-Cotton H., 1980. Experimental studies on reproduction Malacostracea crustacean. Description of vitellogenesis and its control. *Rec. Adv. in Invert. Reprod.*, T. S. Adams & W. W. H. Clark (eds). New York.

Cheng W. W., Guillaume J., 1984. Etude de la nutrition phosphocalcique de la crevette japonaise (*Penaeus japonicus*). CIEM Comite Mariculture F12: 14 p.

Dadd R. H., 1970. Arthropod nutrition. In: *Chemical zoology*, Florkin et Scheer (eds), Vol. 5. Academic Press, New York: 35–95.

Dall W., Moriarty D. J. W., 1983. Functional aspects of nutrition and digestion. In: *The Biol. of Crustacea*, **5**: 215–261.

Deshimaru O., Kukobi K., 1979. Requirement of prawn for dietary thionine, pyridoxine and choline chloride. *Bull. Jap. Soc. Sci. Fish*, **45**: 363–367.

Faure Y., Bellon-Humbert C., Charniaux-Cotton H., 1981. La vitellogenèse secondaire chez *Palaemon serratus* (Pennant); ses relations avec le cycle de mue; son contrôle par l'organe X de la medulla externa (MEX). Premiers résultats. *C.R. Acad. Sci. Paris*, **293**: 461–466.

Favrel P., Van Wormhoudt A., Studler T. M., Bellon C., 1987. Immunochemical and biochemical characterization of gastrin/CCK-like peptides in *Palaemon serratus*. *Gen. Comp. Endocrinol.*, **65**: 363–372.

Fingerman., Dominiczak T., Miyawaki M., Oguro C., Yamamoto Y., 1967. Neuroendocrine control of the hepatopancreas in the crayfish *Procambarus clarki*. *Physiol. Zool.*, **40**: 23–30.

Gabe M., 1967. *Neurosecretion*. Gauthiers-Villars: 1091 pp.

Gibson R., Barker P. L., 1979. The decapod hepatopancreas. *Oceanogr. Mar. Biol. Ann. Rev.*, **17**: 285–346.

Graf F., 1982. Les cellules paracrines des caecums postérieurs de crustacés un nouveau type cellulaire du systéme endocrine diffus du tube digestif. *C.R. Acad. Sci., Paris*, **294**: 319–323.

Guary J. C., Kayana, Murakami, 1974. Lipid class distribution and fatty acid composition of prawn *Penaeus japonicus*. *Bull. Jap. Soc. Sci. Fish.*, **40**: 1027–1032.

Guillaume J., 1986. Besoins énergétiques, protéiniques, vitaminiques et minéraux chez les Crustacés. Collque (eds.) CNERNA, 99–122.

Guillaume J., 1988. L'aquaculture des crevettes: rôle de la nutrition. Actes de colloque, IFREMER, 241–246.

Hirsch G. C., Jacobs W., 1929. Der Arbeitrythmus der Mitteldarmdruse von *Astacus leptodactylus*. *Z. Vergl. Physiol.*, **A12**: 524–558.

Hoyle R. J., 1973. Digestive enzymes secretion after dietary variations on the american lobster. *J. Fish. Bd. Canada*, **30**: 11, 1647–1653.

Kanazawa A., Shimaya M., Kawasaki M., Kaskiwada Ki., 1970. Nutritional requirement of prawn. *Bull. Jap. Soc. Sci. Fish.*, **36**: 949–954.

Kleinholz L. H., 1985. Biochemistry of crustacean hormones. In: *The biology of crustacea, Vol. 9. Integument, pigments and hormonal processes*. D. E. Bliss (ed.). Academic Press, New York: 464–522.

Laubier A., Laubier L., 1988. L'exploitation mondiale des crustacés: bilan et perspectives. Actes de colloque, 8, IFREMER: 231–240.

Loret S., Van de Goor N., Devos D., 1989. Suspensions d'hémocytes et hépatopancréatocytes pour l'étude *in vitro* de la charge en glucose chez un crustacé Décapode. *Océanis*, **15**: 419–429.

Lee P. G., Smith L. L., Lawrence A. L., 1984. Digestive proteases of *Penaeus vannamei*: relationship between enzyme activity, size and diet. *Aquaculture*, **42**: 225–239.

Le Roux A., 1989. Contribution à l'étude du développement larvaire et de la métamorphose chez les Crustacés Eucarides. Thèse doct. d'état. Univ. Rennes.

Lockwood A. P., 1968. *Aspects of the physiology of Crustacea*: 303 pp.

Loizzi R. F., 1971. Interpretation of crayfish hepatopancreatic function based on fine structural analysis of epithelial cell lines and muscle network. *Z. Zellforsch.*, **113**: 420–440.

McLaughlin P. A., 1983. Intermolt anatomy. In: *The biology of crustacea*, D. Bliss & L. H. Mentel (eds). Vol. 5. Academic Press, New York: 1–52.

McWhinnie M. A., Mohrherr C. J., 1970. Influence of the eyestalks factors, intermoult cycle and season upon 14C-leucine incorporation into proteins in the crayfish *Orconectes virilis*. *Comp. Bioch. Physiol.*, **34**: 1415–1437.

New M. B., 1980. Bibliography of shrimp and prawn nutrition. *Aquaculture*, **21**: 101–128.

O'Connor J. D., Gilbert L. I., 1968. Aspect of lipid metabolism in crustaceans. *Amer. Zool.*, **13**: 1274–1276.

Sanders B., 1983. Insuline-like peptides in *Homarus americanus*. *Gen. Comp. Endocrinol.*, **50**: 374–382.

Sedlmeier D., 1988. The crustacean hyperglycemic hormone (CHH) releases amylase from the crayfish midgut gland. *Regul. Peptides*, **20**: 91–98.

Sollaud E., 1923. Recherches sur l'embryogénie des crustacés décapodes de la famille des Palaemoninae. *Bull. Biol. France Belg.*, Suppl.: 1-234, 5 pls.

Sollaud E., 1923. Le développement larvaire des palaemoninae. *Bull. Biol. France Belg.*, **57**: 509–603.

Skinner D. M., 1985. Moulting and its control. In: *The biology of crustacea, Vol. 9: Integument, pigments and hormonal processes*, D. E. Bliss (ed.). Academic Press, New York: 44–146.

Stevenson R., 1985. Dynamics of the integument. In: *The biology of crustacea, Vol. 9: Integument, pigments and hormonal processes*, D. E. Bliss (ed.). Academic Press, New York: 2–43.

Toullec J. Y., Van Wormhoudt A., 1987. Variations quantitatives de peptides apparentés à l'hormone de croissance humaine chez *Palaemon serratus*. *C.R. Acad. Sci.*, **305**: 265–270.

Van Herp F., Bellon-Humbert C., 1982. Localisation immunocytochimique de substances apparentées à la neurophysine et à la vasopressine dans le pédoncule oculaire de *Palaemon serratus* Pennant (Crustacé Decapode Natantia). *C.R. Acad. Sci., Paris,* **295**: 97–102.

Van Herp F., Van Wormhoudt A., Van Venroy, Bellon C., 1984. Immunocytochemical study of crustacean hyperglycemic hormone (CHH) in the eyestalks of *Palaemon serratus* and some other Palaemonidae in relation to variations of the blood glucose level. *J. Morphol.,* **182**: 85–94.

Van Weel P. B., 1960. Digestion in crustacea. In: *Chemical zoology,* Florkin & Scheer (eds), Vol. 5: 97–115.

Van Wormhoudt A., Ceccaldi H. J., Martin B. J., 1980. Adaptation de la teneur en enzymes de l'hépatopancréas de *Palaemon serratus* à la composition d'aliments expérimentaux. *Aquaculture,* **21**: 63–78.

Van Wormhoudt A., Van Herp F., Bellon C., Keller R., 1984. Changes and characteristics of the crustacean hyperglycemic hormone (CHH material) in *Palaemon serratus* (Pennant) Crustacea Decapoda Natantia during the first steps of the purification. *Comp. Bioch. Physiol.,* **798** (3): 353–360.

Van Wormhoudt A., 1987. Régulation hormonale de l'activité digestive de l'hépatopancréas des crustacés. In: *La nutrition des crustacés et des insectes.* CNERNA: 77–87.

Van Wormhoudt A., Porcheron P., Le Roux A., 1985. Ecdysteroïdes et synthèse protéique dans l'hépatopancréas de *Palaemon serratus. Bull. Zool. Fr.,* **110** (2): 192–204.

Vernet-Cornubert G., 1960. Connaissances actuelles sur le déterminisme de la mue chez les Decapodes et étude de quelques phénomènes qui lui sont liés. *Arch. Zool. Exp. Gen.,* **99**: 57–76.

Williamson D. T., 1982. Larval morphology and diversity. In: *The biology of crustacea, Vol. 2: Embryology, morphology and genetics,* D. E. Bliss (ed.). Academic Press, New York: 43–110.

Part IV
Biological basis of fish culture

G. Barnabé

1

Background to fish culture

1. GENERAL

As with all aquatic animals, fish have certain characteristics which suit them to their aquatic existence. They have an elongated fusiform body which is well adapted to moving in a fluid environment and are distinguished from other vertebrates by the absence of a neck, the presence of fins and scales, respiration through gills and many other distinctive characteristics and functions.

Two major groups of fish can be distinguished: the first is made up of rays and sharks and is of no concern to the fish farmer. These are cartilaginous fish (Selachians), differing from the bony fish of Osteichthes which make up the second group. The super-order Teleosteii is part of the latter group and comprises the majority of present-day fish and, with the exception of the sturgeon, all the fish which are currently farmed. All that follows in this part relates to the Osteichthes.

Fish are found from surface waters down to the depths of the oceans, from coasts to the open sea and in all fresh waters. Some are able to breathe air by climbing out of the water, and even up trees, but culture is based on fresh-water, brackish water and marine species which remain in the water.

Although fish inhabit the water in three dimensions, they do not all range through the entire environment. Some species live near to the surface of the open sea and seldom come near to the coast or down to the bed of the sea (pelagic fish); others remain near to the sea bed (benthic fish) and seldom leave the cracks in rocks where they live or remain buried in the sand. Other species seldom leave the sea bed; these are demersal fish. Not all species can be so easily categorized; some species are just as likely to be found around rocks close to the coast as in the open sea at depths of over 100 m, at the mouths of rivers or even in fresh waters. Eels, sea bass and sea bream belong to this category. These are termed ubiquitous fish, and it is worth noting that they are all farmed. In contrast to a fish such as a tuna, they can adapt readily to variable environments and diets.

Here we are only concerned with aspects which relate directly to aquaculture. Problems of systematics and population dynamics, growth studies, sampling, etc., are amply dealt with in other works (e.g. Traité de Zoologie, Grassé, and the physiology textbooks edited by Hoar & Randall, etc.). The bibliography at the end of the chapter aims to help the student in their research.

2. BIOLOGY, PHYSIOLOGY AND ECOLOGY OF FISH

2.1 General anatomy and morphology

The morphology and anatomy of a generalized fish are shown in Fig. 1.

The body is covered with scales, but these do not form the outermost layer of the body; they are embedded in the dermis and covered by a transparent, fragile layer of cells (Fig. 2). This epidermis is covered with mucus and is very different from the cornified layer of dead skin covering the outer surface of mammals. It forms the first barrier against microbial penetration; any handling can damage this layer and open the way to infections which can then spread. Any scale loss leaves a more serious hole but regeneration is possible. Specialized cells secrete mucus which has bactericidal properties and also makes the body of the fish slippery. Under the epidermis are the chromatophores which contain pigments of various colours and can alter the colour of the fish, through expansion or contraction. Fish such as flatfish can alter their colours to camouflage themselves.

Scales are calcareous structures arranged over the surface of the fish like tiles. Because their number is fixed from an early stage they grow in proportion to the growth of the fish. The cessation or the slowing down of growth in winter leads to the formation of a structure known as a winter band or annual ring (Fig. 3). This phenomenon allows scales to be examined to determine the age of a fish and also to work out the growth history of the fish in the preceding years by comparing the radius of the scale and the distance between bands with the length of the fish. This method is largely used with fish captured from the wild. Because, as in warm countries, feeding in culture is continuous, there is no such marked seasonality so the formation of annual rings is less marked. An example of methods used classically in the study of growth can be found in Barnabé (1976).

The skeleton can be seen in X-rays (Fig. 4; Tesseyre 1979). The vertebral column is in the form of an axis made up of vertebrae (the figure shows a deformity which sometimes occurs in rearing where vertebrae are squashed together). Muscles are attached to these vertebrae. In the farmed fish there are few intramuscular bones but this is not true for all fish; pieces of bone show the same marks as scales, and in fish without scales bones and otoliths are used in the study of age and growth. Fish bones are not hollow like those of mammals and there is no bone marrow; blood cells are produced in specialized zones within the kidney.

Some 40% of the weight of the body is striated muscle found on either side of the vertebral column; this muscle is responsible for swimming activity. The muscle is organized into myotomes and appears in the form of Z-shaped blocks on either side of the vertebral column. This is the part of the fish that is eaten. There are few blood vessels and this muscle is usually white, but salmonids (under the lateral zone) and pelagic fish have red muscle with a stronger taste. Physiologically, white muscles convert carbohydrates anaerobically to lactic acid while red muscles work aerobically.

White muscles characterize sedentary fish while red muscles are found in fish which are permanently swimming (pelagic fish). Salmonids are fish which are adapted to slow swimming (maintaining station in a current) and part of their musculature is red. According to Priede & Secombes (1988), the white muscle is not fully utilized, only coming into

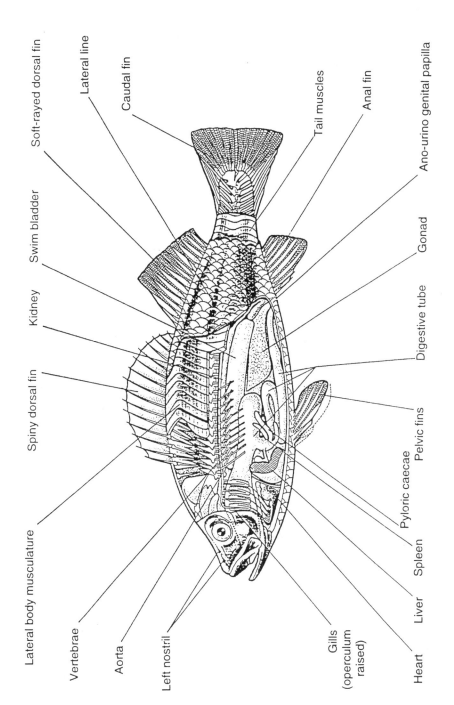

Lateral body musculature

Vertebrae

Aorta

Left nostril

Gills (operculum raised)

Heart

Liver

Spleen

Pyloric caecae

Pelvic fins

Digestive tube

Gonad

Ano-urino genital papilla

Anal fin

Tail muscles

Caudal fin

Lateral line

Soft-rayed dorsal fin

Swim bladder

Kidney

Spiny dorsal fin

Fig. 1. General morphology and anatomy of an Actinopterygien Teleost (perch); lateral view, after Dillon.

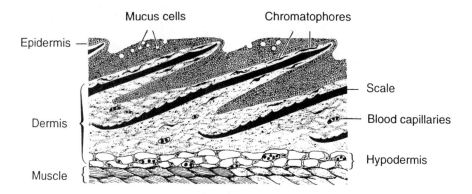

Fig. 2. Diagram of a cross section through the skin of a Teleost fish (from Priede & Secombes 1988). (Figure taken from Laird & Needham (E. Horwood publications).)

action in times of distress when the fish needs to escape through fast swimming when these muscles are operating anaerobically. The rest of the time they act as a store for energy. We shall see that in certain fish these muscles also act as a lipid store (Chapter 2, this part).

Fish mouths can open to a greater or lesser degree depending on the species and on the alimentary regime. During this movement the operculum closes; the mouth produces a suction force and many fish (groupers, bass and also salmonids) literally suck in their prey. Because of this, these predators have a different dentition to that of sharks or barracudas adapted to cutting up prey; these predators without sharp teeth have a mouth which can open wide and which is covered with tiny teeth whose purpose is simply to retain the prey that has been sucked into the mouth. Planktonophagic fish or burrowers capture minuscule prey in a sort of sieve which is formed by extension of the bones on the inner surface of the gill arches; these are the gill rays. A simple examination of these gives an indication of the likely diet of the fish (Fig. 5 et Furnestin, 1966).

2.2 Sense organs

Fish have a pair of nostrils on the snout. These are not connected to the pharynx and are extremely sensitive to odours. Salmonids detect their home streams through olfaction. Fishermen make use of attractant substances which are detected by the fish's sense of smell. Soluble amino acids are often incorporated in the diet of farmed fish (e.g. flat fish) to improve the appetite and the intake of food.

Sense organs permit fish to detect changes in salinity of as little as 0.2% or changes in temperature as tiny as 0.03°C. They are also able to react to falls in barometric pressure which are detected through alterations in the volume of the swim bladder; and can therefore descend to greater depths before storms, which is a phenomenon well known to fishermen. In fish farming, intake of food is also affected by rain and clouds.

The lateral line is a sense organ which is sensitive to low-frequency vibrations. This allows fish to detect movements such as currents, the presence of other fish (which helps in maintaining shoals during the night) or to hold a position in crevices in the rock where there is little light.

(a)

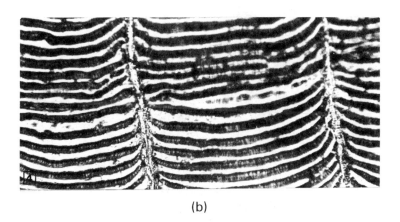

(b)

Fig. 3. (a) Scale from a sea bass *Dicentrarchus labrax* (L.) 8 years old showing 7 winter rings.
(b) Detail of a winter ring.

Many species recognize food particles by sight; the fish eye, well known for its wide angle of vision, is extremely sensitive to the movement of prey and less sensitive to details of shape (see also Chapter 4, this part). This eye is also sensitive at low light levels: larval bass are still able to chase prey at levels as low as 10 lux, and, in the wild, fish hunt by moonlight. They can also see above the surface but, because of the reflection of light at the surface of the water (Chapter 1, Part I), this vision may not extend beyond

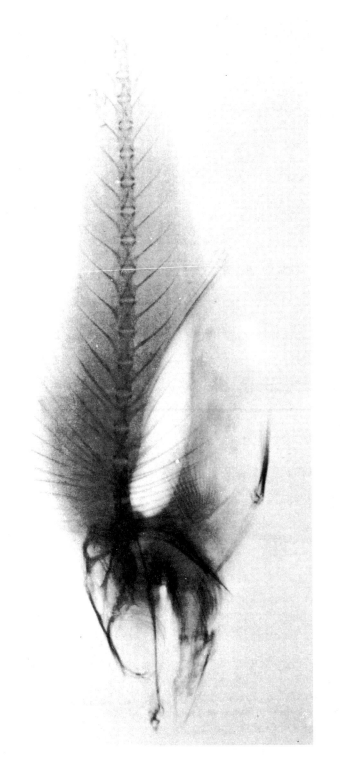

Fig. 4. X-ray of a farmed sea bass, showing an example of compacted vertebrae (above the swim bladder, which shows up as a light area). Tesseyre 1979.

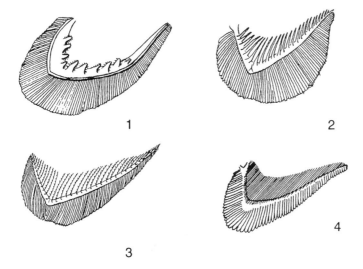

Fig. 5. Gill arches from several species of fish (after Maurin 1966). 1: *Pagellus erythrinus* (Spanish sea bream); carnivore. 2: *Spicara maena* (Picarel); omnivore. 3: *Scomber scombrus* (Mackerel); small fish, plankton. 4: *Sardina pilchardus* (Sardine), planktonophage.

an angle of 48° with the vertical. Many fresh-water species are able to attack aerial insects by leaping out of the water.

2.3 The digestive tube and associated organs

The digestive tube of the fish is shown in Fig. 6: the oesophagus leads to the stomach which is elongated in predatory fish, allowing them to swallow other fish which may be almost as long as the predator. This is particularly noticeable at the end of the larval stage in culture where alevins of various species attack weaker siblings: they ingest the posterior part and continue to swim with the anterior part of the body of their victim sticking out of their mouth. As the stomach breaks down the prey is drawn further into the mouth (this may take several days). For this reason it is important to watch rearing tanks carefully: cannibals are easily recognized by the possession of two heads!

The stomach leads laterally to the intestine which is short in carnivores and long in herbivores and planktonophagic fish (Fig. 7). Pyloric caecae are found at the exit point of the stomach. These are closed sacs which run into the intestine (Figs 6 and 8). Their number varies greatly between species (Fig. 7) and may reach many tens. The major part of protein digestion takes place in this region. There is a large liver (Fig. 6) and gall bladder. Fish can withstand prolonged fasting; up to a year for salmonids.

The tubular intestine forms loops in the body cavity. The spleen, a reddish organ, is found in one of the loops. Fat is deposited around the intestine. Protein is stored in the muscles and there is an intense mobilization leading up to spawning (Chapter 3, this part); as a consequence of this the quality of the flesh (i.e. the muscles) changes dramatically in salmon which have spawned (kelts), drastically reducing their commercial value. Muscle can therefore be described as a true storage organ. The structure of the digestive

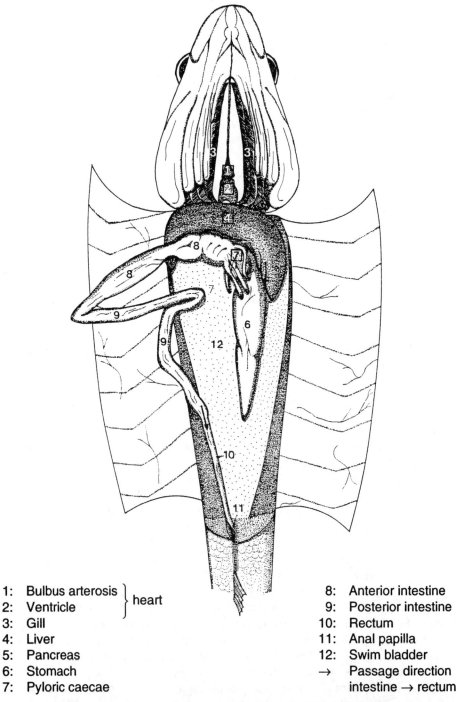

1: Bulbus arterosis ⎫ heart
2: Ventricle   ⎭
3: Gill
4: Liver
5: Pancreas
6: Stomach
7: Pyloric caecae

8: Anterior intestine
9: Posterior intestine
10: Rectum
11: Anal papilla
12: Swim bladder
→   Passage direction
    intestine → rectum

Fig. 6. Diagram showing the digestive system of the sea bass *Dicentrarchus labrax* (Benhalima 1982).

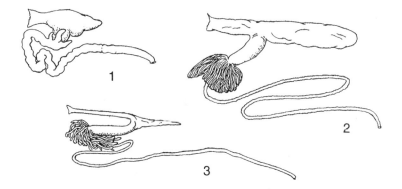

Fig. 7. Digestive tubes from three Mediterranean fish adapted to a different diet (from Maurin 1966). 1: *Pagellus erythrinus* (L), carnivore; 2: *Scomber scombus* L., predominantly a planktonophage; 3: *Sardina pilchardus* sardina Regan, feeds only on plankton.

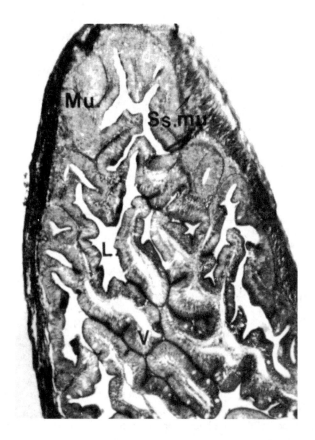

Fig. 8. Histology of the pyloric caecae. Longitudinal section, note the presence of many rangled villi, ramifying through the pyloric caecum. (One step trichrome) (× 35) (Benhalima 1982).
L = lumen, Mu = mucus, Ss mu = sub-mucosa, V = villus.

tube of fish has been well studied, notably for the sea bass, *Dicentrarchus labrax*. Fig. 9 shows a photograph of the ultrastructure of the intestinal epithelium of this species taken from Benhalima (1982).

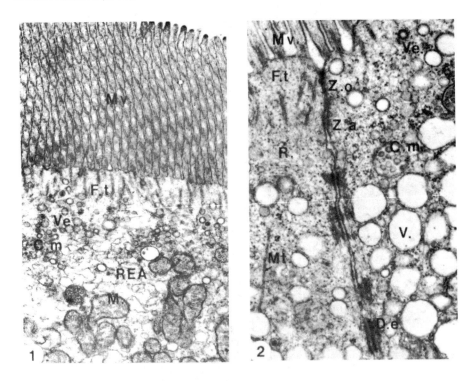

Cm: Multivesicular bodies	R: ribosome
De: desmosomes	REA: agranular endoplasmic reticulum
Ft: microfilament end	V: Vacuole
M: mitochondria	Ve: vesicle associated with pinocytosis
Mt: microtubule	Za: zonula adherens
Mv: microvillus	Zo: zonula occludens

Fig. 9. Ultrastructure of the intestinal cells of *Dicentrarchus labrax* (Benhalima 1982). 1. (× 19 000): Enterocyte. Note the microvilli, the ends of the micro filaments and the vesicles associated with pinocytosis at the tip of the cell. 2. (× 32 000): Enterocytes linked at the tip by a complex junction; zonula occludens, zonula adherens and numerous desmosomes. Note the presence of multivesicular bodies beneath the microfilament ends.

The kidneys appear as long reddish bodies running the length of the vertebral column, lodged between individual vertebrae, above the swim bladder (Fig. 1). Apart from the formation of the blood cells in the haematopoietic tissue, the kidneys are responsible for the production of certain hormones (adrenaline and noradrenaline).

The production of urine, the classic pathway for the excretion of wastes by the kidney, is part of a vast osmoregulatory system which involves several organs.

2.4 Osmoregulation

Fish are immersed in a medium where the level of dissolved salts is most frequently different from that of their internal environment, particularly that of the blood; different mechanisms allow them to regulate their osmotic pressure. This therefore permanently maintains their body fluids at the same concentration of dissolved salts whatever the concentration of the external environment. This is termed osmoregulation and requires energy taken from food. This energy is therefore not available for growth; the salinity of the external environment therefore has a direct impact on the utilization of food (see also Chapter 2, this part). Some fish can live in waters of very different salinities; this is especially true for salmonids which live for part of their life in fresh water and the other part in sea water. Other species are also capable of living in waters of variable salinity (eels, sea bass, sea bream, mullet, catfish etc.). Teleosts are different from other groups of more primitive fish; these, however, will not be discussed in this work.

The composition of the internal environment is regulated by the exchange of ions and water by several organs which are in contact with the internal or external environment; the gills, kidneys and digestive tube.

The 'salinity' of the blood of fish is intermediate between that of fresh water and sea water (fresh water contains 0.2 mM NaCl and sea water 500 mM NaCl) and often approaches 10‰. This means that fish are constantly fighting against an increase in the salinity of their internal environment in sea water and against a decrease in fresh water.

— In fresh water ions from the interior of the fish have a tendency to diffuse to the outside across the gills and water to enter by osmosis. This process is limited in two ways:

(a) The kidney produces huge quantities of urine made up of water with a low concentration of salts to compensate for the water flowing in across the gills.

(b) Ions lost in this urine are compensated for by those taken in by the gills by means of cells termed chloride cells (see below). These ions are supplemented by mineral salts extracted from the food. The fish can extract salts at low concentration from the surrounding water but is unable to survive in pure (distilled) water; there is no true "fresh-water" fish.

— In sea water, the internal environment of the fish is less saline than that of the environment; fish compensate for the osmotic loss of water which occurs at the gills by ingestion of sea water which is taken in through the mouth (drinking) and absorbed by the intestine. This absorption is compensated by an excretion of NaCl through the gills; the kidneys excrete very little urine (Fig. 10).

2.5 Respiration and circulation

The parts of the fish most exposed to the external environment are the gills, which are constantly bathed in a current of water in order to extract oxygen. This function requires a large area of contact because of the relatively low level of oxygen in water (Part I, Chapter 1). The red coloration of the gills comes from the high level of blood vessels to facilitate exchange between the blood and the external environment across the very thin epithelial layer.

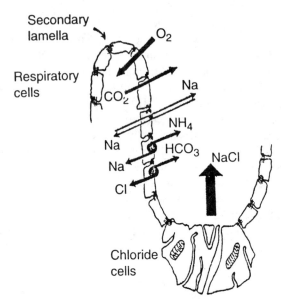

Fig. 10. Diagram showing the pattern of exchange of Na^+ and Cl^- across the primary and secondary lamellar epithelium in trout which have been adapted to sea water (Payan & Bornancin 1979).

The surface of the gills is made up of very fine lamellae laid out in filaments suspended from the branchial arches (Fig. 11); there are two rows of filaments on each gill arch. In fish the heart is situated under the oesophagus (Fig. 1) It is made up of four cavities in series: blood from the veins arrives in the sinus venosus, passes to the atrium, then the ventricle and finally the bulbus arteriosus. The muscular ventricle pumps the blood.

Blood leaves the heart through the aorta, penetrating the branchial arches through four afferent branchial arteries (Fig. 12a). These arteries ramify through each gill filament in an afferent filamentous artery, efferent filamentous artery and vein (Fig. 12b). In each filament there are capillaries situated between the afferent and efferent arteries and a central compartment formed by a huge sinus venosus situated between the filamentous arteries (Fig. 12b). This compartment is connected to the branchial vein but it is also crossed by connections between the arteries (Fig. 12b). These compartments are bordered by the branchial epithelium. This in turn is subdivided into the primary lamellar epithelium, which borders the vein, and the secondary lamellar epithelium, which delimits the arterial circuit.

The first of these is made up of numerous respiratory cells, mucus cells, and a few cells which are rich in mitochondria and are termed chloride cells because of their resemblance to the cells in the stomach which secrete hydrochloric acid. These are involved in the excretion of NaCl but are especially abundant in the epithelium of the secondary lamellae. The passage from fresh water to sea water changes the appearance of the chloride cells. Fig. 10 demonstrates the exchanges taking place at the level of the gills

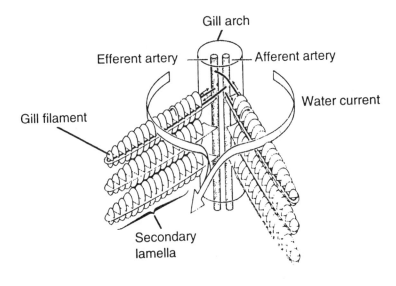

Fig. 11. Structure of the gills (adapted from Priede & Secombes 1988). The diagram shows part
of a gill arch seen from the opercular cavity, looking towards the inside of the mouth.

which are concerned with respiration, excretion of NH_4 and osmoregulation; these functions are not separate but are interdependent.

Respiratory and ionic exchanges take place across this epithelium; the mechanism has
been described elsewhere (Payan & Bornancin 1979–1980; Priede and Secombes 1988).
The transfer of oxygen to the organs is through red blood cells which are nucleated.

The swim bladder shows up clearly in X-rays (Fig. 4) because of the poor absorption
of X-rays by the organ. The formation of the swim bladder in larvae can sometimes cause
problems in culture (Chapter 4, this part) and in its absence the serious malformations
described by Tesseyre (1979) occur. The swim bladder ensures that the overall density of
the fish is low; however, some fish, notably those which live in the pelagic zone, have no
swim bladder.

Respiration of fish in water is made possible by regular movements and the alternation
of the opening of the mouth and the raising of the operculae. The branchiostegal rays
under the lower jaw support a supple membrane which is deformed by each movement of
the operculum and increases the capacity of the buccal cavity. Water, which enters
through the mouth, is expelled across the gills. While the kidney, the liver and the
digestive tube of the fish are analogous to those of other vertebrates, the gills form an
organ which is unique. The gills, with a large area of living tissue permanently bathed in
a current of water generated by the regular movement of the operculum, play several
essential roles. The branchial apparatus is composed of four branchial arches, each made
up of a bony articulated skeleton (the number of bones varies between species).

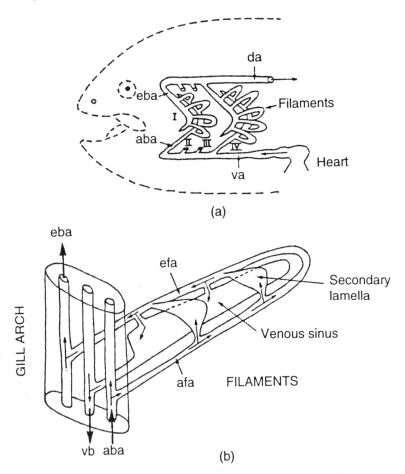

Fig. 12. Diagram of the gill circulation (Payan & Bornancin 1979). (a) Generalized circulation in the four gill arches (I, II, III, IV), (b) Details of the vascular spaces in the filament. aba = afferent arteria; da = dorsal aorta; eba = efferent branchial artery; va = ventral aorta. afa = afferent filamentous artery; efa = efferent filamentous artery.

2.6 Fins
The fins of fish are paired and single appendages (Fig. 1) made up of a membrane suspended by hard or soft rays. The paired fins are attached to bony girdles and are analogous to the limbs of higher vertebrates. The unpaired fins are variable in number. The caudal fin varies in shape and is involved in swimming; however it must not be forgotten that the whole body participates in this activity (Fig. 13).

2.7 The immune system
This system is made up of the white blood cells (leucocytes) which can be subdivided into lymphocytes, macrophages and granulocytes which have different functions. These

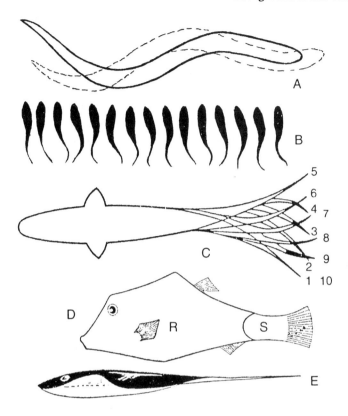

Fig. 13. Different methods of swimming in fish (Bougis & Barnabe 1980). A: *Anguilliform mode*: two successive positions of the body, shown in solid and dotted lines; the fish moves towards the right, the undulation moves towards the tail and is amplified. B: *Carangiform*: successive positions of the body at intervals of 1/25 of a second, a cuprinid swimming in a current of 1 m s^{-1} in order to hold position, although less marked than A, the undulations can be seen clearly. C: *Carangiform mode*: the figures indicate the successive positions of the tail and caudal fin during swimming. D: *Ostraciform mode*: the body of the puffer fish is enclosed in a rigid box and only the beating of the tail assists swimming, with the other fins. E: *Rajiform mode*: As in the anguilliform mode, the undulations increase in amplitude towards the posterior of the fish but the undulations are not in the body but along the edge of the enormous pectorals.

cells allow the fish to fight against invasion by foreign bodies and especially to recognize and eliminate pathogens which the fish has encountered previously.

Macrophages and granulocytes are unspecialized and consume any type of bacteria, producing the same reaction to all types of aggressor.

Lymphocytes, on the contrary, are able to distinguish different types of cells and 'remember' aggressors previously encountered and are thus able to react very rapidly because of previously acquired immunity. The principle of vaccination is based on this reaction (see Part V).

Lymphocytes are subdivided into B and T cells; the B cells manufacture antibodies which are carried by the blood to the tissues and the mucus covering the body and gills. Antibody molecules carried by the B lymphocytes attach directly to a receptor molecule on the pathogen. This recognition initiates the manufacture of antibody molecules of the same type which are liberated into the circulatory system in quantity; these attach to the pathogen and destroy it by various biochemical processes.

T lymphocytes guard normal fish tissues against the development of abnormal cells (infection by viruses or tumour cells).

The immune system is most active at high temperatures. Many infections are linked to specific temperatures. Fish are most sensitive to infections after handling stress, a change in the environment, spawning or a drop in oxygen concentration.

3. FISH CULTURE: OBJECTIVES AND CONSTRAINTS

3.1 Objectives

One of the objectives of aquaculture is to produce the maximum fish (or other marketable species) in a given volume of water in the shortest time and at the lowest possible cost. Achieving this objective requires not only the management and the optimization of growth of the farm stock, but also obtaining them in the first place. It is therefore clear that a knowledge of nutrition and reproduction is the basis of the farming operation.

Fish live in water whose salinity, temperature, pH, oxygen level and other constituents are very variable; different species have different requirements of these various environmental factors. The distinction between fresh-water and sea-water species is commonplace, as is that between warm-water and cold-water species, but the hardiness or robustness of a species is less easy to define.

In addition, information on osmoregulation, digestion and respiration are as essential as that on nutrition and reproduction and underlines the interdisciplinary nature of aquaculture. Fontaine (1989) expressed the worry that there was a danger that too much effort would be dissipated in studying these different aspects rather than getting on with producing fish in places where there are urgent human needs for the products.

3.2 Constraints linked to the biology of the species

For the farmer the selling price of the product depends on the various costs involved in rearing (purchase of fry, labour costs, etc.) but also criteria associated with the biology of the species. Several of these have been identified:

1.   The length of the rearing cycle needed to obtain a marketable product. Many species grow too slowly to be of interest to the fish farmer (small fresh-water fish and certain sparids, for example). Together with the stock level, the length of the rearing cycle governs the number of ponds, tanks or cages needed to obtain a given annual production. The length of this cycle can be altered by temperature, and a rearing cycle which is too long to be profitable in cold water may be successful in warm water.

2.   The conversion rate of the ingested diet to biomass of fish (the relationship between the weight of dry compound diet allocated and the gain in weight of the fish). If the rearing operation is to be profitable the conversion rate must not be high; it is important to remember that any mortality occurring during rearing increases the conversion rate (because a dead fish has been consuming food but the biomass of

this fish will not be part of the final production total). This rate is compatible with profitability for trout, salmon, sea bream, sea bass, turbot, eels (<3 and most frequently <2) but is very high for pelagic fish such as tuna and yellowtails. The poor conversion rate characteristic of these species (between 4 and 8) makes rearing uneconomic unless there is a ready availability of low-price fodder fish, as in Japan.

3.   The maximum density of the fish compatible with good growth. Many species can live at densities suitable for commercial rearing (from 10 kg m^{-3} in cages and tanks for salmon and bass up to a maximum of 400 kg m^{-3} for eels; reported by Jespersen 1989).

4.   The 'hardiness' of the species as expressed in its capacity to survive and grow in different and variable physico-chemical conditions, to survive handling operations (grading, treatments) and various changes in rearing conditions. The hardiness of the trout with regard to handling is a good example; this fish does not show extreme sensitivity to lack of oxygen. However, the truly hardy fish are the tilapias, as carp, although equally robust in ongrowing, are less so during spawning, as are eels. Farmed marine fish (sea bass, sea bream, etc.) are more sensitive than salmonids to handling stress; however, some adult pelagic fish and other species such as haddock in the larval stages are so difficult to handle that, at present, they cannot be reared in captivity. This sensitivity results in high rates of mortality from various causes or abnormalities (lack of pigmentation in turbot, lordoses in bream species).

It can therefore be seen that the realization of the objectives of the fish farmer depends on the biology of the species, but other constraints, e.g. those of the environment, can affect the economics of the rearing enterprise.

3.3 Ecological constraints (constraints linked to the farm site)

In continental Europe, where the culture of fish in fresh water and traditional mollusc culture are long-established activities, most of the available sites are already in use. There is a similar situation in many countries; the most favourable land-based sites are already used and prospects are therefore limited for fresh waters where water bodies are of relatively modest size (rivers, ponds or lakes). The majority of sites still available are in shallow lagoons or in the sea. Happily, Europe and the Mediterranean are rich in lagoons but sheltered marine sites are less abundant (Scotland, Croatia). The exploitation of open sea sites, while difficult, is becoming technically possible, although at present the sheltered coastal zone is the only truly productive zone for marine aquaculture. However, future developments are likely to take place in the open sea as technology progresses.

There is a further fascinating long-term possibility: tropical lagoons are extensive, well protected, often quite deep and with a good water turnover rate, and may be a site for aquaculture in the future. The productivity of these waters is not very great, however, as we have seen (Part I, Chapter 2), but the potential of these waters should be considered in the light of recent discoveries in oceanography. The fertilization of marine waters is a technique that could be further developed.

The natural environmental conditions offered by a site should correspond as closely as possible to the biological needs of the species being cultured. While many factors are relatively similar in Europe in fresh or saline waters, one factor varies significantly from north to south and in lagoons and the open sea: this is temperature.

The growth rate of fish varies with temperature: the optimum is between 10 and 18°C for salmonids, above 20°C for pond fish, 18°C for turbot, between 22 and 23°C for sea bass, around 25°C for sea bream and close to 30°C for penaeid prawns. Sites offering the longest possible time at these optimum conditions will therefore produce better growth rates and better rates of food conversion. All else being equal, sites with the best conditions for rearing a given species produce the fastest growth and the lowest mortality; this leads to the lowest rearing costs.

One complication is that the temperature preferences for growth for a species may be different from those for reproduction (hatchery stage); the sea bass spawns between 10 and 14°C and grows fastest between 20 and 25°C.

With technological developments, changes can be made to the environment or to the biology of the species so that rearing can take place outside the natural range (e.g. production of tilapia or eels in northern Europe (Mélard *et al.* 1989 and Jespersen 1989) or the biological resistance to high temperatures (production of sea bass in the tropics, Barnabé & Lecoz 1987).

Such 'exotic' rearing is only on a small scale and producing minimal tonnages but may nevertheless be extremely profitable. This is particularly true for ornamental fish, the rearing of which is a major activity in places such as Florida.

3.4 Technical constraints and prospects

Technology can affect the constraints linked both to the biology of the cultured animal and to the culture sites.

The amberjack of the Mediterranean has been cultured experimentally, but will not take in dry diet, feeding only on fish (Barnabé 1978). This requirement, the poor food conversion rate and inability to control breeding in captivity, prevents the culture of the species at present.

In cages in warm water, automatic feeders allow food to be distributed many times each day. Their use promotes much higher growth rates than that obtained when food is distributed manually on fewer occasions during the day. This can therefore be considered to be a technical success, as can anti-fouling baths used for cage nets which reduce the need to change nets frequently and thus economize on manpower or the various types of aerators which allow spectacular increases in production to be made.

The management of the farming of tilapia (originating in fresh water) in sea water opens up many prospects for intensive fish culture in tropical lagoons, as in Martinique (Lecoz *et al.* 1989); this result came from the use of an anti-stress compound in the diet.

Technical developments have overcome many other constraints; there is much potential for further progress, from the pumping of deep, cool water to the surface allowing the culture of salmonids in the tropics to the culture of fish in the open sea which will lead to products of higher quality (Sveäl 1988) or rearing eels at very high density in recycled water (Jespersen 1989).

4. CONCLUSION

The choice of species which are actually cultured is the result of a series of widely varying factors that act on each other.

It is also worth noting that fishing is often the starting point for farming activities; collecting alevins or juveniles from the wild for rearing (eels, yellowtails, etc.) and harvesting plankton or artemia cysts for hatcheries. The use of fish meal in dry diets for intensive rearing makes aquaculture and fisheries complementary activities.

The development of technologies requires a knowledge of both the biology and ecology of the cultured species; the term 'biotechnology' is therefore appropriate. The combination of technical and biological expertise will make possible tomorrow that which is impossible today.

REFERENCES

Barnabé G., 1976. Contribution à la connaissance de la biologie du Loup *Dicentrarchus labrax* (L.) Poisson Serranidae. Thèse Doct. Etat, mention Sciences, Univ. Sc. Techn. Languedoc, Montpellier: 426 pp. (Published *in extenso* in 1980 by the Museum National d'Histoire Naturelle, Institut d'Ethnologie, Paris: 9 microfiches).

Barnabé G., 1978. Les potentialités géographiques et techniques du littoral languedocien pur l'aquaculture des Thons. Actes de Colloque, CNEXO (ed.), Paris, 8: 203–207.

Barnabé G., Lecoz C., 1987. Large scale cage rearing of the European sea-bass *Dicentrarchus labrax* (L.) in tropical waters. *Aquaculture*, **66**: 209–221.

Benhalima K., 1982. Structure et développement du tractus digestif du Loup (*Diocentrarchus labrax* L.). Thèse 3e Cycle, Univ. Sc. Techn. Laguedoc, Montpellier: 75 pp, 33 pl.

Bougis P., Barnabé G., 1980. *Je reconnais les poissons marins*, Tome 2: *En plongée*, A. Leson (ed.), Paris: 111 pp.

Fontaine M., 1990. Préface. In: *Aquaculture*, G. Barnabé (ed.). Lavoisier—Tec. & Doc., Publ., Paris: IX–XI.

Furnestin M. L., 1966. Le plancton indicateur halieutique. *Rev. Trav. ISTPM*, **XXX** (2 et 3): 163–170.

Grassé P. P. (ed.). *Traité de zoologie* (Tome 13, Vols 1, 2 and 3). Masson Publ., Paris.

Hoar W. S., Randall D. J. (eds). *Fish physiology* (11 volumes). Academic Press Inc., New York.

Jespersen T., 1989. Eel farming in a indoor recirculating system. In: *Aquaculture: a biotechnology in progress.* De Pauw, Jaspers, Ackefors, Wilkins (eds). European Aquaculture Society, Bredene: 185–196.

Lecoz C., Marguerit P., Marion J. P., 1990. L'elevage en eau de mer du Tilapia rouge à la Martinique. In: *Aquaculture*, G. Barnabé (ed.). Lavoisier—Tec. & Doc., Publ., Paris: 929–938.

Maurin C., 1966. Les poissons planctonophages. *Rev. Trav. ISTPM*, **XXX** (2 et 3): 208–216.

Mélard C., Ducarme C., Philippart J. C., Lasserre J., 1989. The commercial intensive culture of Tilapias in Belgium. In: *Aquaculture: a biotechnology in progress*, De Pauw, Jaspers, Ackefores, Wilkins (eds). European Aquaculture Society, Bredene: 226–232.

Payan P., Bornancin M., 1979–1980. Osmorégulation chez les poissons: aspects physiologiques, biochimiques et enzymatiques. *Océanis*, **5** (5): 799–822.

Priede I. G., Secombes C. J., 1988. The biology of fish production. In: *Salmon and trout farming*, L. Laird & T. Needham (eds), Horwood Publ., New York: 32–68.

Sveäl T. L., 1988. Inshore versus offshore farming. *Aquacultural Engineering*, **7**: 279–287.

Tesseyre C., 1979. Etude des conditions d'élevage intensif du Loup (*Dicentrarchus labrax* L.). Thèse 3e Cycle, Univ. Sc. Techn., Languedoc, Montpellier: 115 pp.

2

Fish nutrition

1. GENERAL

We know that fish consume other living matter from their environment; unlike plants they are incapable of synthesizing organic matter (Part I, Chapter 2). The intake of food supplies their energetic requirements (movement, basal metabolism, etc.), reproduction and growth. Fish are no different from other farmed animals except with regard to temperature. They live in water—a medium with a high specific heat (Part I, Chapter 1) —and fish, like the majority of aquatic animals, are poikilotherms. The temperature of the blood is therefore close (around 1°C) to that of the environment in which they are living. There is thus no requirement to consume energy in order to maintain body temperature at a different level to that of the environment. These animals living in water must adapt to all the other parameters (dissolved gases, salinity, light, pH, pollutants, etc.) which have a bearing on their nutrition.

Fish utilize food in a more efficient manner than terrestrial animals, as is shown in Table 1 (Lowel 1988).

In this chapter we are not giving a detailed account of the processes of digestion, nor shall we deal in detail with fish nutrition which is a specialized subject in itself; our ambition is limited to demonstrating its importance in the culture of fish. For more detailed study the reader is directed to the works of Halver (1989), ADCP (1980, 1983), New (1987), Lowel (1988), and Steffens (1989), etc.

Fish convert food to flesh more efficiently than any farmyard animal. An American catfish increases in weight by 0.84 g when it ingests 1 g of food, while the chicken (the most efficient converter of all the warm-blooded animals) increases in weight by 0.38 g for every 1 g of food consumed. The percentage of edible tissue in a fish is overall greater than in cattle, pigs or chickens. For example, over 90% of the weight of a catfish is made up of flesh (50–70% for the majority of other species). From the qualitative point of view the fish is an excellent source of food for man, except for calcium, vitamin A and vitamin C. One portion of fish supplying all the protein needed will contain 280 calories while the equivalent hamburger would have 750 calories; this underlines the importance of fish in the diet.

Table 1. Comparison of the effectiveness of the use of feed, protein and energy in cat fish, chickens and cattle (adapted from Lowel 1988)

Animal	Composition of the diet of the animal		Efficiency		
	Protein (%)	Energy (kcal ME g^{-1})	Weight gain per g feed consumed (g)	Protein gain per g protein consumed (g)	Protein gain per Mkcal ME consumed (g)
Catfish	32	2.7	0.84	0.36	47
Broiler chicken	18	2.8	0.48	0.33	23
Cattle	11	2.6	0.13	0.15	6

ME = Metabolizable energy.

2. THE DIFFERENT TYPES OF FISH CULTURE

No distinction is made here between fresh-water and marine fish as many species are reared in both environments (sea bass, eels, salmonids, tilapias), but fish are farmed by different processes with different feeding strategies.

2.1 Extensive production
In extensive production the farmed fish feed exclusively on the natural production of the aquatic environment; this may be planktonic or benthic. The farmed fish are therefore the final link in the food chain in an environment which is either entirely enclosed or has a slow rate of water turnover, and the natural productivity of the ecosystem is utilized (Part I, Chapter 2). The term 'extensive fish culture' is used because it takes place in large bodies of water (ponds) where low concentrations of animals obtain all or part of their food from the surrounding environment. In practice, harvesting the fish removes mineral elements from the closed ecosystem; these are replaced by mineral or organic fertilizers.

The subject of the operation of the ecosystem will be described in Chapter 5 of this part; pond culture produces the majority of farmed fish throughout the world.

2.2 Semi-extensive production
This consists of supplementing the natural production consumed by fish in ponds with manufactured diets, wastes from agriculture, or animal feed or human wastes. This supplementary food is used to increase natural production.

In general, the culture of species which are able to utilize supplementary food is extremely profitable. For example, for common carp in fertilized ponds production is around 400 kg ha^{-1}. The addition of cereal by-products increases production to 1500 kg ha^{-1}. The addition of high-quality food increases this figure to as high as 3000 kg ha^{-1} (with records of 10 500 kg ha^{-1}; Jhingran 1989). For American catfish, production is from $3700 \, kg \, ha^{-1}$ in fertilized ponds but an addition of food increase production to 5 tonnes ha^{-1} (Lowel 1988).

Semi-extensive production can be linked to terrestrial rearing. Pond culture can utilize wastes from various domestic animals (chickens, ducks, pigs, cows) to fertilize ponds where fish are cultured: according to Castell (1989), the Chinese are able to achieve a production of $13\,500 \, kg \, ha^{-1}$ in a farm made up of 70 ha of ponds with 1060 pigs, 123 cows and 400 000 ducks (whose livers are exported to France). This type of fish rearing is classified as integrated aquaculture.

Semi-extensive and integrated fish culture are linked to both extensive fish culture (because of the reliance on the functioning of the aquatic ecosystem) and intensive fish culture (because of the supply of feed). These different aspects are discussed in Chapters 5 and 6 of this part.

2.3 Intensive fish production under artificial conditions

In intensive production fish are reared at high density in tanks or cages in which all the food they eat comes from outside. The water serves as a physical support for the fish, provides oxygen, removes metabolic wastes and regulates temperature. This type of aquaculture can be classified as one where food which is not suitable for direct consumption by man (low-value fish, various types of flour or meal) is converted into valuable nutrition for other animals.

The goal of this method of rearing is the maximum production per unit effort and space; this justifies the use of manufactured diets which meet all the nutritional requirements of the fish. Good examples of this type of production are trout, salmon, and Japanese yellowtail which are farmed in raceways or cages. In this system water contributes nothing in the way of food, although some species are able to absorb minerals and certain amino acids from solution across the skin, as larvae (Pavillon 1988) and even as adults (Fiala-Médioni 1988). They therefore play a direct role in nutrition but are not very important.

Production costs are high and feed makes up around 60% of them. Salmonids are fed on dry pelleted diet; Japanese yellowtails on fish of low market value. This type of aquaculture is based on the conversion by the farmed fish of products with a low market value into a more valued product (transformation aquaculture).

The fish are fed in the water and any food which is not consumed almost immediately represents not only an economic loss but may also, if it decomposes, affect the quality of the water adversely. In addition, the method of feeding and the stability of the food in the water are important factors. A difference from integrated or extensive aquaculture is that wastes do not become integrated into the natural trophic web and pose problems of pollution (dissolved organic matter and sedimentation of suspended solids). This will be discussed below.

Depending on their type of diet, it is possible to place fish into the four major categories of herbivores, detritivores, carnivores and omnivores, but in culture this classification is not always clear; many species show adaptations. On top of this, pond fish can receive supplementary feeding (semi-extensive rearing), fish cultured in cages may eat plankton (attracted by artificial lighting in the example of juvenile coregonids reared in cages in lakes), while depending on the stage in the life cycle (larva, juvenile, adult) a single species may be reared extensively or intensively.

3. NUTRITIONAL REQUIREMENTS

Whether the fish are fed naturally or artificially their nutritional requirements are the same: they need proteins, lipids, sometimes carbohydrates, minerals, vitamins and growth-promoting factors, which may come either from the surrounding aquatic environment or from prepared diets.

There are essential differences separating fish from other farmed animals. Their energy requirements are lower than those of warm blooded animals and fish require certain fatty acids which are not essential to warm-blooded animals. Because fish are able to absorb some minerals from the water, their requirements for these are lower; for example, many waters contain enough dissolved calcium for the needs of the fish.

The nutritional requirements of fish do not vary greatly from one species to another. There are exceptions, for example, in the requirements for essential fatty acids and the capacity to assimilate sugars. Quantitative needs established for several species can therefore be used as the basis for the calculation of the requirements of other species; thus, the formulation of diets for intensively reared marine fish is largely (perhaps too much) based on the formulation of trout diet.

3.1 The relationship between environmental factors and nutrition

The speed of biochemical reactions, which is the basis of the metabolic process, depends on temperature and doubles every 10°C (Q_{10} Law) and all poikilothermic species are therefore, overall, most active in the warm season. Their temperature requirements are very variable; some fish are adapted to cool waters (salmonids), to temperate waters (some carps, sea bass) and to warm waters (other carps, tilapia, etc.).

The optimal temperature for growth is therefore characteristic for each species and is not necessarily the same temperature which is best for reproduction of that species. Within the range of temperatures that can be tolerated by a species (i.e. the range over which the species feeds and grows normally) metabolic activity and the requirement for food increases from a minimum up to an optimum temperature. This optimum corresponds both to the best growth and the best utilization of food (lowest food conversion rate); above this optimum the fish converts food to flesh less efficiently even though it eats more and grows faster.

Because of this, in temperate zones which are characterized by seasonal variations in temperature, fish will eat more in summer than in winter and growth (which often ceases in winter) will be concentrated in the warm season. However, in tropical waters growth will continue throughout the year. Temperature has an effect on both the intake and the use of food and is imposed by the environment. The enzymes which take part in metabolic processes are adapted to function over a range of temperatures associated with each species; some may adapt to different temperatures while others will not (Chow 1980). This factor is already taken into account in the choice of sites for farming a particular species.

For trout, Fig. 1 (Cho 1986) clearly shows the effects of temperature, but this factor may also have a direct action on the requirement for protein, for example (Table 2). Many other environmental factors play their part in influencing metabolism; a detailed account can be found in the work of Luquet & Kaushik (1986), information from which is reproduced in Tables 3 and 4.

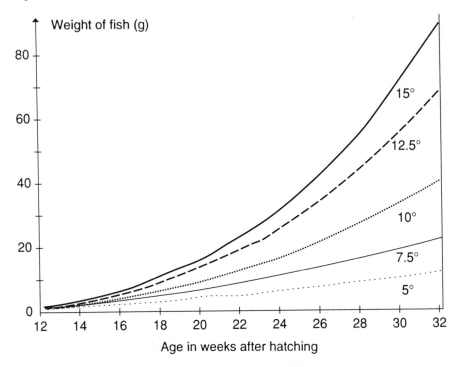

Fig. 1. Influence of temperature on the growth of rainbow trout under similar rearing conditions and with the same diet (after Cho 1986).

To these factors must be added (at least in intensive rearing) the effect of the density of individuals (biomass in $kg\ m^{-3}$) which has variable effects; in larval rearing high densities improve the intake of food; this is also the case in the super-intensive rearing of trout in large on-growing tanks (Table 5) where the water is renewed 25 times each hour and the level of O_2 maintained at $9.5\ mg\ l^{-1}$. In other examples, high densities limit growth, especially in small tanks. In general, high densities are not compatible with good reproduction in captivity; small tanks also have a negative influence on reproduction and growth but the effect varies between species. The composition of manufactured diets and their moisture content obviously has a major influence which we shall discuss later (dry pelleted diets).

Pollutants which are present in the environment can have various effects on the nutrition of fish (Péres & Boge, pers. comm.). These authors report that certain substances are able to alter the properties of the protective mucus produced by fish, and alter the properties of enzymes (organochlorines, mercuric chloride), while potassium bichromate

Table 2. Influence of temperature on the protein requirements of fish (Luquet & Kaushik 1986)

Species	Temperature (°C)	Protein requirements
Chinook salmon	8	40%
	15	55%
Rainbow trout	10	No difference
	16	
Rainbow trout	21	Growth > with 35%
	21	Growth < with 40%
Catfish	20	No difference
	24	

Table 3. Effect of reduction in the oxygen level on certain aspects of protein metabolism for rainbow trout (Luquet & Kaushik 1986)

Oxygen in the water	mg l^{-1}	8.5	5.4
Weight gain	(g)	47.6	7.5
RNA/DNA	liver	4.49	3.71
RNA/DNA	muscle	3.52	2.43

Table 4. Effect of the concentration of non-ionized ammonia on the 'performance' of rainbow trout (adapted from Luquet & Kaushik 1986, after Forster 1979)

Non-ionized ammonia	Weight gain (%)	Food conversion index
0	189	1.49
0.007	221	1.31
0.014	176	1.53
0.027	191	1.45
0.069	182	1. 45
0.103	145	1.67

(Length of trial: 18 weeks, initial weight 100 g.)

Table 5. Influence of the density of fish on growth, conversion index and respiration for lake trout (after Poston, pers. comm.)

Number of fish l^{-1}	Final density $(kg\ m^{-3})$	Weight gain	Conversion index	Respiration $(mg\ O_2\ kg^{-1})$
3.8	51.5	496	1.74	715
7.6	95	458	1.78	673
11.5	137.6	438	1.9	528
15.3	188.7	443	1.79	424
19.3	233.7	442	1.83	391

inhibits the activity of alkaline phosphatase. Toxins can act through the nervous system (loss of appetite brought on by the presence of copper in the water). Detergents can destroy proteins. Our knowledge in this area is still far from complete. Pollutions which wipe out whole stocks are rare but well known; insidious pollution which slows growth or upsets reproduction is more widespread but less well known.

Toxins present in the environment or in food can be accumulated by fish; a study on 16 species captured from the wild by French fishing boats working in the Atlantic found over 20 toxic substances (mercury, lead, cadmium, arsenic, tin, zinc, copper, PCB, lindane, DDT, etc.) but showed that these did not present any great danger as their toxicity is linked to their frequency of consumption by man. Conger eels and dogfish do, however, have high concentrations of mercury; angler fish, arsenic; mackerel, DDT and tin.... The level of contamination in the fish is linked with their place of capture: estuaries are more polluted than the open sea.

3.2 Energetic requirements

Growth is the result of ingested food being converted to body tissue. In order to appreciate the effectiveness of a particular food, nutritionists have developed the concept of energy: energy can be defined as the capacity to carry out work. It is required for the muscular activity involved in movement, and also for all the metabolic processes carried out by the organism (digestion of food, respiration, nervous activity and the osmotic processes necessary to maintain the equilibrium of the body fluids with that of the external environment). Even when resting the processes which are essential to sustain life require energy (cellular exchanges, respiration, circulation, maintenance of osmotic pressure, etc.): this is basal metabolism. It is important to understand that the digestion and assimilation of food consumes energy (around 30% of the energy in the food); this is Specific Dynamic Action or SDA.

The *crude energy* of a food or the crude calorific value is the total energy contained in the food. It is measured by the complete oxidation (combustion) under an atmosphere of oxygen of the substance in water, carbon dioxide gas and other gases in a bomb calorimeter. The increase in the temperature of a water bath surrounding the calorimeter allows the determination of the number of calories contained in a given weight of the substance (1 calorie is the quantity of heat required to raise the temperature of 1 g of water by 1°C,

but the more widely used unit is the kilocalorie, which is equivalent to 1000 calories).
However, not all of the energy in the food is useable by the animal. Different parts of the
diet have different rates of availability of energy and thus the figure for crude energy is
not a useful way to characterize a diet (Fig. 2).

The *digestible energy* of a diet is equal to the total energy minus the energy present in
the faeces (which have been collected and weighed). This energy varies, depending on
the type of diet (see Section 4 below). For trout it has been estimated that 28% of the
ingested energy is excreted as faeces.

The *metabolizable energy* corresponds to the digestible energy minus the losses of
energy in the urine and across the gills which can be measured in a metabolic chamber.
New (1987) gave information on the crude energy and digestible energy of the major
categories of primary foodstuffs used to feed the fish (Table 6).

The *free energy* is therefore that which remains for biological activities (movements,
gamete production) and growth after the maintenance energy requirements (basal metabolism, SDA) have been satisfied.

The metabolizable energy (ME) is therefore lower than the digestible energy: this is
from 3070 cal kg^{-1} ME for 4040 cal kg^{-1} digestible energy (DE) for Norwegian herring
meal (BNA 1973). Norwegian herring meal has a higher energetic value to that of herring

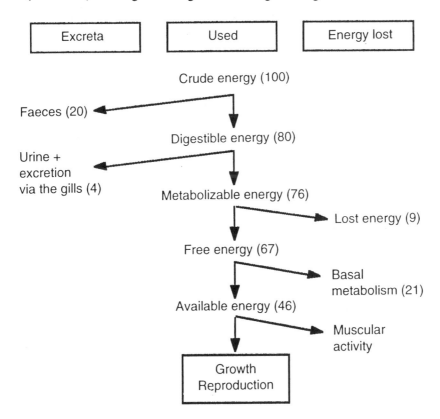

Fig. 2. Use of dietary energy in rainbow trout at a temperature of 15°C (adapted from Cho 1986).

meal from the American coast (3650 ME/2940 DE cal kg^{-1}; BNA 1973). This use of energy is a means of comparison of feeds of different origin; however, diets may vary with respect to other criteria.

Table 6. Calculation of digestible energy for fish (after New 1987)

Feed	Energy (kcal g^{-1})	Digestible energy estimated (kcal g^{-1})
Carbohydrates (other than legumes)	4.1	3.0
Carbohydrates (legumes)	4.1	2.0
Animal proteins	5.5	4.25
Vegetable proteins	5.5	3.8
Fats	9.1	8.0

4. ORGANIC CONSTITUENTS OF THE DIET

While energy is a basis for the comparison of different diets, these are all made from the same organic constituents (carbohydrates, protein and lipids)—materials which are common to all organisms.

4.1 Carbohydrates

Carbohydrates (sugars, starches, cellulose, chitin) are scarce in the aquatic environment and in any case the complex forms are poorly utilized by fish (apart from some herbivorous carp and catfish which can digest them effectively), and do not therefore constitute essential components of the diet. Simple sugars are more digestible, but there are phenomena of intolerance which prohibit the exceeding of certain limits (20% for trout; Billard 1989). Cooking, extrusion and expansion, which break up long chain molecules, improve this for trout (digestibility can increase from 40 to 90%) (Billard *loc. cit.*). The normal practice is therefore to incorporate modified carbohydrates into the diets of fish, as there are unused sources (by-products of cereals) available at low cost. Incorporation of these substances has no effect on the body composition of the trout.

4.2 Proteins

These are the essential component of the tissues of living things. As most fish are predators it is not surprising that they are able to digest these substances more efficiently than terrestrial animals which are more likely to be adapted to use plant cellulose or starch. The protein requirements of various species are given in Tables 6–10, based on New (1987). It can be seen that most of them lie between 35 and 55%, although other workers have found figures of between 25 and 60%. All these figures are higher than those for terrestrial animals. It is important to note the major differences in results obtained by different researchers as proteins are used for energy, not just for the materials for growth in fish. The protein requirements can vary according to the level of other

Table 7. Nutritional requirements of salmonids (New 1987)

Proteins	Starter diet	50%
	Growth diet	40%
	Production diet	35%
Lipids (1)	Starter diet (alevins)	15%
	Growth diet (juveniles)	12%
	Production diet (adults)	9%
Amino acids	Lysine (alevin diet) (2)	5%
	Methionine (alevin diet) (2)	4%
		(in the absence of cysteine)
Available phosphorus		> 0.8%
Available energy		2800–3 300 kcal kg^{-1}
Essential components	Fish meal (3). At least 1% fatty acids n-3 series. Natural or artificial carotenoids.	
Negative factors	Nitrogen 'non-protein' (e.g. urea). High level of carbohydrates (> 20%. Fibre > 6%). Rancid fats.	

(1) Commercial diets for salmon and trout are increasing the lipid level to > 20%.
(2) As a percentage of protein in the ration.
(3) The partial or complete replacement of this expensive ingredient has been the object of much research. Synthetic amino acids can be used to balance plant protein.

Table 8. Nutritional requirements of common carp (New 1987)

Lipids	up to 18% (where protein is high)
Proteins	25–38%
Amino acids	lysine 5.7% ⎱ of the protein in the diet methionine 3.1% ⎰
Available phosphorus	0.6–0.7%
Metabolizable energy	2 700–3 100 kcal kg^{-1}
Essential components	At least 1% of each of the n-3 and n-6 fatty acids. Increased lipids for oogenesis in brood fish.
Negative factors	Non-protein nitrogen although it has been shown that carp can utilize this (this is contested). Rancid fats.

Table 9. Nutritional requirements of tilapia (after New 1987)

Proteins	50%	(alevins up to 0.5 g)
	35%	(0.5–35 g)
	30%	(from 35 g to market size)
Lipids	10%	(alevins up to 5 g)
	8%	(0.5–35 g)
	6%	(from 35 g to market size)
Digestible carbohydrates	25%	
Fibres	8%	(alevins of 10 g)
	8–10%	(from 10 g to market size)
Lysine	4.1 ⎱	of proteins in the diet
Methionine + 50% cysteine	1.7 ⎰	
Dietary energy		2500–3400 kcal kg^{-1}
Essential components		At least 1% of each of the n-3 and n-6 fatty acids
Negative factors		Rancid fats

substances in the diet (e.g. lipids). According to Jackson (1988), salmon require 44% protein and 23% lipids; trout require 44 and 16% respectively.

The metabolic activity of small animals is greater than that of larger animals. Small animals have a faster growth rate in terms of daily increments (growth curves show a continuous decrease in slope with size). The protein requirement of the young is high: trout diets which contain 38–55% protein for alevins contain no more than 37–47% for fish of over 5 g weight. For the American catfish, young fish of 3 g weight have a protein requirement four times higher than 250 g fish (Lowel 1988). The maximum size at which farmed fish are sold is usually well below the size at which growth rate shows a marked decrease unless there are special commercial reasons (for example, the market for large salmon and trout for smoking).

Proteins are assemblages of amino acids united by a peptide bond; these amino acids can be separated into two groups: ten can be synthesized by the fish from other compounds and are therefore not essential. The other ten cannot be synthesized and must therefore be supplied in the diet: these are arginine, histidine, isoleucine, leucine, lysine, methionine, phenylalanine, threonine, tryptophane and valine (for more details see Lowel 1988; New 1987; Halver 1989).

Studies have shown that each species of fish has precise amino acid requirements; these requirements for essential amino acids can be determined by using deficient diets in one or other acid or also, less accurately, by determining the composition of the body of the fish with respect to each acid. Table 11 (from New 1987) shows the requirements for various species which has been put together from the work of several authors. When an essential amino acid (EAA) is deficient the utilization of others can be blocked. In an ideal situation the level of EAA in the diet should correspond to the requirement of the species to avoid waste. A high requirement for proteins can thus be caused by a low level

Table 10. Nutritional requirements of marine fish (New 1987)

Component	Species	%	Type of diet (1)
Proteins	Groupers	40	E
(% MD) (2)	Japanese sea bream	52–60	C
	Yellowtail	53–57	E
	Yellowtail	49–62	C
	European sea bream	40–49	E
	European sea bream	44–55	C
	European sea bass	40–53	E
	European sea bass	50–69	C
Lipid	Groupers	14	E
(% MD) (3)	Japanese sea bream	6	C
	Yellowtail	15–16	E
	Yellowtail	6	C
	European sea bream	8–9	E
	European sea bream	9–11	C
	European sea bass	12–13	E
	European sea bass	9–10	C
Lysine (% protein)	European sea bream	5.0	
Methionine (% protein)	European sea bream	4.0 (Met + Cyst)	
Available phosphorus		Not less than 0.7%	
Dietary energy		2700–3799 kcal g^{-1}	

Essential components	Marine proteins. At least 1–2% n-3 fatty acids (C20 & C22). Carotenoids for the Japanese sea bream.
Negative factor	As for salmon and trout.

(1) E = Experimental diet; C = Commercial diet.
(2) The highest levels are for alevin diet.
(3) The levels of lipids in the commercial diets for the Japanese sea bream and the yellowtail are probably below optimum.

of one or other EAA; the others are present in excess and are not fully utilized (waste food). Other problems, such as the linking of amino acids with other molecules, can limit the availability of EAAs. It is therefore important to consider the qualitative aspects of the proteins in the diet.

Table 11. Amino acid requirements of various fish and level in comparison with several products (New 1987, from various sources)

Amino acid	Requirements, according to species (in percentage of the level of protein)							Amino acids in meal		Amino acids in the flesh of sea bass	
	Tilapia	Carp	Eels	Rainbow trout	Sea bream	Cat fish	Coho salmon	Herring meal	Blood meal	Muscle	Eggs
Arginine	<4.0	4.2	4	3.5	<2.6	4.3	6	6.4	4.2	6	6
Histidine		2.1	1.9	1.6		1.5	1.7	2.3	6	2.3	2.4
Isoleucine		2.3	3.6	2.4		2.6		4.3	1.1	3.7	4.2
Leucine		3.4	4.8	4.4		3.5		7.2	12.7	8.5	9.5
Lysine	4.1	5.7	4.8	5.3	5	5.1		7.5	8.6	10.1	7.7
Methionine	1.7	3.1	2.9	1.8	4	2.3		3.4	1.4	3.7	2.7
Phenylalanine		6.5	5.2	3.1		5		5.4	8.1	6	6.6
Threonine		3.9	3.6	3.4		2.3		4	4.2	4.8	5.2
Tryptophan		0.8	1	0.5	0.6	0.5	0.5	1.1	1.2	1.1	
Valine		3.6	3.6	3.1		3		6	8.7	4.2	5.3

Methionine and lysine are the most frequently limiting amino acids in fish diets and are often added to dry compound feeds. New (1987) shows the amino acid profile of several primary constituents of diets; the composition of hens eggs is given for reference.

Fish meals which are highly digestible represent one of the best sources of proteins for fish diets (although they are expensive). For the bass *Dicentrarchus labrax* (Table 11), the composition of herring meal in amino acids is almost identical to the profile of amino acids for the bass established by Alliot *et al.* (1974) or New (1986).

The incorporation of new sources of protein has allowed the use of up to 33% of protein from unicellular organisms from activated sludge from sewage works where domestic waste is purified (Tacon 1979). This provides a new form of food chain, although the recycling of wastes is not without health problems and this technique therefore remains experimental. There may be benefits in the incorporation of plant protein: for sea bass, the substitution of 30% soya meal, given different heat treatments (from 0 to 20 min at 100°C) for animal meals has been successful. A diet containing such plant material has given results which are as good as those obtained from a commercial diet containing only animal meals (Amério *et al.* 1989). However, soya meal contains trypsin inhibitors and heat treatment for 12 min at 100°C is the only way of achieving results which are as good as those for commercial diets.

Billard (1989) demonstrated the possibility of using up to 80% soya in the diet of trout (double cooking and extrusion at 200°C for 10 min). This shows that in some instances it may be possible to replace fish meal completely. However, this may not be profitable as the cooking process is expensive.

4.3 Lipids
Lipids are a better source of energy for fish than proteins. Table 6 shows the relationship between crude and digestible energy for proteins, lipids and carbohydrates. In spite of their lower price, lipids can only replace proteins within certain limits (while carbohydrates are only tolerated at very low levels).

The basic constituents of lipids are fatty acids, made up of a chain of carbon atoms in which the number runs from 6 to 22 (the notation is written C6 to C22, counting the carbon atoms from left to right). The atoms are joined, but the link may be a double bond. Fatty acids without a double bond are termed saturated; the others are unsaturated. Acids with more than one double bond are polyunsaturated. In works on nutrition, polyunsaturated fatty acids are referred to by the abbreviation PUFA or HUFA (poly or highly unsaturated fatty acids, respectively). Aquatic animals have a high requirement for polyunsaturated fatty acids, higher than for terrestrial animals particularly for the n-3 series.

The designation of a fatty acid is based on the number of carbon atoms, the number of double bonds (:), the number (n) of the first double bond on the chain of carbon atoms. Thus linolenic acid is C18:3n-3 with 18 carbon atoms and three double bonds, the first of which is situated at the third carbon atom. Linoleic acid (C18:2n-6) has two double bonds and the first is found at the sixth carbon atom. Acids with a double bond at the third carbon atom constitute the n-3 series; there is also an n-6 series, etc. Table 12 (Kaushik 1990) lists the principal fatty acids of importance to fish.

In practical terms, lipids can be subdivided into oils, fats and waxes, based on their melting points (oils are liquid at normal temperatures). They may be associated with

Table 12. Important fatty acids in the diet of fish (Kaushik 1990)

Fatty acid type	Abbreviated chemical formula
O_3 fatty acids n-3 (CH—CH$_2$—CH=)	
Linolenic acid	C18:3n-3
Eicosapentaenoic acid (EPA)	C20:5n-3
Docosahexaenoic acid (DHA)	C22:6n-3
n-6 fatty acids	
(CH$_3$—CH$_2$—CH$_2$—CH$_2$—CH$_2$—CH$_2$—CH=)	
Linoleic acid	C18:2n-6
Arachidonic acid	C20:4n-6

proteins (lipoproteins), minerals (phospholipids), etc.; fats (glycerides) are fatty acid esters of glycerol and are the form in which animals store their reserves (for overwintering, reproduction, migration), while waxes are the storage form for plants and some animals (e.g. copepods).

Apart from their energy-providing role, lipids contain essential nutrients which cannot be synthesized by the animal and therefore must come from the food. These include vitamins and trace elements and may be either unsaturated or polyunsaturated fatty acids. The type of essential fatty acid varies between fresh-water and sea-water fish (Table 13); trout requires 1–2% linoleic acid (C18) which it converts to longer chain fatty acids (C20 and C22), but since these acids are two times more effective, only half the equivalent quantity of their precursors is required to be synthesized. Marine fish have a requirement for C20 polyunsaturated fatty acids (ecosopentaenoic acid C20:5n-3) and C22 docosahexaenoic acid (C22:6n-3). The requirement is for 0.5–1% of each, especially during periods of stress. Salinity affects the requirement for essential fatty acids.

The fatty acid requirement also depends on temperature, the stage of development (deficiencies may be irreversible in larvae and juveniles) and, undoubtedly, many other factors. Catfish and warm-water fish are less sensitive to the absence of fatty acids; however, these substances still have beneficial effects on growth. Trout can ingest diets containing 20–25% lipids without any problems occurring.

Deficiencies in fatty acids are essentially a characteristic of sea water animals rather than fresh-water. Marine fish appear to have a greater requirement for HUFA than fresh-water or migratory species, but it is not known why they cannot utilize the n-6 series as effectively as the n-3 series. Cold-water species have a higher requirement for n-3 than warm-water species. In spite of the requirement for fatty acids, levels of PUFA which are too high can be toxic in the feed (this is also true for liposoluble vitamins in mammals and man). It is therefore essential to understand the dietary requirements of a species in order to provide the optimum diet.

The best source of PUFA is lipids from marine animals; this, however is also the most expensive source. Plant oils tend to have high levels of the n-6 series (linoleic series); these cannot therefore be used to replace oils of marine origin (Table 14; Kaushik 1990). It is worth noting that the oils which are essential to fish are also essential to man

Table 13. Fatty acid requirements of different species (Kaushik 1990)

Species	Essential fatty acid	Requirement (% of the ration)
Rainbow trout	18:3n-3	1
	18:3n-3	0.8–1.7
Chum salmon	18:2n-6 or 20:4n-6	1
(Fresh water/sea water)	and 18:3n-3	
Coho salmon	18:3n-3	1.0–2.5
Carp	18:2n-6 and	1
	18:3n-3	1
Eel	18:2n-6	0.50
	18:3n-3	1
Tilapia zilli	18:2n-6 or 20:4n-6	1
Tilapia nilotica	18:2n-6	0.50
Catfish	18:3n-3	1.0–2.0
	AGPI n-3	0.5–0.75
Ayu (Plecoglossus)	18:3n-3 or 20:5n-3	1
Yellow tail	20:5n-3 or 22:5n-3	0.50
Turbot	AGPI n-3	0.50
Japanese sea bream	AGPI n-3	0.40
Striped Jack	AGPI n-3	1.70

(protection against cardiovascular diseases in man, also the functioning of the brain); these oils are not synthesized by marine phytoplankton. Copepods, which are only found in the sea, are a rich source.

The high level of PUFA in artificial fish diets leads to rapid oxidation and therefore rancidity. They should be stored in a cool place, sheltered from the light. ADCP (1980) demonstrated the problems that can occur during storage.

Increasing the lipid content of diet allows economies to be made in proteins, oxygen and also a decrease in the pollution from fish farms (Kaushik 1990); this is a popular development at present.

Table 14. Fatty acid levels of several oils (%)

	Soya	Capelan	Capelan	Whole cod	Cod liver	Whole mackerel	Whole rainbow trout
		Summer	Winter				
C18:3n-3	9.9	0.8	0.6	11.8	23.9	1.3	5.2
C20:5n-3	0	8.9	6.3	12.4	8	7.1	0.4
C22:6n-3	0	5.9	4.2	21	14	10.8	19

4.4 Vitamins, minerals and trace elements

These compounds are essential at low quantities; without them fish will not survive. The role of vitamins is linked to the enzyme systems; the many tasks performed by these compounds are detailed in the general works listed above and specifically by Koenig (1980). Overall, the vitamin requirements of fish are higher than those of mammals. However, details of the requirements are poorly understood and, although Vitamin C deficiency has long been known to cause opercular deformities in sea bass (Halver 1972), one group of workers has looked for the causal mechanism for this disease for many years. Table 15, based on the work of Lowel (1988), summarizes the available information on minimum requirements of vitamins for rearing juvenile fish.

Requirements for minerals and trace elements are shown in Table 16 (from New 1987).

Table 15. Minimum vitamin requirements for the growth of young fish (quantity per kilo diet, compiled by Lowel 1988)

Vitamins	Units	Cat fish	Common carp	Salmonids
A	IU	1000–2000	R	2500
D	IU	500–1000	–	2.4
E	IU	50	R	30
K	mg	R	–	10
Thiamin	mg	1	1	10
Riboflavin	mg	9	8	20
Pyridoxin	mg	3	6	10
Pantothenic acid	mg	20	30–50	40
Niacin	mg	14	28	150
Folic acid	mg	R	N	5
B12	mg	R	N	0.02
Biotin	mg	R	R	0.1
Ascorbic acid	mg	60	R	100
Inositol	mg	N	10	400
Cholin	mg	R	4	3

R: Essential but requirements not known.
N: Need not demonstrated experimentally.

Levels of vitamins and trace elements have long been underestimated for fish; we have shown that for the sea bass *Dicentrarchus labrax*, a supplement of vitamins, polyunsaturated fatty acids and trace elements improves resistance to stress and, particularly, to high temperatures (Barnabé & Lecoz 1987). However, the effects are complex; but explanations may come from recent developments in human medicine. Water soluble vitamins are not toxic in excess (differing in this respect from fat soluble vitamins) and can

Table 16. Mineral requirements of fish (New 1987)

Minerals	Dietary requirement
Ca	0.5%
P (available)	0.70%
Mg	0.05%
Na	0.4–0.3%
K	0.1–0.3%
S	0.3–0.5%
Cl	0.4–0.5%
Fe	50–100 mg kg^{-1}
Cu	1–4 mg kg^{-1}
Mn	20–50 mg kg^{-1}
Co	5–10 mg kg^{-1}
Zn	30–100 mg kg^{-1}
I	100–300 mg kg^{-1}
Mo	Trace
Cr	Trace
F	Trace

therefore be used in environmental stress or disease situations. Vitamin C is used by many fish farmers as a supplement, based on the ideas that high doses can protect against cancer and many other diseases, put forward for man by Pauling. Saroglia & Scarano (1989) estimated Vitamin C requirements for the sea bass to be 200 mg kg^{-1} and showed the need to add a supplement of this vitamin to commercial diets.

The level of trace elements in primary sources of food can vary according to their origin: Fontaine (1989) showed that there was a difference in the level of iodine in two different fish meals used to feed fish which were being studied experimentally for their thyroid activity. The same commercial pellets were always used but the different fish meals used in their manufacture contained different levels of iodine and therefore gave different results. This is also undoubtedly true for many other dietary components which are essential at low levels and have effects similar to those of limiting essential amino acids.

5. WATER CONTENT OF THE DIET AND TYPES OF FEED

Water dilutes the nutritional components of dry diets. It is essential to know the percentage water content of the raw materials and of the dry diet in order to carry out calculations based on dry matter. Humidity plays an important part in determining the constituency of the diet; feeds are designated as dry, moist or paste. The water content also affects the stability and the length of time for which the diet can be stored.

5.1 Moist and semi-moist diets

Moist diets are those which contain 45–80% water. They are made up of fish of low market value which may be minced, kitchen wastes, fresh plant material, etc. These diets may be used in their crude state (salmon) or after mincing or freezing and mincing (salmon, Japanese yellowtail). Some species, white fish, have an insufficient level of fat (2–3%); these must be mixed with fish with a higher fat content (8–20% lipid). Fish which are used directly (forage fish) have a level of water of around 80%. The conversion rate of such diets is high (New 1987), 6–9 for sea fish and fresh meat, up to 10 for fresh plant, and 20–30 for potatoes.

Semi-moist diets contain the same ingredients to which are added various dry products (meals) to lower the water content to between 18 and 45%. Although they are relatively rarely used in Europe these diets form the basis of fry rearing in the North American trout industry. The Oregon moist pellet is the best known example; this contains 60% of dry ingredients (fish meal, wheat millings, yeast). The remainder is made up of the waste products of tuna processing, beef liver and fish from industrial fisheries. The diet is made up by crushing and mixing the components into a paste which is then passed through a mincer. The 'spaghetti' type product emerging from the machine is then fed to the fish. The preparation and storage (by chilling) of this type of diet poses many problems and explains why the use of the diet is declining except for eels which will only feed on a paste diet (Usui 1979). The eel diet is often made up by simply mixing fish meals with water.

A semi-moist diet tested on sea bass (Tesseyre 1979) gave better results than those obtained from commercially available dry diets but the problems associated with their use (see above) could not be overcome.

5.2 Dry compound diets (compressed or extruded)

5.2.1 Compressed dry diets

Dry compound diets for fish are made with presses such as those used for the diets for many terrestrial farmed animals. They are made from powders or meals with a natural humidity of between 7 and 13% which have been mixed beforehand. No more than 12% oil can be added to this mixture, otherwise the granules break up.

The mixture is transported to the press by a continuous screw mechanism. The press consists of a disc perforated with holes which are the same diameter as the pellets to be produced; inside this there are smaller, eccentrically placed discs. These force the mixture of meals through the holes in the grid (2–9.5 mm diameter). During the transition across the press heat is produced which raises the temperature of the mixture. One or two per cent binder is incorporated; under the action of heat and the evaporation of water, this functions to hold the granules together. This increase in temperature can break down vitamins and oxidize unsaturated fatty acids. Because of this it is preferable to add vitamins and lipids after granulation, for example by mixing vitamins in oil (Barnabé & Lecoz 1987) and then adding this liquid to the pellets in a cement mixer. Some suppliers sell pellets which have been treated in this way (oil incorporated before and after granulation); levels of lipids may reach 17%.

The water content (which must be below 12%) and the level of oil determines the cohesion of the granules which may tend to break down to fine particles that accumulate at the bottom of the feed bags and are unavailable to the fish. If these powders are distributed with the granules they act as irritants to the gills of the fish. These granules form the basis of the development of trout culture and salmon culture in Europe; they can be stored at ambient temperature for several months with few problems, either for distribution by hand or for use in automatic feeders.

Dry diets are not necessarily the best type of food for fish; recent studies using electron microscopy (Deplano et al. 1989) have shown that the commercial diet used in the ongrowing of sea bass and sea bream abrades the interior of their digestive tube (Fig. 3).

The conversion efficiency is an expression of the relationship between the quantity of food given and the weight gain (biomass) of the farmed fish expressed in wet weight. This relationship is most frequently between 1.3 and 2 (from 1 to 3). However, it must be remembered that this is an expression of the conversion of dry diet (mean humidity 11%) to wet fish (80% water). The conversion efficiency may sometimes drop to 0.9, but it must be remembered that in terms of dry weight gain this is comparable with natural food chains and is nothing exceptional.

5.2.2 Extruded and expanded dry diets

To improve the quality and handling properties of dry diet there is a new technique, cooking-extrusion, which consists of treating the ingredients with high pressure steam before pelleting; this improves the digestibility of the carbohydrates (starch) which can therefore be included in higher quantities and eliminates certain inhibitors (see proteins, above). On leaving the press the pressure inside the pellet drops suddenly and the water contained in the pellet vaporizes; there is a gelatinization and eventually an expansion of the starch; this is a process similar to that in the making of popcorn. This type of pellet does not break up leaving powder in the bag, and floats or sinks slowly as it is less dense than pellets made by the compression method. The level of oil can reach 25% or more without disrupting the growth and maturation of salmon (Jackson 1988). However, the destruction of vitamins and oxidation of fatty acids is greater and the manufacturing cost 10–15% higher than for the traditional method.

Diets used for aquarium fish which are often in the form of flakes are not described here; information can be found in the work of Spotte (1973). They are more expensive but less polluting and are made by drying a fine layer of paste which is spread on a heated rotating drum.

6. POLLUTION FROM FISH FARMING

In extensive rearing operations in ponds and other still waters the wastes produced by the fish (urine, faeces, carbon dioxide gas) are recycled by detritivores, filter feeders, phytoplankton and bacteria, and contribute to the enrichment and operation of the aquatic trophic webs (Chapter 5, this part). Inputs from outside consist of mineral or organic fertilizers which can be used by either primary or paraprimary producers (Chapter 2,

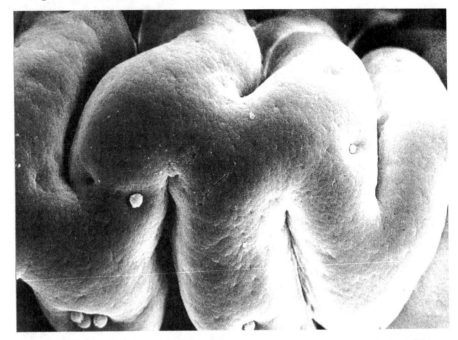

(a)

(b)

Fig. 3. a: View of the intestinal epithelium of a wild sea bass, scanning electron microscope.
Note the uniformity of the surface epithelium. G × 430. Photo Deplano *et al.* 1989. b: View of
the intestinal epithelium of a farmed bass fed on dry manufactured diet, scanning e.m. Note the
abrasion of the epithelium. G × 630. Photo Deplano *et al.* 1989.

Part I). Particulate matter produced as faeces by domestic animals can be consumed directly by fish (carp, tilapia); this is direct recycling.

In semi-extensive rearing operations, the provision of organic compounds and the wastes produced by the fish can also be recycled in the food web as there is usually a lower water turnover rate. The limiting problem in this type of rearing operation, especially at high temperatures, is to maintain the dissolved oxygen at a high enough level for the fish; the nocturnal respiration of the aquatic plants (phytoplankton and macrophytes) contributes to the depletion of oxygen in the environment, even more so than do the fish. In many fish culture units artificial aeration with the use of brushes or paddle wheels is needed, especially in the tropics where the elevated temperature is not compensated by a short night, as happens in summer in temperate zones. In such environments, diurnal variations in oxygen level can be spectacular (from 0 to 100% saturation) because of the production of oxygen by photosynthesis during the daytime.

In intensive culture units (Chapter 6, this part), the quantity of fish per cubic metre is high (10–50 kg) and the surrounding water must bring in all the oxygen necessary for respiration and carry out all the waste products; there is no internal recycling in the system. The wastes exit the system through the outflow of a tank or through the sides and bottom of cages.

The pollution caused by fish farming can be substantial; this can be seen when the figures are compared to the pollution from humans, expressed as an equivalent after treatment of waste water.

Solbé (1988) calculated that the production of one tonne of trout produced pollution equivalent to that produced by:

> 115 000 people in 1 day for BOD
> 44 000 people in 1 day for ammonia
> 312 000 people for 1 day for suspended solids.

'Taking the worse case, the production of one tonne of fish emits a quantity of waste equivalent to the sewage from a town with 300 000 inhabitants' (Solbé 1988).

It must, however, be remembered that this quantity of pollutants is extremely dilute in the rearing system and that in most countries some type of treatment of the water is required before it leaves the farm. The exception to this is cage rearing where it is impossible to treat the wastes.

The use of dry diets reduces pollution by 50% compared with diets made up of untreated fish and slaughterhouse wastes, but a settling pond removes no more than 90% of the pollutants. Low polluting diets are being developed: these use primary constituents with a digestibility of 80–90% and pre-cooked cereals; this can reduce the faecal output by 60%. Reductions in phosphorus, calcium and magnesium are more difficult to make and require that fish be filleted to remove bones before conversion to fishmeal; this is obviously an expensive process (Dekker 1986). Such diets are not in general use.

According to the calculations made by Solbé (1988), based on a study of 25 trout farms in Great Britain, the production of one tonne of trout requires 602 kg oxygen extracted from the water running through the fish farm. The BOD of the water increases by a mean figure of 0.7 mg l^{-1} tonne^{-1}. In spite of treatment by settling there is a production of 1.36 tonnes of suspended solids (in dry weight) from each tonne of fish produced per annum (in wet weight). The production of the various forms of nitrogen is as follows:

> 55.5 kg ammoniacal nitrogen per tonne of fish produced
> 1.8 kg nitrogen in the form of nitrites
> 10 kg of nitrogen in the form of nitrates.

Oxidation by bacteria of the first two forms of nitrogen to nitrate requires the consumption of more oxygen which pushes the total oxygen requirement to 1363 kg O_2 per tonne of production per year. Waste phosphorus has been calculated as 15.7 kg per tonne of fish per year. However, there are major variations in the results presented by different authors (Dekker 1986; Solbé 1988). The range of values reported at the outflow of fish culture units is as follows:

Suspended solids:	293–4015 kg per tonne of fish produced (mean 1066)
BOD	120–600 kg t^{-1} (mean 338)
Ammonia	45–730 kg t^{-1} (mean 182)
Nitrates	4–11 kg t^{-1}
Phosphates	3–6 kg t^{-1}
Total phosphorus	9–75 kg t^{-1}

The effect of these materials on the environment depends on their state (solid or dissolved); solids in suspension increase turbidity, reduce light penetration and therefore limit photosynthesis and the development of the trophic web (Chapter 2, Part I). These materials, which have a high organic content, have an elevated BOD (130 mg kg^{-1}, according to Solbé 1988). The deposition of suspended solids in sheltered zones favours the proliferation of larval chironomids rather than the insects which are characteristic of trout streams. Fish and parasites can escape from fish farms, as can antibiotics and substances used in treatments; we have seen how such products can interfere with the composition of natural trophic webs (Chapter 2, Part I). Dissolved substances which are rich in mineral salts can cause the development of algal growths with the appearance of 'cotton wool' which is unsightly and can trap suspended particles. The build-up of these effects is not always apparent and controls are not always foolproof.

The effects on the environment of the wastes produced by fish culture depend on the biomass of fish and the size of the water body receiving the wastes. The problems are identical for intensive farms in cages, lakes or waters with a low turnover rate (fjords, lagoons); in the former Yugoslavia there have been heavy mortalities of marine fish in cages where the concentration of fish was too high and there was no cleaning of the sea bed under the cages. In Norway, submerged units for circulating water are used around cages to combat this problem.

The productivity of fresh-water lakes is limited by the level of phosphorus; this can therefore be altered by the installation of a fish farm. Solbé (1988) calculated that the production of one tonne of fish (10 kg of waste phosphorus) in a lake of $500 \times 1000 \times 10$ m can lead to the eutrophication of the lake; the concentration goes from 0 $\mu g \, l^{-1}$ phosphorus, which is characteristic of an oligotrophic lake, to 20 $\mu g \, l^{-1}$, characteristic of an eutrophic lake. Thus, cages should be installed with particular regard to the turnover of water in a lake. Indeed, cage rearing may bring about a degree of fertilization which may be seen as beneficial to a lake but will completely change the macrophyte flora. In Norwegian fjords the sedimentation under salmon cages is 22 times greater than the natural sedimentation (see Chapter 6, this part). Such situations illustrate the role of ecological studies in the management of bodies of water.

REFERENCES

ADCP, 1980. *Fish feed technology.* FAO (Ed.), Rome, ADCP/REP/80/11/: 395 pp.

ADCP, 1983. *Fish feed and feeding in developing countries.* FAO (Ed.), Rome, ADCP/REP/83/18: 97 pp.

Amério M., Costa M., Mazola A., Crisati E., 1989. Use of extracted soybean mealin diets for sea-bass. In: *Aquaculture: a biotechnology in progress*, De Pauw, Jaspers, Ackefors. Wilkins (eds). European Aquaculture Society, Bredene: 603–608.

Barnebé G., Lecoz C., 1987. Large scale cage rearing of the European sea-bass *Dicentrarchus labrax* (L.) in tropical waters. *Aquaculture*, **66**: 209–221.

Billard R., 1989. La salmoniculture en eau douce. In: *Aquaculture*, G. Barnabé (ed.). Lavoiser—Tec. & Doc., Publ. (2): 569–613.

BNA (Bureau de la nutrition animale), 1973. *Fiche d'analyse, Hareng (Norvège)*. Group VIII Fiche 10a, BNA (Ed.), Marseille: 2 pp.

BNA (Bureau de la nutrition animale), 1973. *Fiche d'analyse, Hareng (Côtes Atlantiques américaines)*. Group VIII, Fiche 10b, BNA (Ed.), Marseille: 2 pp.

Castell J. D., 1989. An integrated fish farm in China. *World Aquaculture*, **20** (3): 20–23.

Cho C. Y., 1986. Effects of water temperature on requirements and digestibilities of protein and energy in rainbow trout (*Salmo gairdneri*). In: *Environment and nutrition; determining factors in intensive fish farming.* Proceedings of International Symposium. Aquacultura 86 (Ed), Verona: 25–38.

Chow K. W., 1980. Carbohydrates. In: *Fish feed technology.* FAO (Ed.), Rome, ADCP/REP/80 /11: 55–63.

Dekker I. G., 1986. Evolution of low pollution diets. In: *Environment and nutrition; determining factors in intensive fish farming.* Proceedings of International Symposium. *Aquacultura 86* (Ed.), Verona: 57–62.

Deplano M., Connes R., Diaz J. P., Paris J., 1989. Intestinal steatosis in the farm-reared sea bass *Dicentrarchus labrax*. *Dis. Aquat. Org.*, **6**: 121–130.

Fontaine M., 1990. Préface. *Aquaculture*, G. Barnabé (ed.). Lavoisier—Tec. & Doc., Publ. (2): IX–XII.

Fiala-Médioni A., 1988. Matière organique dissoute et production secondaire (adultes). *Océanis.*, **14** (2): 399–407.

Halver J. E., 1972 (ed.) *Fish nutrition.* Academic Press, New York: 713 pp. (new edition 1989).

Jackson A., 1988. Growth, nutrition and feeding. In: *Salmon and trout farming*, L. Laird & T. Needham (eds). E. Horwood Publ., Chichester: 202–216.

Jhingran V. G., 1990. Aquaculture in India. In: *Aquaculture*, G. Barnabé (ed.). Lavoisier—Tec. & Doc., Publ. (2): 1117–1152.

Kaushik S., 1990. Importance des lipides dans l'alimentation des poissons. *Aqua Revue*, **29**: 9–16.

Koenig J., 1980. Les vitamines, les macro et oligoéléments minéraux dans la nutrition et l'alimentation des poissons. *Océanis*, **5** (2): 131–144.

Lowel T., 1988. *Nutrition and feeding of fish.* AVI Book, Van Nostrand Reinhold Publ., New York: 260 pp.

Luquet P., Kaushik S., 1986. Effets de facteurs environnementaux sur le métabolisme et les besoins alimentaires chez le Poisson. In: *Environment and nutrition; determining factors in intensive fish farming.* Proceedings of international Symposium. Aquaculture 86 (Ed.), Verona: 9–24.

New M. B., 1986. Aquaculture diets of post larval marine fish of the superfamily Percoïdae, with special references to sea-bass, sea-breams, groupers and yellowtail: a review. *Kuwait Bulletin of Marine Science*, **7**: 75–151.

New M. B., 1987. *Feed and feeding of shrimp and fish.* Aquac. Develop. and Coord. Prog., UNEP-FAO, ADCP/REP/87/26, FAO, Rome: 274 pp.

Pavillon J. F., 1988. Matière organique dissoute et production secondaire (larves). *Océanis*, **14** (2): 389–397.

Saroglia M., Scarano G., 1989. Studies on the need of vitamin C of sea bass (*Dicentrarchus labrax*) farm in intensive aquaculture. In: *Aquaculture: a biotechnology in progress*, De Pauw, Jaspers, Ackefors, Wilkins (eds). European Aquaculture Society, Bredene: 697 pp (résumé only).

Solbé J., 1988. Water quality. In: *Salmon and trout farming*, L. Laird and T. Needham (eds). E. Horwood Publ., Chichester: 69–86.

Spotte S., 1973. *Marine aquarium keeping.* Wiley & Sons, New York: 171 pp.

Steffens W., 1989. Principles of fish nutrition. E. Horwood Publ., Chichester: 384 pp.

Tacon A. G., 1979. The use of actived sludge-single protein (ASCP) derived from the treatment of domestic sewage in trout diets. Proc. on Finfish Nutrition and Fishfeed Technology, Hamburg 20–23 June 1978. Vol. II, Berlin: 249–267.

Tesseyre C., 1979. Etude des conditions d'élevage intensif du Loup (*Dicentrarchus labrax* L.). Thèse Doct. 3e Cycle, Univ. Sc. Techn. Languedoc, Montpellier: 115 pp.

Usui A., 1979. *Eel culture*. Fishing News Books (Ed.), Farnham: 188 pp.

3

Fish reproduction

1. INTRODUCTION

In any form of rearing, the basic problem is to obtain the stock to initiate the culture cycle. In vertebrates juveniles are produced through sexual reproduction. In this category fish, particularly marine species, are characterized by their high fecundity; it is not unusual for a female to spawn several thousand eggs per kilogram body weight. The control of sexual reproduction is therefore the starting point of aquaculture activities (see Kinne 1977).

The first stage is to spawn fish at the time of year when it occurs naturally in the wild; spermatogenesis and oogenesis depend on external factors and generally show a seasonal pattern, at least in temperate zones.

Reproduction in captivity under similar conditions to those in the natural environment has been a routine operation for many years for pond fish and salmonids (Billard 1971 & 1981; Richter & Goos 1982; Laird & Needham 1988), and has been carried out from 1971 for marine fish (Barnabé 1971; Barnabé & René 1972, 1973); luckily the processes of reproduction are not completely inhibited in captive fish. In general, gonads develop normally up to the final stages of maturation of the gametes; the sequence is only interrupted at the point immediately before the liberation of the gametes from the gonad (reviewed by Harvey & Hoar 1980). It is possible to intervene at this stage and induce spawning through hormonal manipulation.

The fixed seasonal occurrence of spawning, at least in temperate waters, is not always well suited to the later stages of the rearing operation and there have been numerous attempts to advance or retard the date of spawning. The ultimate goal of this type of manipulation is to obtain captive spawning 'on demand', throughout the year; this has happened for the goldfish and for trout and it seems likely that the experimental successes with sea bass can be transferred to the production stage (Barnabé 1974; Girin & Devauchelle 1978; Carillo *et al.* 1989). The methods developed for obtaining this spawning have depended on a thorough understanding of the mechanism of maturation and spawning; this is described below.

2. GAMETOGENESIS

The ova and spermatozoa of adult fish differentiate from the primordial germ cells. In most fish the sexes are separate; although hermaphrodites are found (Sparidae, e.g. sea bream, white bream; Serranidae, e.g. grouper, comber). In these fish the male and female gonads do not mature at the same time (protandrous hermaphroditism in the sea bream where the male gonad matures before the female, and proterogynous in the grouper where the female matures before the male). It is therefore possible to deal separately with spermatogenesis and oogenesis.

2.1 Spermatogenesis

This is the transformation of a germinal cell which shows little sign of differentiation into a male gamete, the spermatozoa within the testis (Fig. 1). Fish have paired testes. In trout the resting spermatogonia are found at the periphery of the tubules (or testicular lobules); these have differentiated from the primary germ cells and are included in a layer of 'nurse' cells, the Sertoli cells, which are somatic in origin. These type A spermatogonia (SA, Fig. 1) are always present and, in salmonids, undergo division between April and June to give type B spermatogonia which subdivide but remain grouped in cysts. Each subdivision is accompanied by a decrease in size. Between July and September the meiosis which follows these two divisions gives rise to primary and secondary spermatocytes. Spermatogenesis is accompanied by a change in volume and weight; this affects the gonado-somatic index (GSI) (see 2.2 below).

Spermiogenesis is the transformation of spermatids into spermatozoa and involves the formation of the flagella, elimination of cytoplasm and condensation of chromatin. Histological sections of the testis show that the flagellae are in the lumen of the lobules with the heads of the spermatozoa still attached to the layer of cells lining the wall of the tubule (Fig. 2). Spermatozoa (Fig. 3) only reach the efferent canals (which secrete part of the seminal fluid) two months later (October–November). The term spermiation is used to denote the condition when sperm is readily expelled through the urinogenital orifice. The first sperm to be emitted has low motility and fertility and only released in small quantities. During the middle part of the period of sexual maturity, the abundance, fertility and motility are at their maximum (100% of ova are fertilized); this lasts for no more than two months (December–January) because of the phenomenon of ageing. This timetable for salmonids can be applied to marine fish; the phenomenon of ageing of spermatozoa has been demonstrated for the sea bass (Billard *et al.* 1977 and Fig. 4) and is undoubtedly found in other species.

The maturation of male fish usually proceeds without interruption in captivity and there are generally few difficulties in obtaining sperm emission from fish kept in tanks.

2.2 Oogenesis

Oogonia are derived from the primordial germ cells situated in the germinal epithelium of the ovary. At an early stage of development they become surrounded by a layer of somatic cells (the follicular cells); this is called the ovarian follicle.

The oogonia is transformed into an oocyte which is characterized by its yolk reserves; this then becomes a huge cell, the egg (which is in fact still an oocyte because the second

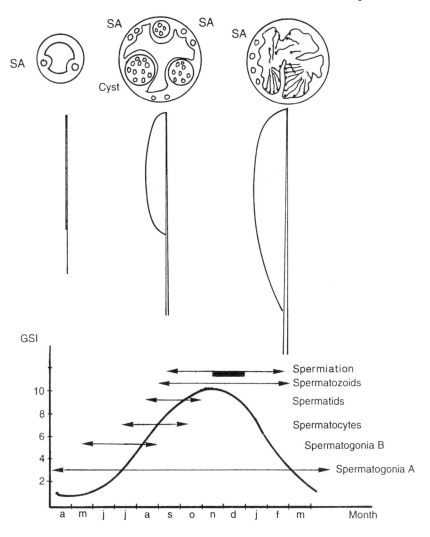

Fig. 1. Diagrammatic representation of spermatogenesis in trout with time (shown in months below the figure). Above: cross section through sperm tubules. Centre: plan of the testes over the same time scale. Bottom: seasonal development of the GSI and categories of cell present in the testis.

polar body is not thrown off until after fertilization). The follicular cells help in building up the yolk (vitelline) reserves of the egg.

The follicular cells differentiate to form the granulosa layer which is separated from the egg by the non-cellular zona pellucida; a distinct theca differentiates from the neighbouring connective tissue. These structures, the granulosa, zona pellucida and theca, are collectively referred to as the follicle or follicular envelope (Fig. 5).

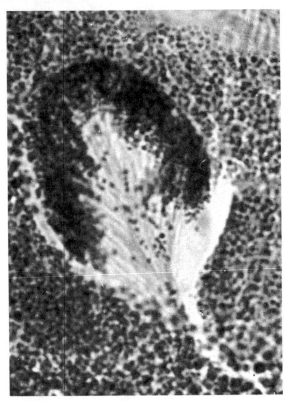

Fig. 2. Section of a mature testis of a *Dicentrarchus* of 37.5 cm, taken 23 December 1969 at Sète. Sperm tubule, enlarged × 1250. The sperm heads are mostly attached to the wall of the tubule; the flagellae in the lumen, giving the appearance of a 'parachute', in sections of fish testes.

The progressive increase in size of the ovaries comes from the growth of the individual oocytes through accumulation of yolk. At the time of spawning, ovaries can represent one-third of the total weight of the fish. The weight of gonads is estimated in relation to the total body weight; this allows the establishment of the gonado-somatic index, the GSI, which is equal to 100 times the weight of the gonads over the whole body weight.

While it has long been known that the maturation of gonads in fish and also their reproductive behaviour are responses to environmental stimuli such as temperature and the length of the day (photoperiod), the relationship is particularly marked for sea bass. The maturation of this fish will be followed, relating the GSI to studies of gonads made with the microscope.

Fish, being poikilotherms, are more active at warm temperatures. Bass consume food more actively between May and November than at other times of the year. This leads to the accumulation of fatty reserves in the mesentery around the digestive tube of non-oily fish. Oily fish accumulate these reserves in the muscle blocks.

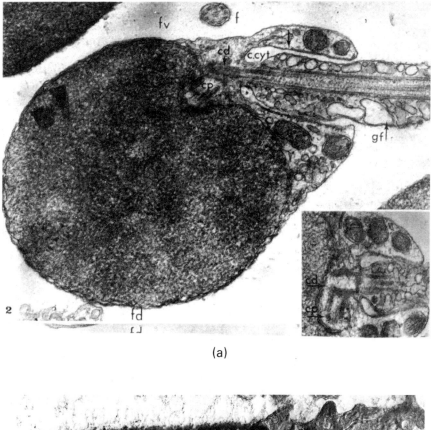

(a)

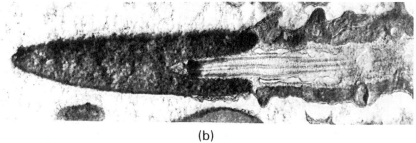

(b)

Fig. 3. (a) Head of a carp spermatozoa × 30 000 (Billard 1969). c.cyt = cytoplasmic canal, cd = distal centriole, cp = proximal centriole, f = section of a flagellum, fd = dorsal surface, fv = ventral surface, gf = shaft of the flagellum. The head contains chromatin, surrounded by plasma membrane and nuclear membrane, in fold. (b) Guppy spermatozoa × 30 000 (Billard 1969).

These fatty reserves are important for spawning; this can be demonstrated using the Nikolski scale (an empirical 4-point scale), following their change over the year. Studies have been carried out with sea bass during sexual maturation and with juveniles whose gonads have not begun to develop (Fig. 6). It is noticeable that in the juveniles the reserves remain almost constant, while in adults they are at their maximum in the warm season and at their lowest in winter.

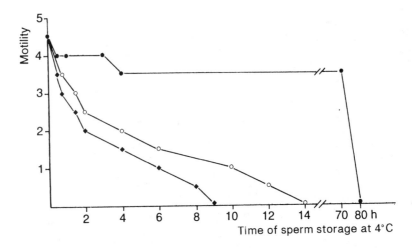

Fig. 4. Development of sperm mobility with time in the bass (*Dicentrarchus labrax*) kept at 4°C from 11th December ● (eight males); ○ 12th January (seven males); ◆ 16th February (eight males).

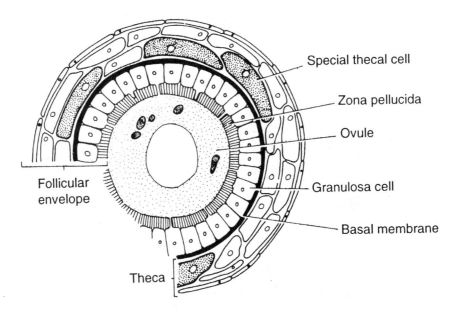

Fig. 5. Diagram of a developing oocyte (Harvey & Hoar 1980).

The weight of the ovary (and that of the testis) has been followed over the course of the sexual cycle for sea bass in the Sète region (Barnabé 1976 and Fig. 7). From June until September the weight of the ovary remains very low, corresponding to a period of sexual dormancy at a time when the bass has its greatest rate of food intake. A slight increase in size is apparent in September but the increase is at its most rapid between

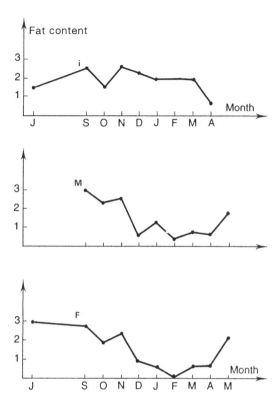

Fig. 6. Seasonal change in fat content of male (M); female (F) and immature (i) bass
Dicentrarchus labrax in the Sète region.

October and December (Fig. 7). The GSI is at its highest in January, which is the height
of the spawning period; the drop in GSI is as spectacular as its rise and the figure is low
by March, the end of the spawning period (<2). This is the period of oocyte atresia where
the eggs which have not been spawned are resorbed. The GSI value is close to 1 during
the period of sexual dormancy, the warm season, which is also the growth season in sea
bass.

The weight of the liver in relation to the whole body weight can be expressed as the
hepato-somatic index, HSI (100 times the weight of the liver over the whole body
weight). Fig. 7 shows that in females this organ shows variations in weight when the
ovary is at its maximum rate of increase or decrease in size. The liver is the 'transit'
organ for substances which have been metabolized by the organism and Figs 6 and 7
show a correlation between the decreases in the fat reserves of the mesentery, the oscillation of the weight of the liver and the increase in the diameter of the maturing oocytes;
the inverse effect occurs during oocyte atresia.

This comparison of the change in the organs, fat reserves and water temperature shows
that gonad development begins at the end of summer but is especially active during the
autumnal drop in temperature (Fig. 8).

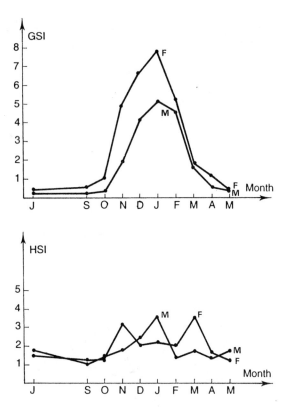

Fig. 7. Seasonal change in the GSI and HSI in male (M) and female (F) sea bass *Dicentrarchus labrax* in the Sète region.

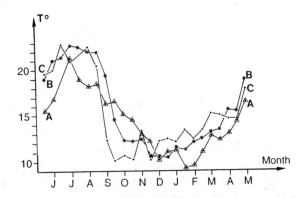

Fig. 8. Seasonal changes in temperature in the natural environment (Sète) during three consecutive years, A, B and C.

The sexual cycle can be fitted into the general life-cycle pattern of the fish; the feeding of the fish during the warm season leads to growth of the body and the deposition of fat in the mesenteries. These reserves are resorbed during vitellogenesis, passing through the liver. This leads to major changes in the HSI between October and December and on accumulation of yolk in the oocytes. Those oocytes which are not spawned break down (atresia), leading to a return of reserves to the mesentery at a time when temperature is low and feeding is limited.

At the cellular level, histological studies (Barnabé 1976; Caporiccio 1976) on oogenesis in the bass have added to the detail of events leading up to spawning: Plate 1 (Fig. 9) shows a section of an ovary from a fish in May. Two types of small oocyte are shown (around 25 and 90 μm diameter) in the ovarian lamellae; these are seen again in June (Plate 2, Fig. 9), but the oocytes are larger (90–100 μm). In September the size has increased further (100–130 μm) although all the photographs show the presence of very small oocytes which are not developing. These stages correspond to those of vitellogenesis defined by Caporiccio (1976).

From September there is a change in the pattern of the development of the oocytes: the diameter of some of them increases rapidly and the follicular cells (nurse cells) become clearly visible (A, Plate 4, Fig. 9); the size of the other oocytes shows little change; the beginning of sexual maturation only concerns a few of the oocytes present in the ovary.

There is a marked increase in the diameter of the maturing oocytes between Plates 5 and 6, taken at the start and end of October, but the increase is truly spectacular between the end of October and the end of November (Plate 7, Fig. 9). The diameter of the oocytes goes from 600 to 700 μm. Spawning ensues one month later; Plate 8 (Fig. 9)

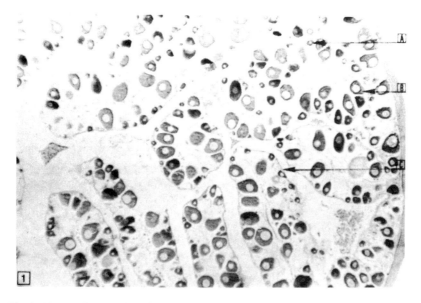

Fig. 9 (Plate 1). Cross section of an ovary from a sea bass (33 cm total length) taken in May (× 75). A: Small oocyte (around 25 μm). B: Pear-shaped oocyte with non-central nucleus: homogeneous cytoplasm. C: Ovarian lamella.

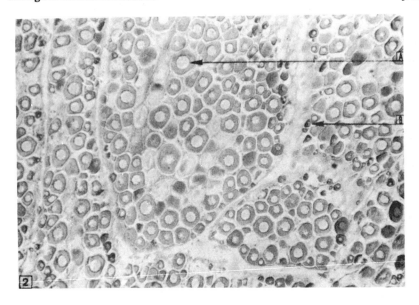

Fig. 9 (Plate 2). Section of an ovary from a 36 cm sea bass, taken in June (× 75). A: Perinuclear nucleoli. B: Small, dark-stained oocyte.

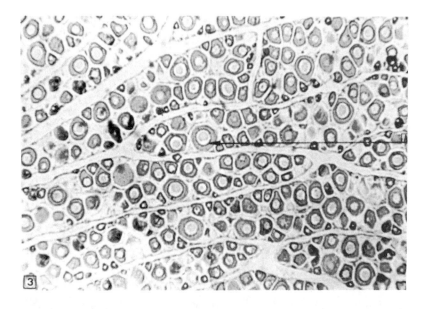

Fig. 9 (Plate 3). Ovary of a female sea bass (36 cm), taken in September (× 75). A: Perinuclear nucleoli.

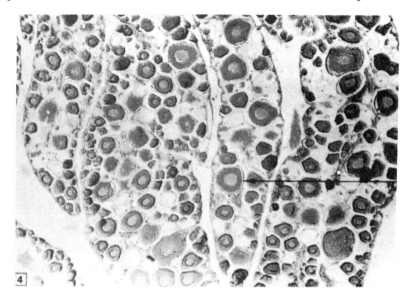

Fig. 9 (Plate 4). Cross-section of the ovary of a sea bass (32 cm) taken in September (× 75).
A: Follicular membrane.

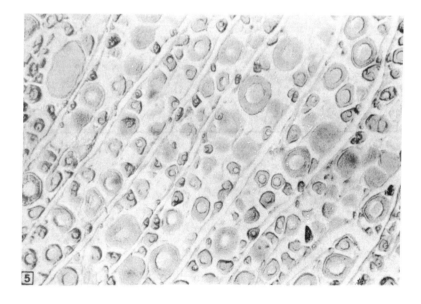

Fig. 9 (Plate 5). Structure of the ovary of a sea bass (32.3 cm) caught in October (× 75).

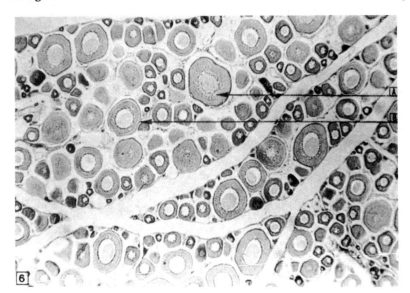

Fig. 9 (Plate 6). Section of the ovary of a sea bass (37.8 cm) taken at the end of October (× 75). A: 1st cytoplasmic inclusions. B: Thickening of the oocyte membrane.

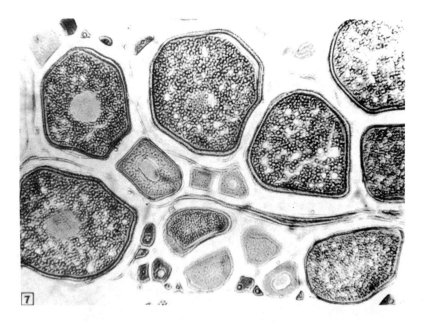

Fig. 9 (Plate 7). Section of the ovary of a female sea bass (60.3 cm) 6 years old, taken in November (× 75).

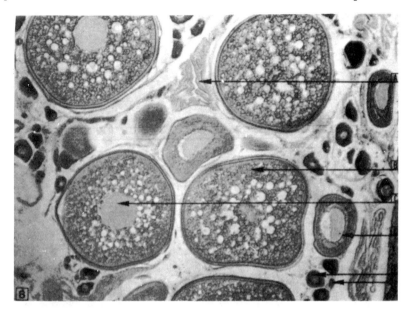

Fig. 9 (Plate 8). Histological appearance of the ovary of a sea bass of 37.5 cm, captured in December during spawning (× 75). A: Empty follicle. B and C: Oocytes in advanced state of maturition. D: Oocyte at the start of maturation. E: Young oocyte (stage 2), resting. F: Very early oocyte stage.

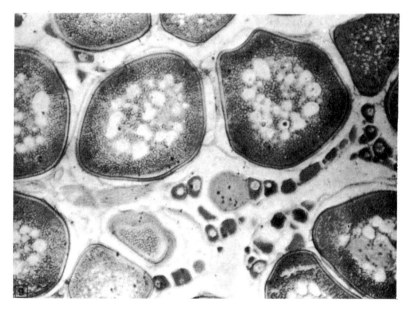

Fig. 9 (Plate 9). Histological appearance of an ovary from the sea bass, *Dicentrarchus labrax*, taken in January (× 75).

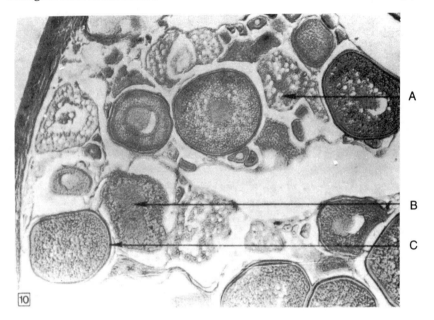

Fig. 9 (Plate 10). Section from an ovary of a fish of 52 cm, taken in February (× 75). A: Advanced stage of oocyte atresia. B: Oocyte undergoing atresia. C: Start of autolysis, marked by a change in the appearance of the membranes.

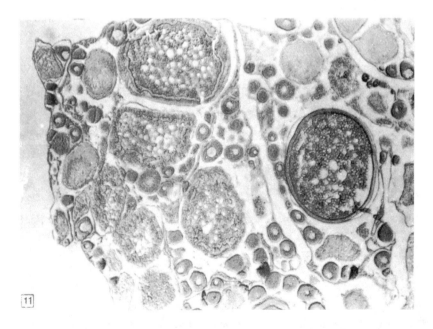

Fig. 9 (Plate 11). Section of an ovary of a 63 cm fish, captured in March (× 75).

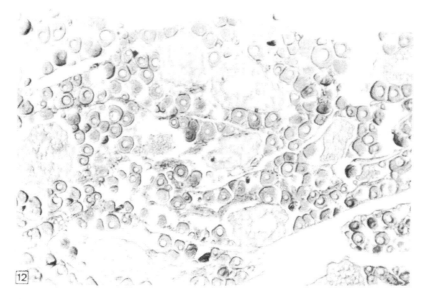

Fig. 9 (Plate 12). Section of an ovary of a 63 cm fish, taken the same day as that in Plate 11 (× 75). Atresia is advanced.

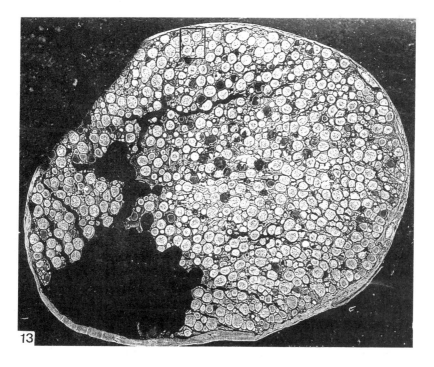

Fig. 9 (Plate 13). Section through the ovary of a sea bass during spawning at the end of December (× 4). The region marked at the top corresponds to the area shown in Plate 8.

shows a section of an ovary of a fish which had been used in an experiment on artificial reproduction (Barnabé & Tournamille 1972). Plate 15 (Fig. 9) shows a cross-section of the whole of this ovary. The empty area is the space left by spawned ova.

Plates 10–13 (Fig. 9) show the oocytes which have not been spawned; these are atretic and in the process of resorption. The release of eggs during spawning is not therefore the inevitable conclusion of the process of sexual maturation; fish may pass the breeding period without spawning and there is a natural phenomenon which is well known to the fish farmer (see 1, above): the blockage of maturation before spawning. This has been described elsewhere (Barnabé 1976). The oogenesis of the sea bass has been the object of several histological studies; Caporiccio (1976) used electron microscopy to show the comb-like microvillae projecting from the follicle into the oocyte (Fig. 10). He showed how the histological development of the oocytes can be matched to a maturation scale (Fig. 11).

As with other teleosts the final stage of maturation is characterized by the migration of the nucleus to the periphery of the oocyte (forming the germinal vesicle, GV) and then the disappearance of the nuclear membrane; the yolk becomes translucent and, in the species which possess them, the lipid reserves coalesce. In some species the oocytes absorb water and increase their diameter, leading to an inflation of the abdomen which is termed hydration; this happens 24–48 hours before spawning. The migration of the germinal vesicle or this coalescence of lipid globules are used as criteria of the state of advancement of sexual maturity (salmonids, pikes and American bass). This technique requires a biopsy; hydration is a more obvious sign of development (Fig. 12).

Ovulation is the point when the oocyte is expelled from the follicle into the ovarian or the peritoneal cavity. This may be accompanied by characteristic behaviour (contractions, resting on the bottom). During ovulation the oocyte envelopes, which are made up of two layers in the zona pellucida and six acellular layers, lose the two extreme layers, breaking the adhesion between the oocyte and the follicle. Helped by the muscles in the ovary wall the oocytes leave the follicles, falling into the ovarian lamellae, and from there pass into the lumen of the ovary. The expulsion of the eggs (oviposition) takes place after the eggs have passed down the oviduct to the genital orifice (most frequent pattern) or passed through the body cavity (salmonids).

Oocyte maturation and ovulation can involve part or all of the stock of oocytes at the end of the vitellogenic stage. There may be a single spawning (sea bass) or sequential spawning (sea bream). Eggs can remain in the body cavity for one to two weeks in salmonids and still remain fertile. For the majority of other fish, the gap between the maturation of oocytes and ovulation is very short (75 h for sea bass) and the eggs are only fertile for a few tens of minutes).

We have seen that gametogenesis is linked to external factors, but also depends on internal factors, such as the build up of the mesenteric reserves in summer and their mobilization via the liver and transfer to the ova in the autumn during vitellogenesis. This has been studied at the cellular level for sole by Nunez-Rodriguez & Le Menn (1988) who identified different structures in the liver cells, associated with the synthesis of vitellogenin and demonstrated their incorporation in the oocyte, while Abraham *et al.* (1988) put forward a scheme for all the interactions between the liver, the ovary and hypophyseal hormone. All these factors are expressed through neuro-endocrine

Cf: follicle cell
D: demosome
Ov: oocyte
R: ribosome

Fig. 10. Contact between the oocyte and follicular cells through microvilli which cross the acellular layer surrounding the oocyte (sea bass oocyte × 31 000, Caporiccio 1976).

complexes which we shall try and summarize. This information is essential for the fish farmer in order to manage reproduction effectively.

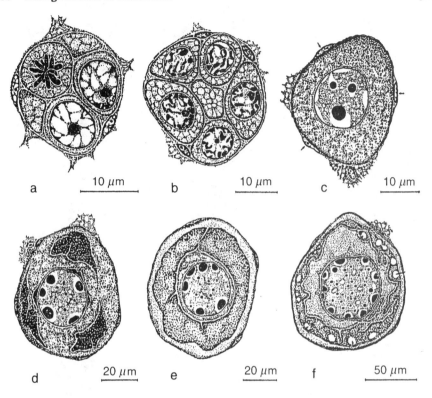

Fig. 11a. Scale of sexual maturity for the ovary of *D. labrax* (Caporiccio 1976).

Stage I: oogonia a);
Stage II: young oocytes b);
Stage III: appearance of follicular cells c);
Stage IV: loss of homogeneity in the oocyte cytoplasm, nucleoli form two groups. Formation of nuclear extrusions d);
Stage V: Formation of three distinct zones in the cytoplasm, e);
Stage VI: First signs of vitellogenesis and start of differentiation of the oocyte envelope f).

3. THE NEURO-ENDOCRINE CONTROL OF GAMETOGENESIS

The reception of environmental stimuli such as day length (photoperiod), temperature and rainfall activates the nervous system and entails the passage of information from receptors (for example, the eyes) to the brain. The pineal organ (or 3rd eye), which is situated on the brain just under the cranium, is also important. This information reaches the hypothalamus and determines the activity of the hypothalamus through chemical messengers called hormones. These hormones, in their turn, cause the hypophysis to liberate a further hormone into the circulatory system: the target organ for this hormone, gonadotropin, is the gonad. Gonadotropin stimulates the production of the sex steroids by the non-sexual cells (the interstitial cells) in the gonad.

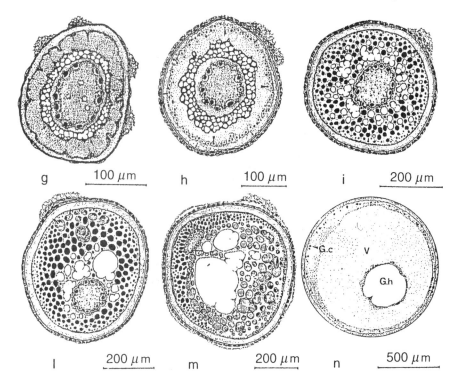

Fig. 11b. Scale of sexual maturity for the ovary of *D. labrax* (continued) (Caporiccio 1976).

Stage VII: arrangement of the first yolk droplets in a crown around the nucleus, and the formation of lobes on the surface of the nucelus g);
Stage VIII: formation of a second type of yolk and the appearance of the 'zona radiata' h);
Stage IX: formation of two types of yolk i);
Stage X: formation of a 3rd type of yolk l);
Stage XI: end of vitellogenesis and migration of the nucleus m);
Stage XII: ovule (o.d. 'oil droplet', Y: homogeneous yolk, cG: cortical granule) n).

3.1 The hypophysis and the hypothalamus

The transition between nerve-borne information (neural) and hormonal control occurs at the hypothalamus–hypophysis interface. The description below is drawn largely from the work of Harvey & Hoar (1980).

As well as the hormone gonadotropin, which is of special interest, the adenohypophysis (the anterior part of the hypophysis) produces somatotropin (growth hormone) corticotropin, prolactin, thyrotropin and the hormone which stimulates the melanocytes.

The neurohypophysis is neural in origin and is a part of the adenohypophysis at the base of the brain, formed mainly of axon and neuron fibres whose cell bodies are in the hypothalamus (Fig. 13, adapted from Harvey & Hoar 1980). These are termed neurosecretory cells. They respond to an electrical signal from the brain by liberating a

Fig. 12. Hydrated females in an anaesthetic bath.

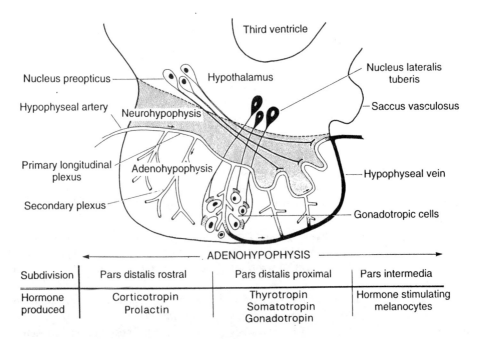

Subdivision	Pars distalis rostral	Pars distalis proximal	Pars intermedia
Hormone produced	Corticotropin Prolactin	Thyrotropin Somatotropin Gonadotropin	Hormone stimulating melanocytes

Fig. 13. Diagram of the adenohypophysis (after Harvey & Hoar 1980).

chemical messenger from the terminal of the axon, ensuring the transition between neural and hormonal information.

The two hypophysial nuclei involved in this process are the *preopticus* nucleus and the *lateralis tuberis* nucleus. The endocrine cells of the adenohypophysis are directly innervated by neurosecretory axons. The chemical messenger liberated at this point is the releasing hormone (RH); this stimulates the production of gonadotropin once released. Gonadotropin is then transported to the gonad by the circulatory system; in the gonad it initiates the production of the sex steroids. These hormones, androgens, oestrogens and progesterones are directly concerned with the development of the gonads.

3.2 Teleost gonadotropins

Many studies have demonstrated the role of the hypophysis in the control of reproduction. Surgical or chemical hypophysectomy shows that this gland is essential for the successful completion of all the stages of the sexual cycle.

The gonadotropic hormone (GTH) has been purified and then injected into several species of fish. The effect of this hormone on maturation *in vitro* has also been demonstrated. More recently, a second hormone has been purified and showed to be essential for vitellogenesis. There is partial species specificity of teleost gonadotropins; while carp or salmon are used as hormone donors (by removal of the hypophysis), they are not always active in other species.

These hormones act on certain tissues which take part either directly or indirectly in the development of the sexual products. The somatic tissue of the gonad secretes the sex steroids; these androgens, oestrogens and progesterones are capable of feedback through the hypothalamo–hypophyseal axis. The thyroid and some tissues in the kidney secrete hormones which are able to alter the action of GTH on the gonad.

The steroid hormones produced by the gonads are known in the male: 11 ketotestosterone and testosterone take part in the control of spermatogenesis in goldfish. In the female, oestradiol regulates the synthesis of vitellogenin in the liver and the mobilization of fat reserves. Progesterones, such as 17α-hydroxy, 20β-hydroxyprogesterone play their part in the control of the maturation of the oocyte (pike, goldfish, carp). Prostaglandins act to initiate spawning behaviour.

The action of gonadotropins on gametogenesis is only partially understood: the maturation of gonads in fish is the indirect result of a slow and regular increase in the rate of secretion of gonadotropins.

Vitellogenesis (the incorporation of yolk into developing oocytes) is regulated by a gonadotropin which has a low glycoprotein content. The formation of yolk vesicles, the first to be produced, is initiated by oestrogens in goldfish; yolk granules are formed under the influence of pregnenolone. The synthesis of the precursors of yolk takes place in the liver and has been shown to be stimulated by oestrogens. We have seen above where the primary material used for the synthesis of yolk comes from. It has been shown elsewhere that low doses of thyroxin stimulate vitellogenesis in immature goldfish. GTH and oestrogens in the plasma increase during the later phases of the cycle. Such steroid hormones are synthesized by the bass at the moment of ovulation (Colombo *et al.* 1978). The steroid 17α-hydroxyprogesterone and 20β-dihydroprogesterone, produced by the follicle in response to secretion of hypophyseal gonadotropin and its transport in the blood, is the most likely mediator for the final stage of oocyte maturation in most species.

The rupture of the follicle and the expulsion of oocytes appears to be independent of hypophyseal control.

4. THE ROLE OF EXTERNAL FACTORS IN THE REPRODUCTIVE CYCLE

Food, temperature, photoperiod, physical and chemical factors (dissolved oxygen, salinity), the density of the animals and the substrate are among the factors which have been implicated in the control of the reproductive cycle.

4.1. Food

The metabolic requirements are met by the food intake; this is the premier factor in the regulation of gametogenesis. Reproduction uses energy which the fish obtains from its food and starved fish will not mature as they do not have sufficient reserves.

In herring, feeding stops during spawning and then begins again, so that fat reserves can be refilled and the gonads can develop again. If feeding stops before spawning, fat reserves become depleted but the gonads continue to develop. A similar phenomenon has been demonstrated for sea bass.

4.2 Temperature

In poikilotherms, the ambient temperature of the surrounding environment plays a determining role on general activity and especially on feeding. Abnormally low temperatures decrease appetite and the likelihood of capture. If the fat reserves are totally depleted the gonads provide the metabolites which allow the animal to survive; this happens with sea bass which are taken from the wild and held at 7°C: no oocyte development is observed. At this temperature the fish do not feed and the gonad is resorbed. The sex glands act as the last store of reserves; but this also acts to the detriment of reproduction (Barnabé 1976).

One species of anchovy only undertakes its spawning migration when the level of lipids in the liver reaches a certain level (>14%). Above this level, it migrates when a temperature drop of 5°C (14 to 9°C) takes place, but if the fat content reaches 22% the spawning migration can be made without encountering a temperature drop (Nikolski 1963).

As well as these indirect effects, temperature has a direct effect; gametogenesis only takes place within a given range of temperature for each species; the geographical distribution of a species is thus made up of a central zone where environmental conditions are suitable for reproduction and a peripheral zone where spawning will not take place even if broodfish are present; many coastal fish enter lagoons and vitellogenesis proceeds. However, spawning only takes place when these fish return to the sea. In the Arcachon Basin (Stequert 1972) or the Etang de Thau (Barnabé 1976), female bass and mullet are effectively sterile.

In some species, gametogenesis takes place when temperatures are decreasing; in others when temperatures are increasing. Often, the last stages (final stages of oocyte maturation and ovulation) are stimulated by a change in temperature which stimulates the neuro-endocrine system (hormonal secretion which controls changes in the gonad).

Temperature has a direct effect on the production of GTH by the hypothalamo-hypophyseal complex. The example of the anchovy spawning migration described above illustrates the interrelationship of various factors in spawning.

4.3 Photoperiod
Where other factors are identical it has been shown that photoperiod plays an extremely important role in gametogenesis. This is true for salmonids and for many species of marine fish.

Where gametogenesis takes place during decreasing photoperiod, an artificially shortened day length brings forward the time of ovulation (e.g. trout which would normally spawn at the end of autumn can be caused to spawn at the beginning of summer). The action of photoperiod is via photoreceptors (the eye, pineal gland), through the nervous system on the hypothalamo-hypophyseal axis.

In addition to the effect of photoperiod, some fish may search visually for a suitable spawning substrate; finding such a substrate may initiate ovulation (pike and tropical species where there is little change in photoperiod).

4.4 Other environmental factors
Zanuy and Carillo (1984) demonstrated the influence of low salinity on the failure of maturation in sea bass; this has also been shown in many other species (e.g. mullet).

If broodfish are kept at high densities, the fecundity of females is reduced (carp, trout).

Physical and chemical properties of the water are also important. In tropical and equatorial waters there are only slight variations in temperature and photoperiod, and humic acids, washed out in the rainy season, may act as a signal for the start of spawning. Aquarists can buy products based on this for spawning tropical fish.

5. APPLICATIONS FOR AQUACULTURE

5.1 Maintaining the reproductive cycle in captivity
In order to maintain a captive breeding stock of fish it is important that diet should be adequate and that temperature, salinity, density and other factors should be optimal. The conditions for holding broodstock are different from those in ongrowing; bass grows best at 22°C but the final stages of gametogenesis will only take place at temperatures between 10 and 15°C (Barnabé 1989).

Carp and most pond fish complete their maturation in broodfish ponds (2000–10 000 m^2) where the density is 100 kg ha^{-1} (500 kg ha^{-1} for pike and some other species). The diet of broodfish and other factors is linked to their maturation (temperature, pH and oxygen levels) and, if satisfactory, should allow maturation to take place naturally (Chapter 5 below); the final stage of maturation and ovulation may need to be induced.

Salmon and trout also mature in large tanks or cages kept specifically for this purpose at densities which remain below 10 kg m^{-3} (Laird & Needham 1988).

5.2 Control of reproduction through external factors

For reproduction in captivity to be successful, an environment suited to the requirements of the species must be provided. The conditions favoured by the species in the wild are therefore imitated (temperature, light, substrate, etc.) and the natural rhythm of gametogenesis is repeated where conditions are most favourable. Fig. 14 shows (diagrammatically) the relationship between external and internal factors in the control of spawning (Bromage *et al.* 1990).

The economic requirements of fish culture impose a shortening of the natural life cycle. This may be aided by changing the timing of the spawning season. The small, fragile larvae can then be reared indoors in hatcheries and nurseries in large quantities during the time of year which is unfavourable for growth outdoors, and then transferred to the ongrowing unit at the start of the favourable season. This can be accomplished by advancing or retarding the timing of the spawning period by changing photoperiod, temperature or both.

The manipulation of these parameters has given results which are sometimes different for different species; there is no general rule, although closely related species (family, genus) are likely to be similar in their responses. We have seen the example of trout where spawning can be brought forward by shortening day length, it can also be retarded by lengthening the day (16–24 h daylight). Similar results have been achieved for marine fish, but the results obtained from out-of-season spawning are often less good than at the normal time of year.

Within the ranges where it has no direct effect, temperature has an influence on gametogenesis which can be used for the profit of the fish farmer through the management of reproduction: the year-round spawning of goldfish can be achieved by holding them at a constant temperature of 10–12°C throughout the year, then transferring them directly to water at a temperature of 18–20°C. The thermal shock brings about the completion of oocyte maturation and ovulation. In some cyprinids (carp species) several spawnings can be achieved each year (up to seven) by raising the temperature. There are limits; for example for the goldfish a constant temperature of 24°C inhibits the development of the gonad even though the secretion of gonadotropin is stimulated.

5.3 The control of reproduction through hormone treatment

We have seen that gametogenesis is not completely inhibited in captivity and that if natural conditions are recreated in ponds or cages, males will complete their sexual maturation and females will complete all but the last stages of oocyte maturation. The block occurs at the end of vitellogenesis in pond fish and also in certain marine fish (see above).

The unlocking of the final stage of maturation (completion of oocyte maturation and ovulation) which does not occur spontaneously in captivity can be brought about through the injection of hormones or products with an active gonadotropin function. Various hormones from different species are capable of stimulating maturation and ovulation. In practice, broodfish which have almost completed maturation are selected for treatment. Thin fish or fish with concave flanks are rejected as they have clearly not matured.

It is also essential to retain enough males to give a sex ratio of two or three males to one female in the spawners. Sometimes there are no secondary sexual characteristics to

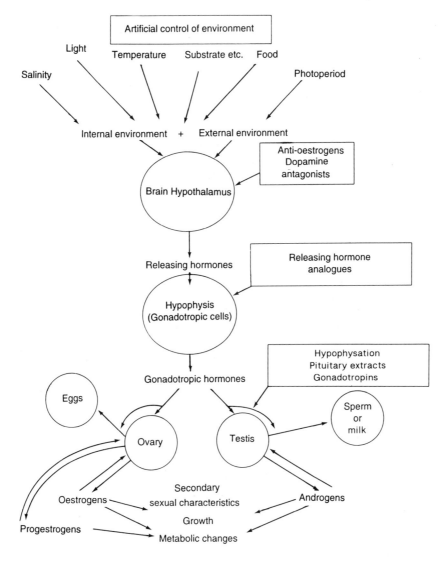

Fig. 14. Control of reproduction through manipulation of the environment or the use of hormones
(adapted from Bromage *et al.* 1990).

aid identification of the sexes; the selection is therefore risky although fully mature males
can be recognized easily; light pressure on the abdomen results in the expulsion of milt.
There are certain characteristics by which mature female carp can be recognized: these
include bulging sides and a reddish cloaca; however assessment of the different stages of
maturity is always subjective, based on colour and shape, and not always reliable. A
plump side may be a reflection of a full digestive system. A reddish, protruding genital
orifice will also be found in fish where the gonads are already regressing.

When the broodfish have been selected, spawning can be induced by three different techniques.

5.3.1 Hypophysation

In pond fish, hypophysation (intramuscular or intraperitoneal injection with crude fish hypophyseal extracts) is the most frequently used technique but gives variable results and is dependent on the collection of large numbers of hypophyses from spawning adults (in order to have significant hypophyseal activity). This technique is used most frequently in carp (Harvey & Hoar 1980) and sometimes in salmonids. The preparation and use of hypophyseal extracts has been described by Marcel (1980).

The hormonal induction of spawning in carp, for example, requires the injection of dried, crushed carp hypophyses; the powder is suspended in water and the solution injected via the intramuscular or intraperitoneal route at a dose of 0.3 mg hypophyses per kg broodfish injected. Spawning normally occurs 12 h later at 20°C.

5.3.2 The injection of substances with a gonadotropic activity

The injection of substances with a gonadotropic function has been practised for 50 years in fish and this has given rise to an abundant literature, reviewed by Pickford and Atz (1957). Unfortunately, few products have a proven activity on a wide range of species. Only one substance has this property; this is human chorionic gonadotropin (HCG).

HCG is extracted from the urine of pregnant women and is most effective in marine fish, although the mechanism by which it acts is not completely elucidated. HCG is used to obtain spawning in many other species of farmed fish. The timing of spawning can be fixed to within a few hours (within the natural spawning period which may extend for three months of the year) by using HCG. This helps in the management of the use of ponds and the production of plankton, rotifers and artemia as food. If eggs can be obtained regularly in known quantities the farm can be operated more efficiently than when the timing and quantity of egg production is left to chance.

The management of spawning in females (males continue to spermiate over a long period) has long been an objective of marine farmers. In many marine species preparations from fish hypophyses have no effect (Barnabé 1976).

5.3.3 Injection of releasing hormone analogues

It is possible to synthesize peptides which have the same properties as releasing hormones (known as luteinizing hormone releasing hormone, LH-RH or gonadotropin releasing hormone, GnRH). This has opened up a new way of managing the reproduction of marine fish. This is reflected in the fact that the numbers of sea bass and sea bream fry only became significant in 1986–87, two years after the publication of the precise methodology for the use of an analogue of LH-RH (des-Gly[10], D-Ala[6]) from the Sigma company (Barnabé & Barnabé-Quet 1985). These synthetic products (analogues) can be used to modify the spawning process: they act at the hypothalamus–hypophyseal interface in the brain, 'upstream' of the hormone system.

With this product there are no problems about specificity (as occur with gonadotropin) as the substance injected ($10 \mu g \ kg^{-1}$) causes the liberation of the appropriate gonadotropin from the hypophysis of the recipient fish. A further advantage is that the products are active in a wide range of species and there are no signs of rejection.

We have shown (Barnabé & Barnabé-Quet 1985) that the injection of synthetic LH-RH into the sea bass (10 μg kg^{-1}) not only stimulates spawning more effectively than HCG but is also capable of acting on oocyte maturation, causing oocytes to mature normally in temperatures outside the normal range for the species (at temperatures between 15 and 20°C for example). These peptides also stimulate the production of other hormones (growth hormone, thyroxine). This benefits the general quality of the gonad products; indeed, it is known that low doses of thyroxine stimulate vitellogenesis in immature goldfish.

It is possible to intervene in gametogenesis by short circuiting the natural stimuli (when they are known) and/or by injecting substances which act as hormones or induce hormone activity (Fig. 13). Unfortunately, the establishment of the whole of gametogenesis solely by biochemical means is possible in theory but far from easy in practice: it involves much stressful handling and also the continued use of implants which liberate infinitesimal quantities of hormones into the body of the fish over long periods of time.

It is not yet possible for a practising fish farmer to manage gametogenesis by totally artificial means. However, rapid progress is being made and there are many published works pertaining to different groups or species of fish (see references).

6. CONCLUSION

The use of LH-RH has been the basis for a recent improvement in the production of larvae of marine fish because of the effects of inducing maturation and improving the quality of the gametes and the survival of eggs and larvae.

Captive marine broodfish, presumably because of the greater availability of food than in the wild, produce bigger eggs than wild fish; this gives rise to bigger larvae (next chapter). This change in larval size alters larval feeding; this has important consequences for commercial production.

These new developments, which have come from physiological research, are not the only ones likely to give rise to further improvements of traditional methods. It is already possible to control the sex of the fish and early maturation experimentally. Sperm can be stored at low temperature for several days, and indeed for several years if frozen with suitable diluents. We have not touched on the potential developments in the genetics and molecular biology of reproduction: these are for the future rather than the present.

REFERENCES

Abraham M., Hilge V., Akert K., Sandri C., 1988. The teleost oocytes and the follicle cells in relation to transport and incorporation of vitellogenin. In: *Reproduction of fish*. INRA (Ed.), Paris: 38–43.

Barnabé G., 1971. Premières inductions hormonales de la ponte chez le Loup *Dicentrarchus labrax* L. Rapport CNEXO: 10 pp.

Barnabé G., 1974. Compte rendu sommaire de la campagne 1972–1973 de reproduction contrôlée du Loup à Sète. In: *Colloque sur l'Aquaculture*. CNEXO (Ed.), Paris: 205–213.

Barnabé G., 1976. Contribution à la connaisssance de la biologie du Loup *Dicentrarchus labrax* (L.) Poisson Serranidae. Thèse Doct. Etat, mention Sciences, Univ. Sc. Techn. Languedoc, Montpellier: 426 pp.

Barnabé G., 1990. L'élevage du Loup et de la Daurade. In: *Aquaculture*, G. Barnabé (ed.). Technique & Documentation (Lavoisier) (Publ.), Paris: 675–720.

Barnabé G., Tournamille J. C., 1972. Expériences de reproduction artificielle du Loup *Diocantrarchus labrax* (Linné 1758). *Rev. Trav. Inst. Pêches Marit.*, **36** (2): 185–189.

Barnabé G., René F., 1972. Reproduction contrôlée du Loup *Dicentrarchus labrax* (Linné) et production en masse d'alevins. *C.R. Acad. Sci. Paris*, **275D**: 2741–2744.

Barnabé G., René F., 1973. Reproduction contrôlée et production d'alevins chez la Dorade *Sparus auratus* Linné 1758. *C.R. Acad. Sci. Paris*, **276D**: 1621–1624.

Barnabé G., Paris J., 1984. Ponte avancée et ponte normale du Loup *Dicentrarchus labrax* (L.) à la Station de Biologie Marine et Lagunaire de Sète. In: *Actes du Colloque 'L'Aquaculture du Bar et des Sparidés'*, G. Barnabé & R. Billard (eds). Publ. INRA, Paris: 63–72.

Barnabé G., Barnabé-Quet R., 1985. Avancement et amélioration de la ponte induite chez le Loup *Dicentrarchus labrax* (L.) à l'aide d'un analogue de LHRH injecté. *Aquaculture*, **49**: 125–132.

Billard R., 1969. Ultrastructure comparée de spermatozoïdes de quelques poissons téléostéens. *Comparative Spermatology*, **137**: 71–79 (4 pl).

Billard R. (ed.), 1978. International symposium on reproductive physiology of fish, Paimpont, 1977. *Ann. Biol. Anim. Bioch. Biophys.*, **18** (4): 759–1106.

Billard R. (ed.), 1981. *La pisciculture en étang*. I.N.R.A. Publ., Paris: 434 pp.

Billard R., Dupont J., Barnabé G., 1977. Diminution de la motilité et de la durée de conservation du sperme de *Dicentrarchus labrax* (L.) (Poisson Téléostéen) pendant la période de spermiation. *Aquaculture*, **11**: 363–367.

Borg B., Ekström P., Van Veen T., 1983. The parapineal organ of teleosts. *Acta Zoologica*, **64** (4): 211–218.

Bromage N., Jones J., Randall C., 1990. Broodstock care and the effects of hormonal and environmental factors in the induction of spawning. In: *Mediterranean aquaculture*, Flos, Tort & Torres (eds). E. Horwood Publ., New York: 88–101.

Caporiccio B., 1976. Etude ultrastructurale et cytochimique de l'ovogénèse du Loup (*Dicentrarchus labrax*, L.). Thèse 3ᵉ cycle, USTL: 111 pp, 35 pl.

Carillo M., Bromage N., Zanuy S., Prat F., 1989. The effect of modifications in photoperiod on spawning time, ovarian development and egg quality in the Sea Bass *Dicentrarchus labrax* L. *Aquaculture*, **81**: 351–365.

Colombo L., Colombo P., Arcarese G., 1978. Gonadal steroidogenesis and gametogenesis in Teleost fish. A study on the sea bass *Dicentrarchus labrax*, L. *Boll. Zool.*, **45** (suppl. II): 89–101.

Girin M., Devauchelle N., 1978. Décalage de la période de reproduction par raccourcissement des cycles photopériodique et thermique chez les poissons marins. *Ann. Biol. Anim. Bioch. Biophys.*, **18** (4): 1059–1065.

Harvey B., Hoar W. S., 1980. *La reproduction provoquée chez les poissons: théorie et pratique*. IDCR, Ottawa, Canada: 48 pp.

Kinne O., 1977. *Marine ecology. Cultivation*, Pisces (Vol. III), 2: 968–1035.

Laird L., Needham T. (ed.), 1988. *Salmon trout and farming*. E. Horwood Publ., Chichester (UK): 270 pp.

Marcel J., 1980. Préparation et utilisation de broyats hypophysaires pour l'induction de la reproduction des poissons. In: *La pisciculture en étang*, R. Billard (ed.). INRA Publ., Paris: 163–172.

Nikolski G. V., 1963. *The ecology of fishes*. Academic Press, London: 352 pp.

Nunez-Rodriguez J., Le Menn F., 1988. Vitellogenesis in a teleost fish: the Sole (*Solea vulgaris*). In: *Reproduction in fish*. INRA (Ed.), Paris: 31–34.

Pickford G. E., Atz J. W., 1957. *The physiology of the pituitary gland of fishes*. Zool. Soc., New York: 613 pp.

Richter C. J., Goos H. J., 1982. Proceedings of the International Symposium of Reproductive Physiology of Fish, Wageningen (Holland), 1982. Pudoc Publ., Wageningen: 256 pp.

Stequert B., 1972. Contribution à l'étude du Bar *Dicentrarchus labrax* L. des réservoirs à poissons de la région d'Arcachon. Thèse 3ᵉ cycle, Fac. Sci. Bordeaux: 149 pp.

Zanuy S., Carillo M., 1984. La salinité: un moyen pour retarder la ponte du Bar. In: *Actes du Colloque 'L'Aquaculture du Bar et des Sparidés'*, G. Barnabé & R. Billard (eds.). INRA Publ., Paris: 73–80.

4

Hatchery rearing of the early stages

1. THE FUNCTIONS OF THE HATCHERY–NURSERY

Whatever the species being reared the basic form of an aquaculture unit is similar in principle: obtaining mature broodstock and inducing spawning have already been described. The fertilization of eggs, their incubation, the rearing of the fragile, tiny animals which hatch from the eggs (larvae) are processes which usually take place in the same structure, the hatchery or in some cases combined hatchery and nursery. There are several biological reasons for this:

— The main achievement in the hatchery is the rearing of the first, most fragile, stages of development of the seed for the farm. Marine fish and many fresh-water species and also molluscs and crustaceans only accept live prey as food; these must be provided in the hatchery. It is thus necessary to manage a type of trophic web (Chapter 2, Part I). The hatchery is central to the problems and the prospects of the 'new' aquaculture.

— In the hatchery, huge numbers of larvae (of very small size) are reared in controlled conditions in small volumes of water. Environmental conditions can be optimized (physical and chemical characteristics, absence of predators) and food supplied to give a survival rate of between 20 and 60% for marine fish and more than 80% for fresh-water fish. Estimates of survival in the wild are below 5% for the eggs of larval marine fish (Bannister *et al.* 1974; Riley 1974).

— The feeding of the larvae of marine species (fish, crustaceans and molluscs) pose many problems as these tiny animals require a large number (hundreds and even thousands of individuals each day) of living prey. The size of the prey given must increase as the size of the mouth of the larva increases. The pioneers of aquaculture established the requirements of the larvae and ways of meeting them.

— Up to the present it has not been possible for technology to develop a manufactured or inert food suitable for the larval stages of planktonic fish similar to that developed for salmonids. It is therefore necessary to supply live plankton as prey. Molluscs require microalgae and crustaceans, and fish need zooplankton. In the hatchery–nursery it is therefore necessary to set up a food chain of planktonic organisms for larval rearing.

— The situation is different for salmonids because of their large, more robust eggs which have massive reserves of yolk; by the time the last of the yolk has been absorbed the larvae are able to feed on diet which is manufactured by industrial companies (dry compound diet). It has long been possible to manage the production of juvenile salmonids; this explains why at present these fish provide the bulk of the production of intensive fish culture.

— The eggs of marine and pond fish are tiny (about 1 mm diameter); tiny larvae hatch from these eggs (the grouper is 1.5 mm long when it hatches). This small size, the absence of significant yolk reserves and their dietary requirements explain why the rearing of larvae is a recent, incomplete development.

— For fish, the hatchery and nursery are associated. For molluscs, there are specific nurseries where the larvae from hatcheries are grown to the spat stage required by the farmers. In this chapter we are only concerned with fish; however, the production of algae and zooplankton is also necessary for other groups. Here we shall use the terminology given by Blaxter (1988): the 'embryo' is the term given to the young fish up to hatching. The term 'larva' covers the development up to metamorphosis and a 'juvenile' has a similar appearance to an adult.

2. SPAWNING, EMBRYONIC DEVELOPMENT AND EGG INCUBATION

Males are placed together with females which are ready to spawn. Eggs and sperm are released directly into the water, simultaneously. Of the many spermatozoa which surround each egg, only one penetrates through a hole, the micropyle, and ensures fertilization.

Immediately after fertilization, the egg absorbs water and the chorion hardens. A perivitelline space forms and the embryo begins development in the form of a blastodisc at the animal pole (Figs 1 and 2). The blastodisc surrounds the yolk, except for one gap, the blastopore. The embryonic axis forms through convergence of tissues and becomes segmented and separated from the yolk; the cephalic (head) zone elongates, followed by the rest of the body. The heart is functional and the eyes form before hatching (Figs 1 and 2).

At the stage immediately after fertilization or when the first cell division is taking place, pressure or temperature shocks can be used to induce triploidy or tetraploidy. For example, trout eggs can be treated for 2 h at −1.5°C or at +26–32°C, depending on authors. Other treatments (inactivation of sperm by ultraviolet treatment, the use of catecholamine B or colchicine) are used with the same aim.

Egg incubation is usually not a constraint in aquaculture but different techniques are employed depending on whether the eggs are benthic or pelagic.

2.1 Incubation of benthic eggs

Such eggs include those of salmonids and many pond fish.

— The large eggs (around 4 mm) of salmonids are incubated in a single layer on grids or in cylindro-conical incubators where the water flows upwards from the base. Dead eggs (white eggs) must be picked out and removed to prevent the rapid spread of *Saprolegnia* filaments which would otherwise occur. This group of fish is characterized by low incubation temperatures (10–12°C), long incubation time (300 degree

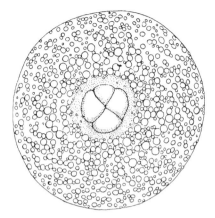

Fig. 1a. Blastomeres stage 4 (28 h).

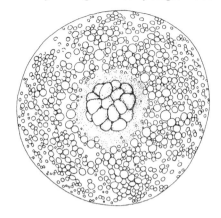

Fig. 1b. Blastomeres stage 16 (36 h).

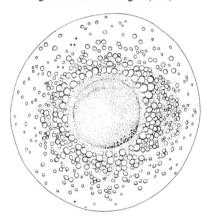

Fig. 1c. Blastula ($3\frac{1}{2}$ days).

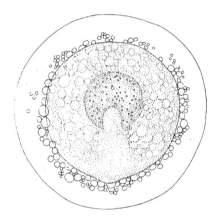

Fig. 1d. Outline of the embryo (seen from above) (6 days).

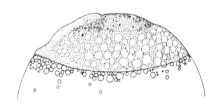

Fig. 1e. Outline of the embryo (lateral view, 6 days).

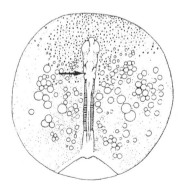

Fig. 1f. Embryo seen from above, optic vesicles being formed (8 days).

Fig. 1. Chronology of the development of the trout *Oncorhynchus mykiss* at 10°C (after Vernier 1969).

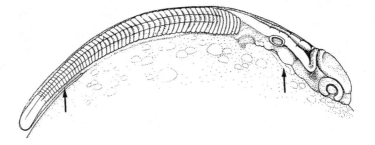

Fig. 1g. Embryo seen from the side, 1st gill slit open (12 days).

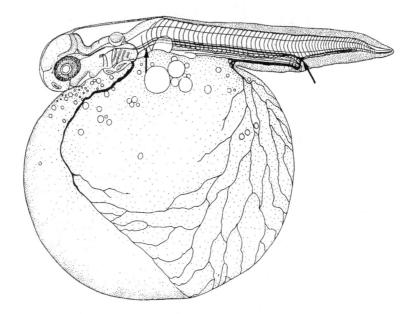

Fig. 1h. Vascularization of the yolk, pigmentation of the chorion (16 days).

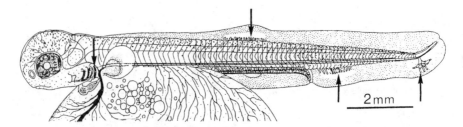

2 mm

Fig. 1i. Outline of the different fins (25 days).

Fig. 1. Chronology of the development of the trout *Oncorhynchus mykiss* at 10°C (continued).

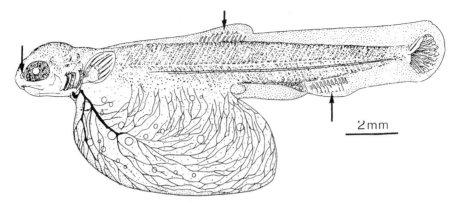

Fig. 1j. Larva 2–5 days after hatching (39 days).

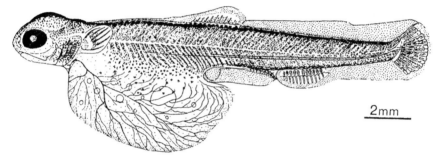

Fig 1k. Larva during yolk resorption (42 days).

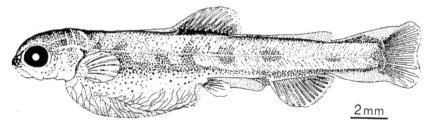

Fig. 1l. Appearance of characteristic arrangement of pigment (59 days).

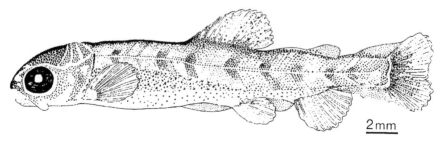

Fig. 1m. Yolk sac resorbed (85 days).

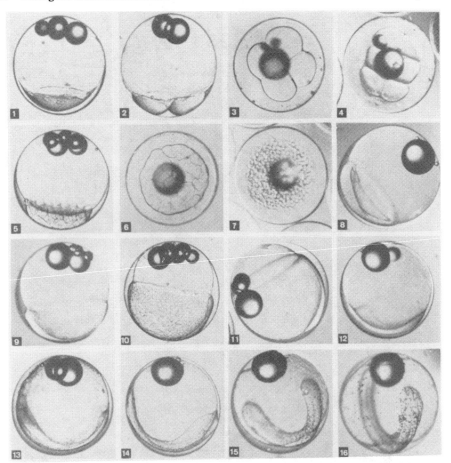

Fig. 2. Embryonic development of *Dicentrarchus labrax* (L.) (after Barnabé 1976). 1: Bass egg (diameter = 1 180 μm) $\frac{1}{2}$ hour fertilization, seen from the side. 2: The first division occurs 1 h 30 after fertilization, egg seen from the side. 3: Stage 4 egg, seen from below, 2 h after fertilization. 4: 2 h 30 after fertilization, the 6 blastomere stage is reached. 5–6: Eggs at the morula stage 4 h 30 after fertilization, seen from the side and above respectively. 7: Blastodisc (germinal disc) seen from above after 9 h of incubation. 8: Egg (from the side) 22 h after fertilization. 9: Yolk enveloped by the germinal disc, 28 h after incubation. 10: This process of surrounding continues, this stage is reached 34 h after incubation. 11: After around 40 h, the outline of the embryonic axis is visible (shown in profile). 12: The same stage as above, oblique view. 13: Differentiation of the embryonic shield is clear after 45 h incubation, seen from the side.

days, or 30 days at 10°C for trout) and saturation of dissolved oxygen. Water flows through at a rate of 500–600 l h^{-1} per 100 000 eggs. Details of the techniques used are given by Ingram (1988) and Billard (1990). Large-capacity incubators with an ascending current are sometimes used (Fig. 3).

— The eggs of various carp species, catfish and pike are incubated in Zoug or similar type jars: these jars, which have no base, are of around 8 litres capacity and supplied with water through the neck which is pointed downwards, as with incubators with an ascending current. The water flow is adjusted so that the eggs are in a constant, gentle

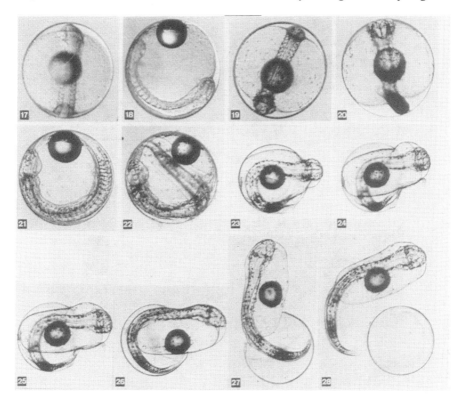

Fig. 2. Embryonic development of *Dicentrarchus labrax* (L.) (continued). 14: Embryonic morphogenesis leads to the appearance of the embryo seen from the side; 50 h have elapsed since fertilization. 15: The embryo occupies half the circumference of the egg after 57 h. 16: As well as the appearance of pigment, the embryo elongates, 69 h of incubation. 17: Metamerization is progressing in the median part of an 80 h embryo, seen from above. 18: The 18 picture shows the same stage shown in profile. Note the thickening of the cephalic zone. 19: Metamerization continues towards the caudal zone; after 85 h incubation. 20: Elongation and individualization of the embryo are clearly visible. 95 h. 21: View from the side of a more advanced individual (97 h). Somites are clearly visible; the primordial fin outline appears. 22: The embryo curves inwards, always in the same manner, in the egg; the primordial fin forms in the 104 h embryo. 23–28: Hatching occurs when the chorion is dissolved by an enzyme secreted by a gland situated on the head of the embryo. This occurs after 110 hours of incubation. In spite of contorsions it takes around 1–2 h for the embryo to free itself from the eggshell (112 h incubation) at 15°C.

current of water which is saturated in oxygen. Incubation time varies with species but is of the order of 60–70 degree days. The water passes up through the eggs (2–3 litres of eggs per bottle) and flows out of the jar at the top; the flow rate is 0.5 l min^{-1}. Sometimes hatched larvae (which are pelagic) are taken off via the outflow for incubation in the troughs or jars where yolk resorption takes place.

These jars are cylindro-conical in shape, of 300–500 l capacity and also fed with water entering through the base. From 300 000 to 500 000 larvae are kept in them until the yolk is absorbed (after around 3 days).

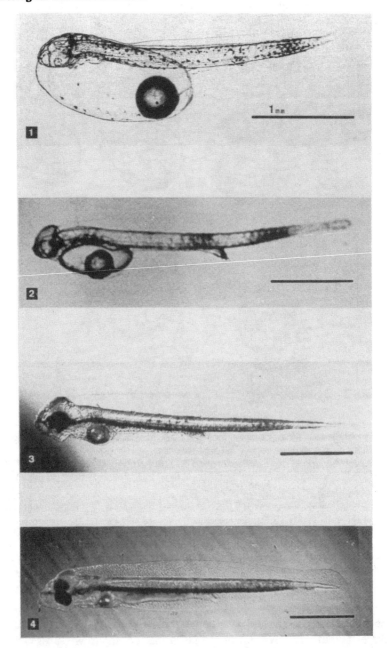

Fig. 2. Larval development of *D. labrax* (L.) (continued). 1: Newly hatched larva (L = 3.25 mm). Note the yolk sac with the oil droplet. 2: 3 day larva (L = 4 mm) yolk reserves have been resorbed. 3: 6 day larva (L = 4.4 mm). The eyes are pigmented, the mouth is opened, the yolk is used up. 4: 9 day larva (L = 4.9 mm). The mouth is open, food is invisible in the digestive vesicle.

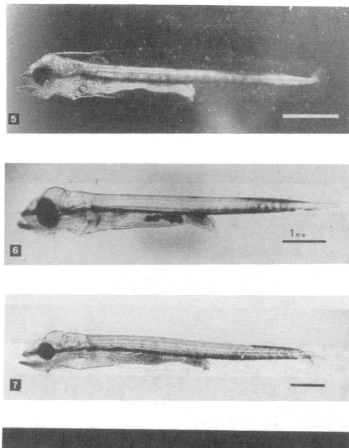

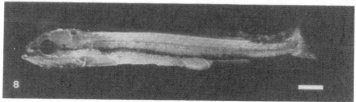

Fig. 2. Larval development of *D. labrax* (L.) (continued). 5: 12 day larva (L = 5.5 mm).
Resorption of lipid reserves is complete. 6: 20 day larva (L = 7.7 mm). The villi of the digestive
tube are visible, as is the ingested prey (*Fabrea salina*). The swim is apparent. 7: 28 day larva.
Formation of the rays of the caudal fin. 8: 34 day larva (L = 12 mm). 9: 36 day larva
(L = 13 mm).

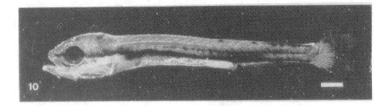

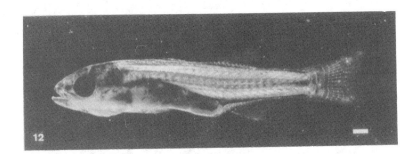

Fig. 2. Larval development of *D. labrax* (L.) (continued). 10: 40 day larva (L = 15 mm). 11: 52 day larva (L = 16 mm). 12: 64 day larva (L = 22 mm). 13: 75 day larva (L = 30 mm). The morphology is similar to that of the adult.

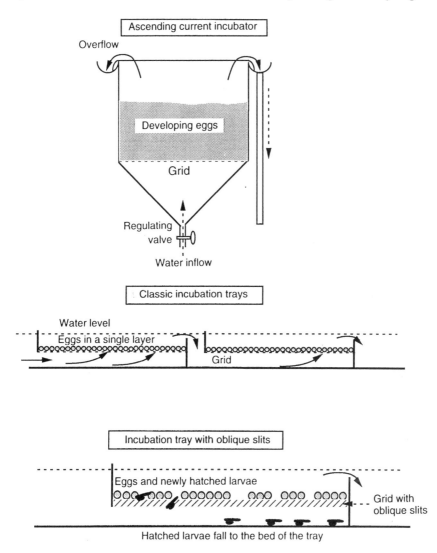

Fig. 3. Types of incubators used for salmonids.

Hatching can be synchronized by interrupting the water flow (which has the effect of lowering the oxygen level). Details of the methods used have been described in works edited by Billard (1980), Marcel (1990) and Jhingran (1990).

2.2 The incubation of pelagic eggs

These eggs are almost all from marine fish. Several systems are used but the only way that they can be incubated successfully is by keeping them in mid-water (Fig. 4). Salinity plays an important part because of its effect on the density of the water (sea bass eggs are in hydrostatic equilibrium at salinities between 34 and 37‰) but air or water fed into the

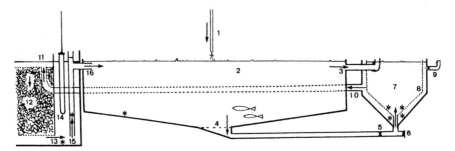

Fig. 4. 1: Water inflow. 2: Spawning tank (4 m diameter, 10 m³ capacity). 3: surface outflow with elbow to regulate water height. 4: Outflow at the base. 5: Sluice. 6: Outflow. 7: Incubator (in diameter capacity 0.5 m³) and eggs. 8: Filter mesh of the incubator (450 μm mesh) 9: Overflow. 10: Pipe for return of water to the filter. 11: Inflow of water to the biological filter. 12: Filter medium (polyethylene fibres). 13: Filtered water passes to return pipe. 14: Heater. 15: Exhaust. 16: Return of filtered water to the spawning tank.

* aeration points, arrows show the direction of water circulation.

tank through the base are also important. In general, cylindro-conical baskets made from bolting silk or other fine mesh material (400–600 μm) with a volume of several tens to hundreds of litres are used to retain the eggs. The hatching rate can exceed 98% when conditions are good.

The duration of incubation varies according to species (1–5 days after fertilization) and is modulated by temperature. The chronology of incubation has been described for many species. The main requirements are for a medium with near constant environmental conditions as appropriate for each species, rich in oxygen. The embryo lives on its yolk reserves and requires no exogenous food. It has been noted that the incubation of the eggs of marine species is facilitated if salinities are increased; this brings the larvae into the surface waters. However, eggs can survive a range of salinities, 7‰ for sea bass. Some authors claim that the density of the eggs depends on the salinity in which the female parent was kept before spawning; others disagree. The egg has a high water content (up to 92% of the weight). This water is hypotonic with the external environment (less dense) and therefore contributes to maintaining the egg in the pelagic zone. The eggs have one or more lipid globules which play their part in flotation and as a source of food.

Egg quality is difficult to define and predict. However, it is certain that quality varies and this affects the survival (or lack of it) of the larvae. This quality is determined by the nutrition and environment of the broodstock (see chapters on reproduction and nutrition).

3. HATCHING AND YOLK ABSORPTION

On a biological and technical level hatching marks the transition from the stage of the passive embryo within the egg to the free-living larval stage. This has less significance than the stage of transition from endogenous to exogenous feeding when the larva has consumed its yolk reserves. In pike, hatching can take place 80% of the way through embryonic development; some authors consider that this is not at a particularly well-defined stage. However, for the farmer, hatching marks an important stage as the different behaviour of eggs and larvae leads to a need to change the rearing environment. Blaxter (1988) reviewed the different stages of development of eggs and larvae. Immediately prior to hatching the embryo becomes extremely active and the chorion is softened by enzymes secreted by a gland situated on the head of the embryo.

The time taken for the resorption of the yolk depends on temperature and the size of the egg, which can vary from 0.75 mm in diameter for groupers to 3.5 mm for a halibut. The first resorbs its reserves in less than 3 days (Barnabé 1974; Barnabé et al. 1976) at 25°C; the second in 50 days at 5.3°C (Blaxter 1988). It can be seen from this that their rearing will be different.

The huge eggs of salmonids require approximately 300 degree days of development (or 30 days at 10°C), giving rise to a larva 15–20 mm long with a large yolk sac attached, allowing it to survive for almost a month. At the time of transition to exogenous feeding this small benthic fish (25–50 mm long) already has a well-developed feeding capacity, distinguishing it from the minute planktonic larvae of other species of fish. Amerio et al. (1987) compared the feeding requirements of the early stages of sea bass and trout and showed them to be very different.

In species with small eggs (marine and pond fish) the animal emerging from the egg is very different from the adult fish. It is tiny (1.7–3 mm for sparids, 3.5 mm for sea bass or tuna), and has limited yolk reserves (yolk sac 0.97–1.37 mm long in Sparids; Kentouri 1985). Neither the mouth nor anus is open, the eyes have no pigment but the otoliths are visible in the otic vesicle and neuromasts are present at the anterior of the trunk. A primordial fin is present and, like the rest of the body, is transparent.

The organs differentiate during the days following hatching; the progress of the differentiation of the digestive system is understood for sea bass (Benhalima 1982, Diaz et al. 1989; Anon. 1990). The stage during which the swim-bladder forms is critical; during the larval life there are many other progressive or abrupt changes, the length increases several times and the weight over one thousand times in marine fish larvae (from less than 1 mg to 1 g).

The characteristic stages of development can be demonstrated by the techniques used in the rearing of carp; hatched larvae are transferred to cylindro-conical vessels such as Zoug jars of around 300–500 litres capacity (1000 larvae per litre), supplied with water through the base for three days at 20–24°C. No food is supplied. The absorption of the yolk is accompanied by the differentiation of the digestive tube, the sense organs, the pectoral fins, etc., so that by the end of the yolk absorption the fish are capable of feeding themselves. They are then transferred to other structures. Larvae with other behavioural patterns (pike which are attached to the bottom) are placed in troughs during yolk absorption and cover the whole of the trough's area (15 larvae cm^{-2}).

In marine fish with eggs which float near to the surface in salt water (above 35‰ for sea bass), incubation and yolk resorption can take place in a variety of different containers: currents are created in the water through aeration or the inflow of new water, and these bring about the homogeneous distribution of the planktonic eggs and larvae. While eggs can be incubated at high densities (up to 15 000 eggs l^{-1}—Barnabé, unpublished data), newly hatched larvae should be maintained at a lower density. Larvae which will remain in tanks for 45 days and be fed there should be at a density of 100 larvae l^{-1} in intensive rearing and 5–10 larvae l^{-1} in extensive.

For all species, the trickiest part of the operation is the adaptation of the larvae to exogenous nutrition. If this is not accomplished the larvae will starve and die after a varying degree of time.

4. LARVAL REARING

Before detailing the processes developed recently by man for larval rearing, it is necessary to review some relevant aspects of their biology.

Swimming activity of the larvae is progressively established (Barnabé 1976a; Kentouri 1985): at first the fish lie passively in a horizontal position with their backs upwards facing the light, showing little movement. Periods of active swimming then alternate with periods of gliding (an extremely efficient plan from the point of view of the energy budget), even in still water. The action of the water on the neuromasts invokes changes in behaviour; swimming and gliding actions are governed by water movement; when the movement ceases the neuromast activates a resumption of swimming activity through reflex action (feedback action).

This stereotyped swimming pattern causes the larva to become displaced in a horizontal plane (sea bass) at the moment when the eyes begin to function. Fish larvae are sensitive to light levels of the order of a few lux (shade). The larvae move towards the light (phototaxis) and tend to swim against the current (rheotaxis). These locomotor activities which are brought on by physical factors allow the fish visually to explore a 'tunnel' of water, the depth of which depends on the visual field and the duration and speed of swimming. Several authors have calculated the volume of water explored by the larvae of various species: these are very similar for fish of equal size. They vary from $2\,l\,h^{-1}$ for a 6-day-old larval sea bass to almost $9\,l\,h^{-1}$ for the same fish at 25 days (Barnabé 1978). This larva reacts to prey situated at least 1 cm in front of it (within a field extending laterally through around 40°).

Hunting activity is initiated when a suitable sized prey organism comes into view; it is a response to a successive stimulation of different cells in the retina. This releases a sequence of genetically programmed movements (a behavioural sequence) characterized by a succession of different phases (Fig. 5). Because different retina cells are excited successively only prey moving in relation to the larva are capable of invoking attack; and then only towards living, moving prey. A further criterion, hardness, is detected when the prey is in the mouth. Items which are too hard are rejected. In the wild these adaptations ensure that the larvae only attack zooplankton and avoid detritus.

In terms of digestive capability, small marine larvae differ from the 'huge' larvae of fish such as most salmonids. While coregonids (salmonids with small eggs) have been reared on microparticles (see below), marine species require living prey, in spite of limited experimental success using microparticles alone (Adron et al. 1974; Barnabé 1976b).

It appears that live prey carry with them essential, unidentified components (enzymes or precursors? intact proteins required for nutrition?). Artemia nauplii, for example, contain an essential liposoluble substance, for others free amino acids are an essential source of energy for larvae and these are extremely soluble. Those in frozen zooplankton are dissolved within the ten minutes following immersion. The addition of digestive enzymes

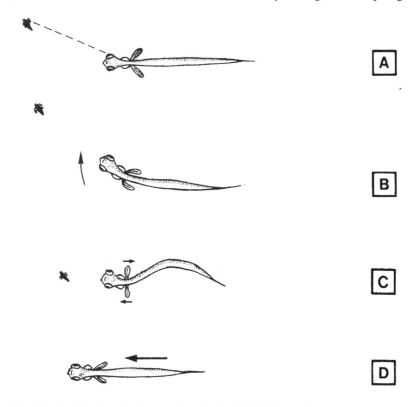

Fig. 5. Hunting behaviour. A: Location of prey in the field of vision. B: Movement of orientation towards the prey. C: Aim and positioning in an S shape. D: Relaxation and preparation.

to artificial diets has not produced much success and the problem of feeding marine larvae with artificial diets remains to be solved, in spite of many trials: the use of live prey remains the common practice.

There have been very few studies on the role of dissolved organic matter in the diet of fish. However, it has been shown that the chorion of herring eggs is permeable to organic compounds of low molecular weight. Studies using autoradiography have shown that 3- to 5-day-old sea bass larvae have taken up amino acids marked with radioactive carbon which has been metabolized and incorporated into proteins (Pavillon & Vu 1981).

The biological process used for larval rearing can be divided into three categories:

1. Extensive rearing out of doors (fry ponds, cages, enclosures). This is used particularly for fresh-water species reared in ponds and lately for marine fish. It is based on the management of the planktonic trophic web in an ecosystem (Billard & Marcel 1980; Marcel 1990).

2. Intensive rearing in an enclosed hatchery is the main technique used for salmonids and marine fish; there are few examples of this sort of rearing for pond fish. This type of rearing is developing with the use of artificial diets for certain larvae and new techniques of rearing and of feeding.

3.   The culture of plankton which can take up most of the space in a hatchery and/or
      the harvest of plankton for feeding the larvae: this is not solely associated with the
      culture of fish and has been described in Chapter 3, Part III).

4.1 Extensive outdoor rearing

The production of the natural environment is used to provide living zooplankton which
are essential to the larval fish. For the majority of pond fish, larval rearing (after the stage
of yolk sac resorption) is carried out in fry ponds which have a small area (100–1000 m^2)
and a mean depth of around 1 metre. These ponds can be drained and are treated according to the method shown in Fig. 6 (Marcel 1989). They are fertilized, for example with 8
tonnes of stable manure, 2 tonnes of pig manure and 150 kg of superphosphates per
hectare in four applications over four weeks of the rearing season. The operation of the
pond ecosystem will be described in the next chapter; here discussion is restricted to the
rearing of the early stages.

     Pond fish larvae are small and therefore have small mouths. Because of this rotifers
are a more suitable food than crustaceans (Fig. 7, Marcel 1989). In Eastern Europe, an
insecticide, Dipterex or Flibol E, is dissolved in the pond water at a rate of $2 \, g \, m^{-3}$
48 hours before the introduction of the larvae. This is non-toxic for the larvae but eliminates the crustaceans in the plankton (daphnia, copepods). During the first week, rotifers,
which can withstand fifty times that concentration of insecticide, are the dominant mem-
bers of the zooplankton. The crustaceans return eventually but by this time the fish larvae
are larger. A supplement of artificial food (30% fish meal, 30% blood, 20% soya and
20% yeast) can be given daily at a rate of 1 kg per 100 000 alevins. This ration is
increased by 10% each day but stopped if the temperature of the water drops below 18°C.
Survival rates can reach 50–60% and the density of the larvae varies between 2 and 6
million per hectare depending on whether or not insecticides have been used. In other
places, from the 10th day, an artificial diet made up of powdered soya and cereals is
given at a rate of 20 kg per million individuals. A detailed account of the production of
eggs, larvae and juvenile carp has been published by FAO (1986a and b).

     The production of the fresh-water striped bass (*Morone saxatilis*) in the USA for
stocking leisure fisheries also utilizes the technique of fertilizing ponds where larvae are
reared. Here the size of the crustacean plankton is well suited to the size of the larval
mouth; the object is to optimize the production of the plankton. Geiger *et al.* (1985)
showed that the application of fertilizer based on cotton seed flour is more effective than
using chicken wastes, both in terms of survival and biomass of alevins produced (63%
and 49 kg ha^{-1} as against 46% and 26 kg ha^{-1}). According to Nichols (1983) the use of
liquid fertilizer (ammonium polyphosphate, 13–34–0) in these ponds saves on manpower
as it can be distributed by tractor. It is more effective and cheaper.

     In Martinique, Tilapia larvae are reared in sea-water ponds. The transition from fresh
water to sea water of the broodstock takes place over 15 days without any drop in
primary production. The enclosure is treated in the same way as a pond and fertilized
(40 kg chicken manure per 15 days for 1000 m^2); this is where the juveniles are then
reared. The regulation of primary production through the addition of fertilizer, dilution or
changing the water in the tanks is easy. The herbivorous zooplankton which develop
naturally in brackish or marine waters are made up of rotifers (*Brachionus plicatilis*) and

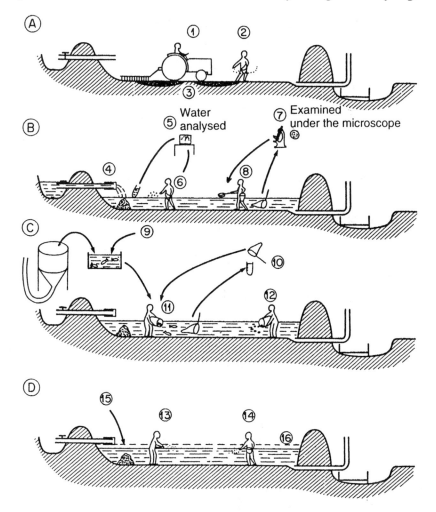

Fig. 6. Preparation and utilization of fry ponds for carp. A: Eight to ten days before larval stocking the bottom of the pond is levelled and quicklime applied (1, 2 & 3) and the pond partially filled (4) with filtered water. B: Six days before stocking (i.e. the day the broodstock spawn), controlled application of minerals and organic dressing takes place (5, 6). The types of zooplankton are studied under the microscope (7) and larger type destroyed by insecticide (8). C: Fry with yolk sac resorbed are taken from the hatchery to the pond (temperature of transport water and that of pond being the same ± 1°C); or the biomass of rotifers present in the pond is quantified (filtration of 100 l of water should give 3 ml rotifers (live value) (10). Larvae are released and zooplankton reintroduced (large forms, daphnia for preference) (11, 12). D: After several days of rearing, the pond is completely filled (15), more fertilizer is added and, eventually manufactured food given (13, 14).

small copepods (species vary between locations). Collecting these is the starting point of such a system. Four ponds, $10 \times 10 \times 0.5$ m (50 m³) in size give a routine daily production of 0.5–1 kg of zooplankton, two-thirds of which are rotifers (250–500 million individuals) and one-third copepods (15–35 million).

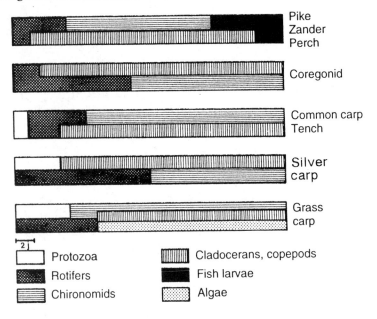

Fig. 7. Dietary preferences of larvae of different species during their first month of exogenous feeding (after Marcel 1989).

Scandinavian workers have been studying larval life in large volumes (mesocosms) for over 15 years (reviewed by Oiestad 1982). For turbot, Paulsen & Andersen (1989), working in concrete tanks of 2200 m^3 volume, achieved a survival of 62% to the 30 day stage with a density of 63 larvae m^{-3}. They concluded that this technique might be used successfully as an alternative to intensive rearing. In intensive rearing turbot often suffer heavy mortalities between the 8th and 10th days. A similar technique is being tested for the halibut (Oiestad & Berg 1989). The culture of the larvae of this species is proving to be extremely difficult. In these trials, plankton from elsewhere are introduced into the rearing tanks. The extensive technique of production of marine larvae has been developed for sea bass and sea bream in Greece (Divanach & Kentouri 1989), using fine mesh cages (mesh size 10–20 μm) and fertilizing the water (0.2–1 mg nitrogen per litre). In subtropical waters, trials in the rearing of the red drum *Sciaenops ocellatus* have produced promising results.

Progress in extensive larval rearing is being helped by the studies of marine ecologists working in mesocosms whose objective is the understanding of the aquatic ecosystem, pollution, etc. The study of the larval stage of marine fish is difficult in the small volumes used in the laboratory and even more difficult in the open sea. It is, however, a major area of study in mesocosms; the rearing of larval fish is only one of the components of such a system but it is a good model of the food chain. The consumption of food, choice of prey by the larvae, interactions between populations of prey and larvae provide vital information for fish farming (Oiestad 1982) and often theoretical and applied studies are carried out in parallel. The potential of the extensive approach to larval rearing has been reviewed by Von Walhert & Von Walhert (1989).

It is worth noting that the best larval survival rates have been achieved in large volumes in such studies (93% for herring larvae to day 39 according to Oiestad & Moksness (1981), but at a density of 2.4 larvae m^{-3}. This high survival rate is attributed largely to the absence of predators which, in the wild, are a major cause of mortality (Hunter 1984). It should also be appreciated that these high survival rates in large volumes have been achieved at prey densities which are much lower than in intensive rearing and that the density of prey in the natural environment is also much lower (Part I, Chapter 2). The space factor should therefore be taken into account. This method may be a lower cost alternative to the intensive hatchery–nursery system and is expanding wherever sites are available and adaptable to the requirements of larval life (Naas et al. 1987; Divanach & Kentouri 1989).

It is difficult to estimate the survival rate of larvae in the open sea; a mortality rate of 94% of the stock of larvae of Norwegian herring (spring spawning) has been reported during the transition to external feeding. Hunter (1980) found that mortalities were 29% per day for eggs and 2–10% for larvae. Bailey and Houde (1989) reviewed the work in this field and showed that mortality through predation is at its greatest when in the smallest larvae. The vulnerability of larval anchovies was studied by Hunter (1984); only 20% of 6 mm larvae responded to attacks by a predator with a flight response, while 85–100% of 33 mm larvae respond. This agrees with observations made during the rearing of the Japanese sea bream (see below). We have also seen (Part I, Chapter 2) how the organization into swarms of larvae or their prey can affect the trophic relations and even sampling; however, natural variability is also significant.

For information on the ecology and physiology of larvae as well as their feeding behaviour in the wild, the reader is referred to the following works: Blaxter (1974); Bardach et al. (1980); Lasker & Sherman (1981); Lasker (1984); Hoar & Randall (1988) and the review of Bailey & Houde (1989).

4.2 Intensive larval rearing in tanks and ponds

4.2.1 Salmonids
Salmonid rearing usually begins with the placing of fertilized eggs onto grids (Fig. 3), submerged in elongated trays. The spacing of the mesh in the grid is chosen so that eggs are supported but do not fall through, but newly hatched larvae can slip down through the spaces. This allows the larvae to be separated from the empty egg shells. When the hatching period is over the grids are removed and the larvae left to complete the resorption of the yolk in the same trays (which are similar to miniature raceways). First feeding may be carried out in these same containers using dry diet, in the form of tiny particles (size 0–1, or 0.3–0.6 mm and 0.6–0.8 mm). The current should be strong enough to force the fish to swim and to take away uneaten food particles through the outflow of the tray. Food is given at frequent intervals, preferably by automatic feeders; the daily feeding time may be extended through the use of a 20/24 h photoperiod. Different types of tanks are used; the usual type are shallow (20–30 cm deep) and rectangular in shape.

Water usually flows through once only and is seldom recycled. Because larval salmonids ingest and digest dry diet readily from first feeding onwards, all salmonid hatcheries can be classed as intensive, many producing millions of individuals. More detailed information can be found in the works of Ingram (1988) and Billard (1989).

4.2.2 Marine and pond fish

The techniques used in the rearing of marine fish and those few species of pond fish which are reared intensively differ because of the extremely small size of the larvae. The characteristics of the rearing system vary according to species. Here we shall review the system used in rearing bass.

Physical characteristics

Rearing is carried out in cylindro-conical tanks, 1–2 m in diameter (depending on the hatchery), with a volume of 2–8 m^3 and a depth of from 1 to over 2 m (Fig. 8). The vessels are constructed from moulded plastic (often black in colour) and arranged in series. They are supplied with salt water and compressed air. The compressed air is introduced to the tank through a porous diffuser at a point above the base of the cone. This aeration ensures that convection currents are set up in the water in the rearing tanks. The flow of water and air can be regulated. All the pipework is plastic. The water flows out through a large cylinder of plankton netting (at least 0.5 m^2), of 80–400 μm mesh. The perforated outflow pipe is located in the centre of this cylinder.

Although there are variations in the physical and chemical characteristics of the water used in actual rearing operations, variables fall within the following ranges:

— Temperature rises from 15–18°C at the start of rearing (the day when hatching occurs is taken as day 0) to around 20°C at day 10, and 22°C at 50–60 days (Fig. 9). Other farmers may begin rearing at 14°C and allow the temperature to rise to 20°C by day 18 and then maintain this temperature. Turbot are reared at between 16 and 20°C; at the time when external feeding is initiated the temperature is normally held at around 18°C (Person Le Ruyet 1987). All stages of sea bream prefer temperatures around two or three degrees above the equivalent stages for bass; 19–20°C for spawning and egg incubation, 18–24°C and even 26°C for larvae.

— The salinity is generally that of the seawater pumped from the natural environment (30–37‰). It can however be decreased to 7‰; this allows the use of freshwater plankton (Barnabé 1983) and ensures good development of the swim bladder (Barnabé 1983; Barnabé & Guissi, in press). The curves shown in Fig. 9 are an indication of the mean; in practice large variations are of little importance in the days following hatching before feeding begins. After feeding has begun, variations in salinity reduce the appetite of the larvae.

— The level of dissolved oxygen should remain below saturation. At the above temperatures it is between 5 and 8 mg l^{-1}.

— Levels of toxic nitrogen compounds such as ammonia and nitrites should not exceed 0.1 mg l^{-1} in the tanks where larvae are reared, although levels of several mg l^{-1} are of

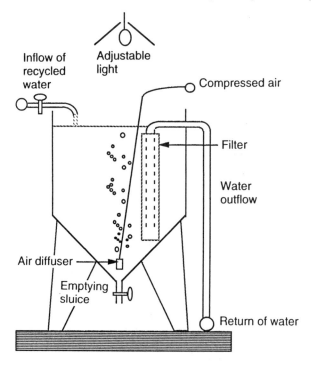

Fig. 8. Diagram of a tank used for larval rearing.

little consequence. High levels of dissolved oxygen increase the sensitivity to nitrogen compounds.

— pH is that of sea water (8–8.2); any value between 7.8 and 8.3 is compatible with larval rearing.

— For the first 10 hours after hatching larval bass are kept in darkness; the work of Weppe & Joassard (1986) and Ronzani-Cerquieira (1986) has shown that if incubation and yolk resorption take place in darkness or low light levels (70 lux), survival is better than at high light levels (700–2000 lux). These high levels are associated with the absence of a swim bladder and numerous other deformities (Tesseyre 1979).

Japanese authors were the first to report swim bladder deformities in *Pagrus major* and to show the important role played by light, access to the surface of the water for the larvae, and diet. Some authors advocate gentle aeration of the rearing tanks in the 12 days following hatching: in a tank of 86 m³, eight diffusers supplying 0.2–0.6 l min⁻¹ are enough; the slight, ascending currents generated by these aerators allow the larvae, whose swimming ability is poor, to take their first gulps of air from the surface 5–10 days after hatching. After this, the control of the volume of the swim bladder appears to be under the influence of light through secretion–excretion by specialized cells. In the wild, the vertical migration of larvae is considered to be an adaptive strategy both for following the migration of zooplankton and for evading predators (Hunter 1980). Larvae of the Japanese sea bream migrate to the surface at dusk and drop towards the bottom at dawn.

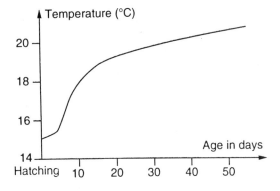

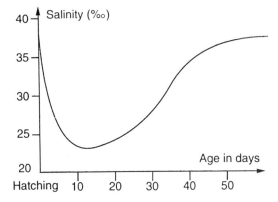

Fig. 9. Change in temperature and salinity in intensive culture of larval sea bass.

When sea bass are reared under extensive conditions under natural day lengths the problem of a failure of the swim bladder to develop does not arise. Other factors are also indicated as having an effect: for example, fluctuations in salinity, as has been shown several years ago (Barnabé 1984). However, other authors have obtained differing results. Chatain (1986) found that air was only taken in from above the water when there was a film of oil covering the surface. Guissi (1988) and Marino *et al.* (1989) showed that the filling of the swim bladder can take place without access to the surface, confirming the results obtained by Doroshev & Cornacchia (1979) for a closely related species.

Work carried out by Equipe Mereq (1986) showed that the intensity of light should be increased progressively between the tenth and twentieth days and the duration of the light should be increased from 12 h to 24 h per day (Fig. 10). The maintenance of darkness is only appropriate for bass. For all other cultured species, moderate light levels (<200 lux) are used at the start of the rearing cycle and the natural day length cycle is maintained but with a long day length (16–18 hours of light). When larvae are reared in large volumes (mesocosms) and in 'green water', they are able to select their preferred light intensity. Japanese workers using these rearing techniques shade part of the rearing tanks.

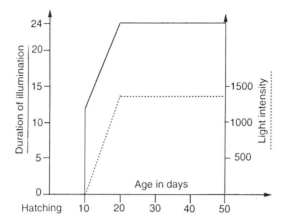

Fig. 10. Change in light intensity and photoperiod during the rearing of larval sea bass.

The turnover of water in the tank is around 30% of the volume of the tank each hour between hatching and day 20, reaching 70% h^{-1} and even 100% h^{-1} at 45 days for bass, depending on the farm. In European hatcheries water is most frequently recycled via mechanical and biological filters. Two filters are used in series: the first is a sand filter and acts as a mechanical filter, the second contains biolite or chabazite, a porous volcanic rock, and acts as a biological filter. Recycling ensures the maintenance of constant physical and chemical conditions in a way that is not possible in open, flow-through systems. A description of the processes involved in filtration has been given by Petit (1990). New water must be added to compensate for losses through evaporation and to prevent the build up of nitrates but the amount is restricted to a few percent of the overall volume. The new water is often introduced after the mechanical filter where there is a loss of water.

Larval density
In production hatcheries the starting larval density is from 100 larvae per litre for bass (2–8 m^3 tank volume) to 30–60 for turbot and the European sea bream. Although they are highly productive, these intensive rearing tanks are very unstable and sensitive to changes. In identical rearing conditions in the former Yugoslavia (same time, same initial egg density, same water circulation, fed to satiation, identical handling operations) rates of survival of 34, 18.7 and 12.2% were obtained after 40–50 days (Berg 1985). Their advantage is to allow a better utilization of live prey and space in the hatchery. The lower the pollution the less is the likelihood of a breakdown which might annihilate the stocks; this is the great risk in intensive larval rearing.

Larvae obtained from these systems are transferred to larger tanks at an age of 40–50 days. These tanks are similar to those used in the culture of salmonids—square or cylindrical with flat bases. The transition to dry diet (weaning) takes place in these tanks.

The Japanese bream is reared in large volume tanks (50–100 m^3 at densities of 20 000–40 000 m^{-3} during the pelagic phase (<7.5 mm in length) and then at 10 000–15 000 m^{-3} (cannibalism begins at this stage). Fig. 11 (adapted from Foscarini 1988) summarizes the

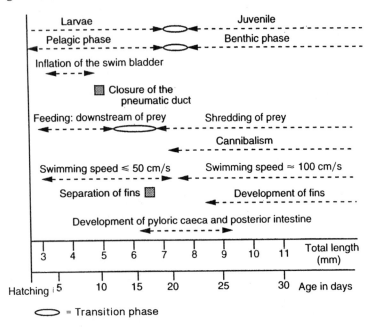

Fig. 11. The main morphological behavioural changes during the 1st month of life of the Japanese
bream (adapted from Foscarini 1988).

biological events during the larval stage; these set the constraints in rearing and then densities per unit volume.

Feeding larvae on living prey
This has been described elsewhere for sea bass and sea bream (Barnabé 1990; Iizawa 1983), turbot and sole (Person le Ruyet *et al.* 1990), Japanese bream (reviewed by Foscarini 1988) and pond fish (Gillet 1980), and will be summarized here. The sequences of prey used (Chapter 3, Part I) are few as man only knows how to raise (and often poorly) two types of plankton in intensive conditions in hatcheries:

1. The rotifer, *Brachionus plicatilis*, reared by Ito (1960) is given from day 4 to day 25 for sea bream, and from day 12 to day 15 for turbot and bass (when it is used at all for this species). The Japanese farmers use this particular prey species up to an age of 30 days for *Pagrus major*. The density of rotifers in the rearing tanks should be maintained at 5 ml^{-1} up to the 10th day of rearing and 3 ml^{-1} after this, with figures of 10–20 ml^{-1} for turbot, reaching 60 ml^{-1} by day 8 (Person Le Ruyet 1987). This author produced a table comparing the biological characteristics of different species during their larval stages; this has been extracted in Table 1 here. In order to meet the nutritional requirements of the larval fish, rotifers and also artemia are often treated before being distributed to the larvae (Chapter 3, Part I).

The consumption of rotifers by larval *Pagrus major* is shown below.

Age (days)	Number of rotifers ingested per day
4	22
9	58
13	156
18	285
21	427
24	747
27	1213
28	1665
29	2151

Table 1. Comparative information on larvae in intensive rearing. Adapted from Person Le Ruyet (1990)

	Turbot	Sole	Sea bass	Sea bream
Mean weight of hatching (mg)	0.2	0.6	0.4	0.3
Mean weight at 1 month (mg)	60–75	50–75	10–25	20–30
Mean weight at 3 months (g)	1.5–2	1–1.5	0.5–1	0.5–1
Critical { end of yolk resorption	***	—	**	***
phases { metamorphosis	*	*	**	**
Duration of larval life (days)	30–40	13–15	40–50	40–50
Dietary sequence best known:				
— Unicellular algae	—	—	—	D2–D5
— Enriched *Brachionus*	D2–D12	—	D2–D15	D2–D15
— *A. salina* nauplii	D8–D25	D2–D25	D10–D25	D10–D25
— *A. salina* metanauplii	D15–D30	D20–D30	D20–D30	D20–D30
Mean age at weaning (days)	30–40	30–40	30–50	30–50
Granules { dry commercial	*	*	***	***
{ rehydrated (experim.)	**	**	*	*
Mean survival at 1 month	≤ 5–40	60–80	30–80	≤10–30
Mean survival to weaning	60–80	60–80	30–80	30–80

2. Recently hatched *Artemia salina* nauplii are given from day 15 to days 35–40 for sea bream or from day 25 in turbot. Artemia nauplii are given to sea bass larvae from the first day of feeding (days 6–10) following the regime developed by Balma and René (pers. comm.). Frozen artemia cysts are bought in; it is easy to obtain the nauplii from these. Thus the rearing of larval bass is relatively simple and increasing numbers of fry

are being produced (tens of millions are currently being produced each year). The nutritional quality of the nauplii (especially their fatty acid content, reviewed by Versichelle *et al.* 1989) varies depending on the origin of the cysts. This brings about variable, unpredictable, mortality rates in the larval bass. Monitoring the quality of these cysts has greatly improved the success of the rearing of marine fish larvae; much of the work has been carried out by the Artemia Reference Centre in Ghent, Belgium (see also Watanabe *et al.* 1983).

The number of larvae consumed by juvenile sea bass has been established by IFREMER (Equipe MEREA, pers. comm.) and is summarized in Table 2. Other authors have produced equations allowing the calculation of the daily prey consumption at different ages and temperatures, although problems of environment and the behaviour of the larvae can affect these calculations. Because of this, the distribution of fresh prey is usually based on the estimation of the prey remaining in the tank.

Table 2. Consumption of live prey in intensive rearing (2 m^3 tanks, $T = 19°C$, 100 larvae l^{-1}, survival at 45 days: 49%)

Prey	Nauplii	Metanauplii
Time over which distribution takes place (days)	10–20	15–45
Duration of distribution (days)	10	30
Total number of prey per juvenile	605	12 000
Daily maximum distribution	55	450

The use of metanauplii (A2) requires that Artemia nauplii must be kept in tanks and fed; because of this it is not widespread. Usually large numbers of nauplii are used instead of the metanauplii. In spite of the fact that various dry diets for Artemia nauplii are sold, many operators only use newly hatched Artemia cysts. For the purposes of calculation a single nauplius (A0) is taken as equivalent to 0.5 metanauplii (2 days old), as the weight of an A0 nauplius is around 16 μg and that of an A2 is 30 μg.

Mixed rearing operations (combining living and inert diet)
Developments in the manufacture of inert diets either as finely divided dry particles or microvesicles (an aqueous solution of lipid-coated particles) give an alternative to the use of living prey. There have been many studies on both Japanese and European sea bream, sea bass and turbot which have indicated that this is a promising method of feeding: Barnabé and Guissi (in press) achieved a survival rate in excess of 60% in a pilot trial (500 litre tank, 100 larvae l^{-1}) using microvesicles made up in the laboratory. Corneillie *et al.* (1989) achieved a survival rate of 51% using Japanese microparticles ('Nippai') alternating with live prey but at a density of only 10 larvae l^{-1} for *Sparus aurata*. Cavalier (pers. comm.) in 1986 showed that 'Kyowa' microparticles were superior to others on the market. This was confirmed by Person Le Ruyet *et al.* (1989). However, the price of this diet means that it is no longer competitive.

These particles (of 250–400 and 400–700 μm in diameter) contain 57% protein, 23% lipid and 4% water. They sink very slowly (0.22 cm s^{-1}). Each gram contains 11 or 5 million particles, depending on the chosen size. Teshima *et al.* (1982) described the main methods by which these microparticles could be manufactured experimentally. The method developed by Lim & Moss (1981) consists of encapsulating living particles and may be the way for the future, but at present this is limited by the cost of the biochemical processes involved.

Several other studies have shown that mixed feeding is more effective in terms of survival (although not growth) than feeding with live prey alone.

The use of microparticles alone
This method used alone gives reduced survival for marine fish larvae for the reasons described above (Adron *et al.* 1974; Barnabé 1976b). Encouraging results have been obtained on an experimental and pilot scale for coregonids (Champigneulle & Rösche 1988, cited in a review by Champigneulle 1988) with different diets: the diet giving the best survival contained 50% dry yeast, 35% powdered beef liver, 5% cod liver oil, 5% mineral salts and 5% vitamins. The other diet is a commercial preparation sold for aquarium fish ('Tetra'). The feeding of larval common carp on microparticles has also progressed to the pilot stage.

We have shown that it is possible to deceive the larvae of marine fish into accepting an inert diet if it appears to move in the current (Barnabé 1978) but the problems of digestibility described above have not been resolved in spite of a great deal of research on the enzymes of the larval fish.

Weaning
The transition from a diet based on living prey to one based on dry diet takes place progressively and is associated with changes in behaviour and digestion. It is necessary to wait until the digestive system of the larva is fully operative and weaning is not usually completed until day 45 for marine fish. If weaning is completed before this there are likely to be many mortalities and reduced growth rate in the surviving fish. The numbers of live prey given are progressively reduced and inert diet is increased to compensate. Mortality may be high at this stage (up to 50%) with ordinary dry compound diet in spite of the precautions taken. Because of this, many farmers use commercially available frozen plankton (Japanese bream) or moist diet. This type of diet may be used for at least two months.

It is now possible to buy dry diets for weaning; one example is 'Sevbar' which is prepared by the cooking–extrusion process (see chapter on 'Nutrition') which have been supplemented with oils and vitamins at the factory. Although the level of survival is not improved significantly in comparison with other commercially available diets, it provides an alternative to the use of frozen plankton. These diets are distributed frequently using an automatic feeder so that the larvae, used to encountering food particles frequently in the natural environment, can be gradually trained to accept a limited number of meals each day given at fixed distribution points. Weaning is facilitated if diet can be distributed on the surface of the vessel; water currents are often used to aid this. Weaning techniques have been described elsewhere for sea bass and sea bream (Barnabé 1989) and for flat fish (Person Le Ruyet 1989). The diagrams in Fig. 12 have been taken from the work of the latter author and provide a résumé of the principal stages of a hatchery for flat fish. For sea bass, sea bream and other mid-water fish, there is no need for the intermediate tank.

Weaning is also important for fish larvae (e.g. milk fish, *Chanos chanos*) which are captured in the wild rather than bred in captivity. A Far Eastern fish, milkfish, spawn in the sea but juveniles migrate towards the coast and lagoons where they are captured in specialist fisheries. In Japan, yellowtail juveniles are captured on floating seaweed in the open sea and are used in cage culture. In the Mediterranean, Italian fishermen capture mullet, sea bass and sea bream from the coasts of Turkey to Morocco; these are then ongrown in the 'vallis' of the Adriatic coast. Most of the elvers captured in the estuaries and rivers of Europe are destined for immediate human consumption; the price reached £130–150 per kilogramme in 1990. The annual catch in France varies between 500 and 1300 tonnes depending on the year (mean weight 0.3–0.4 g) or 2.5–3 billion individuals for each 1000 tonnes. This consumption of alevins is unusual but not unique; certain other young fish are captured in very fine mesh nets.

In all these cases the collection of juveniles from the natural environment exploits the characteristic behaviour or ecology of the species: the search of floating objects for young yellowtails, the arrival of milkfish at the coast and the collection of the migrating larvae of Mediterranean fish or eels for ongrowing in 'fixed engines'. This harvest of juveniles from the wild is not without its problems, the most common of which is insufficient natural recruitment. Although aquaculture based on the supply of juveniles from the wild is well developed, it is often limited by the supply of juveniles and would benefit from the guaranteed supply of larvae from aquaculture.

Larvae which are captured in the wild are placed in acclimation tanks; these may be similar to those found in the hatchery or those in the nursery depending on the size of the larvae. The rearing operation differs little from that of larvae from the hatchery but it is often difficult to persuade fish from the wild to take dry diet. The techniques of the hatchery–nursery must therefore be adopted for these juveniles brought in from the wild. Usually, the wild juveniles are fed on minced fish or shellfish or frozen zooplankton depending on local availability. Weaning onto dry diet is carried out over 4–7 days by the progressive substitution of dry diet for the moist food.

Larval survival

The information on larval survival reported above (Table 1) has been obtained from many different studies on feeding. Improved survival has also come from the improvement in the quality of eggs since the use of LH-RH for the induction of spawning became widespread. This is clearly demonstrated from the explosion in the numbers of sea bass produced from 1985 (the year the technique became available) onwards (Billard & Barnabé 1981, unpublished; Billard *et al.* 1983; Barnabé & Barnabé-Quet 1985). The survival rate of bass from a hatchery varies between 20 and 60% depending on the batch, the competence and care of the farmer, etc. There are fewer figures for sea bream and turbot while for sole larval rearing is simple (and has been carried out since 1905); the problems with this species come in the later stages of rearing.

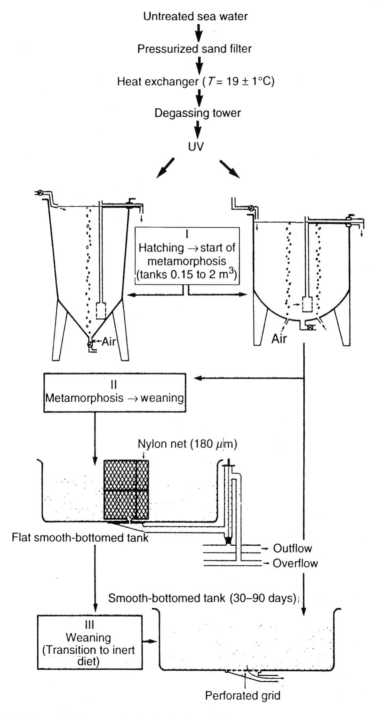

Fig. 12. The principal stages in the hatchery, according to the intensive method used at IFREMER.

5. EFFECTS OF POLLUTION ON LARVAE

In hatcheries bacteria have no effect on algal cultures and may be favourable for the production of rotifers and bivalve larvae but lethal in the culture of turbot larvae (Nicolas *et al.* 1989). Rotifers may act as a vector for pathogenic bacteria such as *Vibrio*.

Fish embryos and larvae are more sensitive to pollution than adults. Well-established rearing procedures allow the production of large numbers of fish larvae which is of much importance in studies on pollution, since adult trout are used in the control of the quality of drinking water.

A study carried out with fish from the coast of California (Sherwood & Mearns 1981) showed that fin erosion (associated with skin tumours and the dilation of the liver) occurred in 72% of juvenile sole (Table 3). These deformities are linked to the exposure to pollutants which contaminate the sea bed rather than waste domestic water.

The chemoreceptive organs of the larvae are blocked by contact with hydrocarbons. Many studies report the effects of heavy metals; these are associated with various deformities and malfunctions. It has been shown that the decline in the numbers of the striped bass (*Morone saxatilis*) around the coasts of America is due to an increase in water pollution. Concentrations of 0.02–0.05 ppm methyl mercury cause various abnormalities in the killifish (*Fundulus heteroclitus*) (Weis *et al.* 1981). The European sea bass is also extremely sensitive to the same pollutants (Table 4) and even more so to detergents which are toxic at levels of $0.0034 \, \text{ml} \, \text{l}^{-1}$ (Beji 1981). The youngest larval stages are the most sensitive; salinity increases sensitivity to detergent.

Table 3. Development of the erosion of fins and skin tumours in sole as they move to the benthic habit captured at 137 m depth on the Palos Verde (California) (adapted from Sherwood & Mearns 1981)

Date	Number of specimens	Size range and median	% erosion of fins	% tumours
19–04	538	55(35–65)	0.93	0
11–05	256	55(45–65)	2.7	0.78
21–06	42	65(45–75)	9.3	4.8
23–08	10	85(75–95)	50	0
19–10	52	105(85–105)	44	1
05–12	403	105(85-115)	72	7.7

6. CONCLUSION

The techniques of the hatchery–nursery system have improved markedly over the last few years through the availability of new biological and physiological information. This has allowed several marine species to be produced on a commercial scale (Japanese sea bream, 20 000 tonnes; European sea bass and sea bream, 6 000 tonnes (on top of the extensive production in the valli); turbot, around 200 tonnes). This total control over the

Table 4. Effect of different concentrations of copper on embryonic development and the hatching rate of eggs of *Dicentrarchus labrax* (adapted from Cosson & Martin 1981)

Control	Normal development	100
5 μg l^{-1}	Normal development	74
10 μg l^{-1}	Normal development	79
50 μg l^{-1}	Interruption at the metamerization stage	18
100 μg l^{-1}	Interruption at the stage of pigmentation	1
500 μg l^{-1}	Eggs die in 48 hours	0
1000 μg l^{-1}	Eggs die in 24 hours	0

rearing of marine species is a new activity for man but is an extension of a process started at the beginning of the century.

In spite of these advances, our knowledge of the larval life of many fish is sketchy as, in the wild, the larvae are found in the open ocean and are difficult to keep in captivity. The progress made through aquaculture has not only led to economic production but has also helped in the understanding of the wild fish.

Examples of this are in the study of the effects of pollution and in studies of plankton production. Curiously, one area where knowledge of the biology of larval stages could be most helpful is little developed in Europe; this is in the problem of the determination of stock recruitment. The absence of the relationship between the stock of spawners and the number of larvae or surviving juveniles (recruitment) is a major stumbling block. The methods of aquaculture could be of major assistance in understanding factors affecting recruitment (Von Walhert & Von Walhert 1989).

On leaving the nursery the juveniles are placed in ongrowing units; their weight has increased hundreds or thousands of times since the start of rearing as an egg or a larva and the problems also change their dimension.

REFERENCES

Adron J., Blair A., Cowey C., 1974. Rearing of plaice (*Pleuronectes platessa*) larvae to metamorphosis using an artificial diet. *Fish. Bull.*, **72** (2): 353–357.

Amerio M., Bertolinelli M., Fiorentini L., 1987. Alimenti naturali e mangami comosti integrati per gli stadi larvali di trota (*Salmo gairdneri* R.) e branzino (*Decentrarchus labrax* L.). *Zootec. Nutriz. Anim.*, **13** (5): 55–562.

Anon., 1990. 1970–1990: 20 ans de recherches sur les bases biologiques de l'aquaculture. Univ. Sc. Tech. Languedoc, Montpellier, Lab. Biol. Anim.: 25 pp.

Bailey K. M., Houde E., 1989. Predation on eggs and larvae of marine fishes and the recruitment problem. *Advances in Marine Biology*, **25**: 1–83.

Bannister R. C. A., Harding D., Lockwood S. J., 1974. Larval mortality and subsequent year-class strength in the Plaice (*Pleuronectes platessa*). In: *The early life of fish*, J. H. S. Blaxter (ed.). Springer Verlag, Berlin: 21–37.

Bardach J., Magnuson J., May R., Reinhart J. (ed.), 1980. *Fish behavior and its use in the capture and culture of fishes*. ICLARM, Manila, Philippines: 512 pp.

Barnabé G., 1974. La reproduction du Mérou *Epinephelus gigas*: Observations préliminaires de terrain. *Aquaculture*, **4**: 363.

Barnabé G., 1976a. Contribution à la connaissance de la biologie du Loup *Dicentrarchus labrax* (L.) Poisson Serranidae. Thèse Doct. Etat, mention Sciences, Univ. Sc. Techn. Languedoc, Montpellier: 426 pp.

Barnabé G., 1976b. Elevage larvaire du Loup (*Dicentrarchus labrax* (L.), Pisces Serranidae) à l'aide d'aliment sec composé. *Aquaculture*, **9**: 237–252.

Barnabé G., 1978. Etude dans le milieu naturel et en captivité de l'écoéthologie du Loup *Dicentrarchus labrax* (L.) (Poisson Serranidae) à l'aide de nouvelles techniques. *Ann. Sc. Nat., Zoologie, Paris*, 12e serie, **20**: 423–502.

Barnabé G., 1983. Mass rearing of sea-bass (*Dicentrarchus labrax* (L.)) larvae with plankton collected from oxidation ponds. *Journal of World Mariculture Society*, Sp. Publ., **3**: 83–91.

Barnabé G., 1984. Aliment composé pour animaux aquatiques et procédés et dispositifs pour élever des larves de poisson. Brevet USTL no. 84/17155, BNPI, Paris: 11 pp.

Barnabé G., 1990. *Aquaculture*, G. Barnabé (ed.). Technique & Documentation (Lavoisier), Paris, 2 Vol.: 1308 pp.

Barnabé G., Boulineau F., René F., 1976. Chronologie de la morphogenèse chez le Loup ou Bar *Dicentrarchus labrax* (L.) (Pisces Serranidae) obtenu par reproduction artificielle. *Aquaculture*, **8**: 351–363.

Barnabé G., Barnabé-Quet R., 1985. Avancement et amélioration de la ponte induite chez le Loup *Dicentrarchus labrax* (L.) à l'aide d'un analogue de LHRH injecté. *Aquaculture*, **49**: 125–132.

Barnabé G., Guissi A. Combined influence of salinity and diet on the larvae of the European Sea-Bass *Dicentrarchus labrax* (submitted for publication).

Beji O., 1981. Recherche de matériels biologiques marins, méthodologie pour la réalisation de bio-essais normalisables. In: *Convention de recherches avec le Ministère de l'Environnement et du cadre de vie*, Lapailleur & Barnabé, No. 79–31: 65 pp.

Benhalima K., 1982. Structure et développement du tractus digestif du Loup *Dicentrarchus labrax* (Linné 1758). Thèse 3e Cycle, Ecologie appliquée et générale, Univ. Sc. Tech. Languedoc, Montpellier: 65 pp, 33 Pl.

Berg L., 1985. Yougoslavie; développement de la culture du Bar. Rapport préparé pour le projet 'Developpement de la culture du Bar'. FAO, Rome, Réf. : FI TCP/YUG 2301 (Ma), Document de travail 1, mars 1985, 1 microfiche.

Billard R. (ed.), 1980. *La pisciculture en étang*. INRA Publ., Paris: 434 pp.

Billard R., 1990. La salmoniculture en eau douce. In: *Aquaculture*, G. Barnabé (ed.). Technique & Documentation (Lavoisier) Publ., Paris: 569–613.

Billard R., Marcel J., 1980. Quelques techniques de production de poissons d'étangs. *La Pisciculture française*, **16** (1): 9–16 and 41–50.

Billard R., Weil C., Barnabé G., 1983 (1985). Induction de l'ovulation et stimulation de la spermiation par le LHRH ou un analogue de LHRH associé ou non au pimozide chez quelques espèces de poissons téléostéens. Colloque: *Bases biologiques de l'aquaculture*. Montpellier Déc. 1983. Actes de Colloques IFREMER, No. 1: 321–332.

Blaxter J. (ed.), 1974. *The early life of fish*. Springer Verlag., Berlin: 765 pp.

Blaxter J., 1988. Pattern and variety in development. In: *Fish physiology*, W. Hoar and D. Randall (eds). Academic Press, New York, Vol. 11: 1–58.

Champigneulle A., 1988. A first experiment in mass-rearing of Coregonid larvae in tanks with a dry food. *Aquaculture*, **74**: 249–261.

Chatain B., 1986. La vessie natatoire chez *Dicentrarchus labrax* et *Sparus aurata*. 1. Aspects morphologiques du développement. *Aquaculture*, **53**: 303–311.

Corneillie S., Ollevier F., Carracosca M., Peci F., Rendon A., 1989. Reduction of the use of Artemia nauplii by early feeding of sea bream larvae (*Sparus aurata*) with dry food. In: *Aquaculture 89*. Short communications, Abstracts. European Aquaculture Society, Bredene, Belgique: 73–74.

Cosson P., Martin J. R., 1981. The effects of copper on the embryonic development, larvae, alevins and juveniles of *Dicentrarchus labrax* (L.). *Rapp. P.-v. Réun. Cons. Int. Explr. Mer*, **178**: 71–75.

Diaz J. P., Connes R., Divanach P., Barnabé G., 1989. Développement du foie et du pancréas du Loup *Dicentrarchus labrax*: I. Etude de la mise en place des organes au microscope électronique à balayage. *Ann. Sc. Nat., Zool.* 13e ser., **10**: 87–98.

Divanach P., Kentouri M., 1990. Elevage larvaire en conditions intensives. In: *Aquaculture*, G. Barnabé (ed.). Technique and Documentation (Lavoisier) Publ., Paris: 911–928.

Doroshev S., Cornacchia J., 1979. Initial swimbladder inflation in the larvae of *Tilapia mosambica* and *Morone saxatilis*. *Aquaculture*, **16**: 57–66.

FAO, 1986a. La Carpe commune. 1—Production massive d'œufs et de post-larves. Coll. FAO formation, **8**: 87 pp.

FAO, 1986b. La Carpe commune. 2—Production massive de carpillons en étangs. Coll. FAO formation, 9: 85 pp.

Foscarini R., 1988. A review: Intensive farming procedure of red sea bream (*Pagrus major*) in Japan. *Aquaculture*, **72**: 191–246.

Geiger J. G., Turner C. J., Fitmayer K., Nichols W., 1985. Feeding habits of larval and fingerlings Striped Bass and zooplankton dynamics in fertilized rearing ponds. *Prog. Fish-Cult.*, **47** (4): 213–223.

Gillet C., 1980. Ecloserie et production de larves. In: *La pisciculture en étang*, R. Billard (ed.). INRA Publ., Paris: 175–188.

Guissi A., 1988. Influence des facteurs écologiques sur les populations larvaires de Loup *Dicentrarchus labrax* et de Daurade *Sparus aurata* en élevage intensif. Thèse Doctorat, Univ. Sc. Techn. Languedoc. Montpellier: 230 pp.

Hoar W., Randall D., 1988. *Fish physiology. The physiology of developing fish* (Part A). Academic Press, San Diego: 546 pp.

Hunter J., 1980. The feeding behaviour and ecology of marine fish larvae. In: *Fish Behaviour and its use in the capture and culture of fish*. ICLARM Conferences Proceedings, Manilla: 287–330.

Hunter J., 1984. Feeding ecology and predation on marine fish larvae. In: *Marin fish larvae*, R. Lasker (ed.). Univ. of Washington Press, Washington: 33–77.

Iizawa M., 1983. Ecologie trophique des larves du Loup *Dicentrarchus labrax* en élevage. Thèse 3ᵉ Cycle, Univ. Sc. Techn. Languedoc, Montpellier: 140 pp.

Ingram M., 1988. Farming rainbow trout in fresh water tanks and ponds. In: *Salmon and trout farming*, L. Laird and T. Needham (eds). E. Horwood Publ., Chichester (UK): 153–189.

Ito T., 1960. On the culture of mixohaline rotifer *Brachionus plicatilis* O. F. Muller in sea water (in Japanese). *Rep. Fac. Fish. Pref. Univ. Mie.*, **3** (3): 708–740.

Jhingran V. G., 1990. L'aquaculture en Inde. In: *Aquaculture*, G. Barnabé (ed.). Technique & Documentation (Lavoisier), Paris: 1117–1152.

Kentouri M., 1985. Comportement larvaire de 4 Sparidés méditerranéens en élevage. Thèse de Doctorat d'état, Univ. Sc. Tech. Languedoc, Montpellier: 492 pp.

Lasker R., Sherman S., 1981. The early life history of fish: Recent studies. *Rapp. R.-v Réun. Cons. Int. Explor. Mer.*, **178**: 607 pp.

Lasker R. (ed.), 1984. *Marine fish larvae*. Washington Sea Grant Program Publ. University of Washington Press. Seattle: 131 pp.

Lim F., Moss D., 1981. Microencapsulation of living cells and tissues. *J. Pharm. Sc.*, **70**: 351–354.

Marcel J., 1990. La pisciculture en étang. In: *Aquaculture*, G. Barnabé (ed.). Technique & Documentation (Lavoisier) Publ., Paris: 615–652.

Marino G., Boglione C., Cataudella S., Saroglia M., 1989. Development anatomy and interpretation of swimbladder inflation mechanisms on sea-bass (*Dicentrarchus labrax*). In: *Aquaculture 89*, Short communications, Abstracts. European Aquaculture Society, Bredene, Belgique: 311–312.

Merea, 1986. Maîtrise de la qualité des alevins de Loup *Dicentrarchus labrax* produits en élevage intensif. *La Pisc. Fr.*, **85**: 17–23.

Naas K., Berg L., Klungsoyr J., Pittman K., 1987. *Natural and cultivated zooplankton as food for halibut larvae*. ICES, CM., F., 17: 20 pp.

Nichols W., 1983. Application of liquid fertilizer to hatchery ponds. *Prog. Fish-Cult.*, **45** (4): 223–225.

Nicolas J. L., Ansquer D., Besse B., 1989. Effect of bacteria in culture: Detrimental or beneficial. In: *Aquaculture 89*, Short communications, Abstracts. European Aquaculture Society, Bredene, Belgique: 187–188.

Oiestad V., 1982. Application of enclosures to studies on the early life of fishes. In: *Marin mesocosms*, G. Grice and M. Reeve (eds). Springer Verlag, New York: 49–62.

Oiestad V., Berg L., 1989. Growth and survival of Halibut (*Hypoglossus hypoglossus* L.) from hatching to beyond metamorphosis carried out in mesocosms. In: *Aquaculture, a biotechnology in progress*. European Aquaculture Society, Bredene, Belgique: 233–240.

Oiestad V., Moksness E., 1981. Studies of growth and survival of herring larvae (*Clupea harengus* L.) using plastic bag and concrete basin enclosures. *Rapp. P.-v Réun. Cons. Int. Explr. Mer.*, **178**: 144–149.

Paulsen H., Andersen N., 1989. Extensive rearing of Turbot larvae. In: *Aquaculture, a biotechnology in progress*. European Aquaculture Society, Bredene, Belgique: 241–248.

Pavillon J. F., Vu Tan T., 1981. Un aspect de l'absorption des substances organiques dissoutes chez les organismes marins et d'eaux saumâtres: données autoradiographiques aux premiers stades de développement chez *Artémia* sp. et *Dicentrarchus labrax*. *Océanis*, **6**: 705–708.

Person Le Ruyet J., 1987. L'élevage des poissons marins en écloserie. *Océanis*, **13** (1): 5–23.

Person Le Ruyet J. (ed.), 1990. L'élevage des poissons plats. In: *Aquaculture*, G. Barnabé (ed.). Technique & Documentation (Lavoisier) Publ., Paris: 721–775.

Person Le Ruyet J., Alexandre J. C., Le Ven L., 1989. *Early weaning of sea bass larvae using a commercial microparticule*. EAS, Special Publication No. 10: 205–206.

Petit J., 1990. L'approvisionnement en eau, le traitement de l'eau et le recyclage en aquaculture. In: *Aquaculture*, G. Barnabé (ed.). Technique & Documentation (Lavoisier) Publ., Paris: 47–182.

Riley J. D., 1974. The distribution and mortality of sole eggs *Solea solea* (L.) in inshore areas. In: *The early life of fish*, J. H. S. Blaxter (ed.). Springer Verlag, Berlin: 39–52.

Rosch R., 1988. Mass rearing of *Coregonus lavaretus* larvae on a dry diet. *Finnish Fisheries Research*, **9**: 345–351.

Ronzani-Cerqueira C. V., 1986. L'élevage larvaire intensif du Loup *Dicentrarchus labrax*: influence de la lumière, de la densité en proies et de la température sur l'alimentation, sur le transit digestif et sur les performances zootechniques. Thèse de Doctorat, Univ. Aix-Marseille 2: 165 pp.

Sherwood M., Mearns A., 1981. Fate of post-larval bottom fishes in a highly urbanized coastal zone. *Rapp. P.-v Réun. Cons. Int. Explr. Mer*, **178**: 104–111.

Teshima S., Kanazawa A., Sakamoto M., 1982. Microparticulate diets for the larvae of aquatic animals. *Min. Rev. Data File Fish. Res.*, **2**: 67–86.

Tesseyre C., 1979. Etude des conditions d'élevage intensif du Loup (*Dicentrarchus labrax*). Thèse 3e Cycle, Univ. Sc. Techn. Languedoc. Montpellier: 115 pp.

Vernier J. M., 1969. Table chronologique du développement embryonnaire de la truite arc-en-ciel *Salmo gairdneri*, Rich, 1836. *Ann. Embryol. Morpho.*, **2** (4): 495–520.

Versichelle D., Leger P., Lavens P., Sorgeloos P., 1990. L'utilisation d'Artémia. In: *Aquaculture*, G. Barnabé (ed.) Technique & Documentation (Lavoisier) Publ., Paris: 241–259.

Von Walhert G., Von Walhert H., 1989. Extensive rural mariculture in the tropics—experiences and issues. In: *Aquaculture, a biotechnology in progress*. European Aquaculture Society, Bredene, Belgique: 97–102.

Watanabé T., Kitajima C., Fujita S., 1983. Nutritional value of live organisms used in Japan for mass propagation of fish. A review. *Aquaculture*, **34**: 115–143.

Weis J., Weis P., Ricci J., 1981. Effect of cadmium, zinc, salinity and temperature on the teratogenecity of methylmercury to the killifish (*Fundulus heteroclitus*). *Rapp. R.-v Réun. Cons. Int. Explr. Mer*, **178**: 64–70.

Weppe, Joassard, 1986. Preliminary study: effects of light on swim-bladder's inflation of cultured sea bass (*Dicentrarchus labrax*) larvae. Communication presented at the Congress PAMAQ: Pathology in Marine Aquaculture (unpublished).

5

Extensive culture of fish in ponds

1. GENERAL

In contrast to intensive culture where the water serves simply to act as a physical support for fish, extensive culture makes use of the natural trophic webs to produce fish which are then eaten by man.

Extensive pond culture is particularly developed in China, India and central Europe. The greatest developments are in China where there are more than 0.7 million hectares of ponds (33% of the world's ponds) and 67% of the world's production of pond fish. Production ranges from 1.5 to 15 tonnes per hectare per year (mean 3.2 t ha^{-1}).

Table 1 (Marcel 1990) shows the production from different European countries. In France it is less developed than intensive aquaculture but still results in the production of 8000 tonnes of fish each year.

One essential difference distinguishes Chinese and Indian pond culture from that carried out in Western countries: there are no aesthetic barriers to the use of animal wastes (manure). This means that fish can be produced cheaply, in line with the low price likely to be obtained in the market. The differences between the two systems are summarized in Table 2 (from Wohlfarth & Schroeder 1979).

Table 1. Production through pond fish culture in different countries (Marcel 1989)

Country	Production (t)	Area (ha)	Weight ha^{-1} (kg)
France	5 000	27 500[*]	180
Hungary	20 000	14 000	1 400
Israel	13 000	3 700	3 500
Germany (West)	6 500	18 000	360
Poland	20 000	30 000	600
Czechoslovakia	16 000	42 000	380
Former Yugoslavia	20 000	20 000	1 000

[*] Out of 150 000 hetares sampled (300 000 hectares with various lakes and reservoirs).

Table 2. Rearing pond fish in China and South East Asia

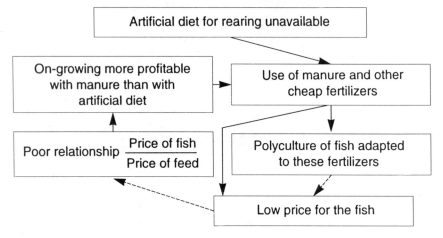

(a) Rearing pond fish in China and South East Asia

Artificial diet for rearing unavailable

On-growing more profitable with manure than with artificial diet

Use of manure and other cheap fertilizers

Poor relationship $\dfrac{\text{Price of fish}}{\text{Price of feed}}$

Polyculture of fish adapted to these fertilizers

Low price for the fish

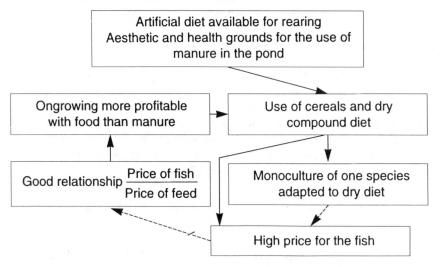

(b) Rearing pond fish in Europe and the USA

Artificial diet available for rearing Aesthetic and health grounds for the use of manure in the pond

Ongrowing more profitable with food than manure

Use of cereals and dry compound diet

Good relationship $\dfrac{\text{Price of fish}}{\text{Price of feed}}$

Monoculture of one species adapted to dry diet

High price for the fish

Fish are farmed without artificially made diets, but with the addition of animal waste products to the pond mainly in Asian countries where supplementary nutrients are costly and/or difficult to obtain. In China and India the ecosystem of ponds is managed highly efficiently for the production of fish (Marcel & Doumenge 1990; Jhingran 1990). Pond management is also highly developed in the temperate climate of Eastern Europe.

This type of aquaculture, where the natural productivity of waters is transformed to products which are consumed by man, is termed extensive aquaculture or sometimes production aquaculture.

2. BIOLOGICAL PRODUCTION OF PONDS

Some fish feed exclusively on plankton. These planktonophages are able to filter and concentrate particles in suspension in the water (Chapter 1, this part). In the sea, such fish include herring, anchovies and sardines; in fresh water there is the silver carp. Other fish such as the tilapias are able to feed on plankton but can also feed on material from the bottom. The common carp (*Cyprinus carpio*) is an efficient grazer on the bed of ponds. Other fish, such as the grass carp (*Ctenopharyngodon idella*), consume large quantities of rooted or floating aquatic plants.

The processes of production in aquatic ecosystems have been described in Part I, Chapter 2. Extensive aquaculture in shallow ponds follows the basic principles with a few characteristics of its own; the shallowness of the water allows photosynthesis through the water column; phytoplankton, various macrophytic algae or flowering plants are the primary producers which compete with each other to colonize the environment. These bodies of water are different from larger, natural waters because it is possible for man to influence the type of production to a limited extent (over areas of a few tens of hectares) to produce species which are regarded as useful.

2.1 Physical characteristics of the ponds used in extensive fish culture

The mean depth of these ponds goes from 1.5 to a maximum of 3–4 metres and their area is most frequently between 10 and 100 hectares. They are usually sandy around the edges and muddy in the centre.

The ponds can be emptied (Fig. 1) and are fished out each year (in central Europe this takes place in the autumn). The temperature of the water follows that of the air but there is stratification in the summer, with a drop in oxygen near the bottom and an accumulation of phosphates and ammonia (Fig. 2). At the end of the summer the surface waters become mixed with deeper waters; this occurs even in the absence of wind because of the cooling of the upper waters which takes place at night. There is also an equilibrium between carbon dioxide and the oxygen in the air if the pond is well aerated; the consequence of this is a pH range between 7.5 and 10 with a mean of 8.2 (Korinek *et al.* 1987). In 'normal' conditions a pond should have a dissolved oxygen level >3.5 mg l^{-1} (Figs 3

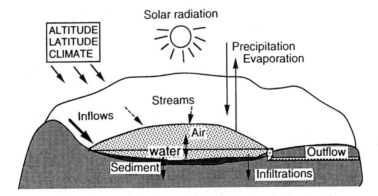

Fig. 1. The pond and its relationship with the environment (adapted from Balvay 1980).

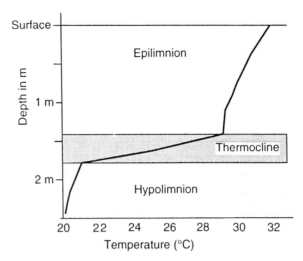

Fig. 2. Stratification in a pond in summer in calm weather (adapted from Boyd & Lichtkoppler 1979).

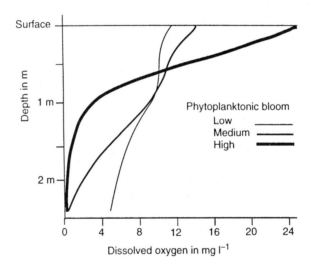

Fig. 3. Variations in the level of dissolved oxygen as a function of depth, depending on the intensity of the phytoplanktonic development (adapted from Boyd & Lichtkoppler 1979).

and 4) and an ammonia level <0.02 mg l^{-1}. The fish most frequently reared in ponds are cyprinids which have relatively low oxygen requirements. Although these species convert food most efficiently when oxygen levels are above 3 mg l^{-1}, they are able to survive levels of 1 mg l^{-1} which are lethal for salmonids..

Emptying and resuspension of sediments by currents induced by winds ensures the renewal of the natural fauna. The annual emptying of the ponds ensures the control of aquatic populations with long life cycles such as fish.

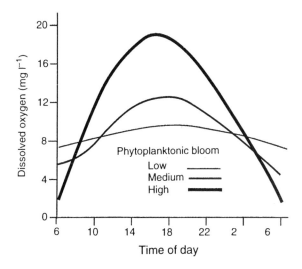

Fig. 4. Variations in the level of dissolved oxygen during a day/night cycle, depending on the intensity of the phytoplanktonic bloom (adapted from Boyd & Lichtkoppler 1979).

2.2 Liming, other chemical treatments and pH

In Chapter 1, Part I, it was explained that the total hardness of water is a measure of the total concentration of divalent cations in the water expressed in milligrams per litre calcium carbonate (calcium and magnesium are the two ions present); total alkalinity expresses the total quantity of bases measurable in the water, expressed in milligrams per litre equivalent of calcium carbonate. The hardness and total alkalinity are normally similar, but in certain waters the calcium hardness is lower than the alkalinity because the alkaline anions are associated with magnesium, sodium or potassium rather than calcium. Sodium and potassium are soluble and magnesium is more soluble than calcium carbonate.

Ponds are often constructed in places where agriculture is barely possible, in particular in acid soils. However, in nature the most productive waters have a calcium (Ca) level of between 80 and 100 mg l^{-1}. It is therefore necessary to use lime to improve productivity. The quick lime used to supply calcium has a disinfectant action when applied to a dried-out pond (with the addition of water quick lime becomes slaked lime: $CaO + H_2O = Ca(OH)_2$).

Liming increases the alkaline reserves in the water and therefore helps maintain a stable pH (Chapter 1, Part I, and Boyd et al. 1978), in particular when photosynthetic activity displaces the CO_2–carbonates equilibrium, exhausting the carbonates, threatening lethal pH levels (>9 in salt water and >10 in fresh water). Calcium is also important in other ways, increasing the pH of silty sediments and the availability of phosphorus, and increasing the concentration of the bicarbonate ion in water. This action is important in water purification ponds and may also be used in mollusc culture ponds in salt water. Martin (1987) details the use of various forms of calcium in pond fish culture.

Lime acts as an effective flocculant for microalgae (at doses of 400–800 mg l^{-1}; Barnabé 1981) and also of mineral matter in suspension. It is therefore essential to avoid using it in excess or at the wrong time.

High pH values occur during the summer (Fig. 5); they come from an excessive development of plants (phytoplankton or macrophytes) and the exhaustion of carbon reserves (Chapter 1, Part I) because of the use of carbon dioxide during photosynthesis. These values may cause the elimination of plankton and fish and also animals, such as prawns, which are buried in the mud. Because of this, pH should be controlled to within acceptable limits (Fig. 5). There are several possible methods.

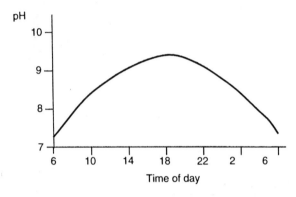

Fig. 5. Change in pH over a 24 hour cycle, depending on the intensity of the development of phytoplankton (adapted from Boyd & Lichtkoppler 1979).

— A drop in pH can be brought about by adding ammonium (NH_4) which is transformed to nitrites (NO_2) in the water with the production of two H^+ ions before being oxidized to nitrate (NO_3) by bacteria: $NH_4 + \frac{3}{2}O_2 = NO_2^- 2H^+ + H_2O$.

— The ammonium sulphate used in south Carolina at a level of 224 kg ha^{-1} decreases the pH by 2–3 units in 24 hours but the development of an intense plankton bloom returns the pH to its original value within 2–3 weeks. The introduction of ammonia thus ensures nitrogen fertilization (see below).

— Alum (aluminium sulphate) is also used to limit the increase in pH. Each molecule of alum liberates six H^+ ions when combining with water. It takes 1 g of alum to neutralize the alkalinity of 1 g of calcium carbonate. This treatment does not combat the cause of the phenomenon; it merely provides a chemical neutralization.

— The use of gypsum ($CaSO_4.2H_2O$) increases the concentration of calcium but does not increase pH. The calcium precipitates carbonates in the form of calcium carbonate in alkaline waters; this results in a drop in pH. If the type of gypsum used in agriculture is used (80% $CaSO_4.2H_2O$), 1 g of gypsum is equivalent to 2.15 g alkalinity.

2.3 Manure and fertilizers

The increase in primary production through the addition of fertilizer can be said to be intentional eutrophication. The limits are imposed by the survival of the fish stock. Several types of fertilizer may be used.

2.3.1 Direct application of organic manure and minerals

There are various published accounts of such treatments using varying amounts of fertilizer. It appears that there is a degree of empiricism in these treatments which is not
surprising in view of the differing characteristics of the ponds (Wurtz-Arlet 1980).

Valdeyron (1989) recommended annual applications of nitrogen and phosphorus of
between 15 and 40 kg ha^{-1} as P_2O_5 (phosphoric acid) which corresponds to 75–200 kg
phosphate of lime. The use of nitrogenous fertilizers remains rare and is limited to
5–10 kg per hectare per year of nitrogen (N). When organic manures are used the
volumes are:

— animal wastes: 10–20 m^3 ha^{-1} yr^{-1} given at the rate of 1–2 m^3 ha^{-1} week^{-1}.
— fresh animal droppings: 1–3 t ha^{-1} yr^{-1}, at a rate of 0.5–1 t ha^{-1} per application;
— dehydrated manure: 100–300 kg ha^{-1} yr^{-1}, at a rate of 100–300 kg per application.

Where organic manure is used in well managed ponds the rate of application can reach
10–20 kg ha^{-1} day^{-1} of dry matter (the manure is distributed weekly).

— Potassium-rich fertilizers which favour the development of phytoplankton are used at
   a rate of 30–40 kg potassium per hectare per year.
— The treatment of fry ponds has been studied in detail: 12–15 tonnes of liquid manure
   ha^{-1} or 6 t manure are applied before water is added to the pond. The pond is
   disinfected with 0.5 t quick lime ha^{-1}.

Manure made from dried chicken droppings is marketed, but fresh chicken manure is also
used effectively. Organic manure brings about a rapid development of zooplankton; pig
manure, for example, favours the rapid development of cladocerans.

2.3.2 Intensive agriculture (with fertilization) during the periods when ponds are dried out

Ponds are left dry between cultivation cycles and are frequently cultivated. The remains
of the plants enrich the pond and this increases the production of fish; parasites and
predators are killed, aquatic vegetation is eliminated. When the ponds are flooded again,
tubificid worms reappear, hatching from cysts which can survive as long as the water
level in the soil does not fall below 20%.

2.3.3 Other types of fertilization and methods of increasing production

Additional food supplied to the fish (cereals and dry diets) are not all eaten; wastes are
therefore produced. A large part of these are recycled within the pond and become
responsible for generating primary production. This is also true of excretory products
(NH_4, urine and faeces). Carp and gudgeon which remain on the bed of the pond recycle
these nutrients directly; they also find their way to zooplankton and bacteria (Chapter 2,
Part I). Depending on the species cultured, the fish themselves participate in recycling of
wastes.

The production of ponds is increased where movement is initiated in the water: Costa-
Pierce (1989) showed that ponds where there was little water movement became stratified
(Fig. 2), and dominated by green algae in the phytoplankton, whereas ponds with good
mixing of the water are dominated by diatoms (brown algae). Although there is no

difference in the level of chlorophyll a, the production increases from 2.1 t tilapia ha^{-1} to 5.1 t ha^{-1} in ponds where the water is agitated (the tilapia are fed with chicken feed with a 25% protein composition). This example demonstrates our ignorance of ways of optimizing the food chains for the production of fish.

2.3.4 Dynamics of nutrient salts in ponds

When fish are harvested, plankton lost through the outflow or aquatic plant cut-back, there is a loss of nutrient salts from the ponds; these are replaced by fertilizers. From this point of view, a pond cannot be considered to be a closed system.

Phosphorus is lost from the pond as a constituent of the harvested fish at a rate of around 0.5 g m^{-2} yr^{-1}, but after decades of fertilization the level of phosphorus in the mud and silt at the bottom of the ponds supplements that added during fertilization. Studies in lakes have shown that the buffering action of the sediments is important for this element and the annual rate of phosphorus application (in the form of superphosphate) can thus be dropped from 1 to 0–0.3 g m^2 yr^{-1} without any change in the production of the fish in Eastern European ponds. The increase of the release of phosphorus from the sediments may in some circumstances be a better method of fertilization than the addition of manure. Phosphorus is rare when phytoplankton is abundant, and vice versa. Maximum levels are found when the water is clear (up to 0.7 mg l^{-1}) and there is no development of benthic vegetation.

The concentration of ammoniacal nitrogen varies from one to several hundreds of microgrammes per litre and the minimum level is found during phytoplankton blooms. Mean levels fluctuate between 30 and 300 μg l^{-1} in an unpredictable manner and daily consumption has been estimated at 200–300 μg of ammoniacal nitrogen by various authors; recycling is therefore very rapid.

Running waters are rich in nitrates because of the runoff from agriculture; these nitrates often end up in ponds. In water with a low turnover rate, such as ponds, levels range from several milligrams per litre in the absence of phytoplankton to low or undetectable levels in the presence of microalgae. Nitrite levels are usually very low in ponds.

2.4 Primary production

Photosynthesis takes place through the whole depth of the water column, although the water column may sometimes be very clear because of the actions of filter feeders such as Daphnia (filtration can reach a rate of 100% of the volume per day). The general processes of primary production in the aquatic environment have been described elsewhere (Part I, Chapter 2). The active primary production in ponds is accompanied by major fluctuations in the level of dissolved oxygen; this reaches a maximum which may often exceed 10 mg l^{-1} in the late afternoon (around 18.00 hours) on sunny days with a minimum (1–3 mg l^{-1} in the absence of aeration) caused by the respiration of all the biomass. This minimum may result in complete anoxia and lead to total mortality.

While qualitative variations in phytoplankton are not always linked to the season, nevertheless a seasonal change in species has been observed and characterized for central European ponds by Korinek et al. (1987). At the beginning of spring, flagellates are the dominant group (*Chlamydomonas, Cryptominas*), preceding the chlorococcales. The planktonic alga *Oscillatoria* dominates in April and *Spirogyra* in May; if the water remains clear macrophytes (*Potamogeton*) colonize the whole of the water body in summer. Where huge numbers of phytoplanktonic organisms proliferate suddenly (blooms) half of the light is absorbed at a depth of 11–17 cm and phytoplankton causes shading. Blooms of *Aphanizomenon flos-aquae* are linked with a superabundance of phosphorus. The presence of macrophytes accelerates the rate at which ponds age; these plants cause shading and their decomposition has a negative effect. Their biomass (dry weight) can go from 0.6 to 9.5 kg m^{-2}.

Variations in phytoplankton production are followed by measuring the level of chlorophyll a (usually 1–100 μg l^{-1}) although more simple methods are available:

— The nett production of O_2 can be estimated by measuring the difference between the concentration of O_2 in a transparent bottle and the concentration in a black bottle after a given period of exposure at the same depth. This varies between 0.1 and 10 g O_2 m^{-2} day^{-1}.
— The depth at which a Secchi disc disappears. A Secchi disc is a white disc, 30 cm in diameter, which is dropped into the water. The depth at which the disc disappears is a measure of abundance of chlorophyll a (Part I, Chapter 2).
— Chinese fish farmers dip their hands and forearms into water to assess the density of algae! This may seem rather unscientific but it should be remembered that these people produce the majority of farmed pond fish.

Copper sulphate ($CuSO_4$), used at small doses, eliminates certain unwanted species of algae. It is toxic for developing *Aphanizomenon flos-aquae* at levels of 100–160 μg l^{-1}; at these levels it is not toxic for fish and does not accumulate in flesh. It favours the development of certain planktonic microalgae (chlorophycaeae and diatoms). Balvay (1980) showed how copper could be used to alter the pattern of production in ponds.

2.5 Secondary production

The abundance of zooplankton is linked to the density of the fish which consume them: if this density is low, cladocerans dominate (*Daphnia*), making up 70–95% of the biomass. Their turnover rate is between 5 and 7 days in summer, which indicates that the dominant zooplankton can be turned over more than twenty times during a growth season of 200–240 days. The mean density of daphnids is 100 l^{-1}. In the Mèze lagoon system, production of daphnids during the high season (May–October) has been estimated at 5 tonnes (wet weight) per hectare per month in a small pond of 5000 m^2, comparable with a fertilized pond. Other types of zooplankton include cyclopoid copepods (*Cyclops*), calanoids (*Eudiaptomus*), as well as the rotifers, *Brachionus, Asplanchna, Keratella*.

When there is a high density of fish (1 m^{-2}), zooplankton is composed of small species: cladocerans such as *Bosminia* and *Ceriodaphnia*, cyclopoid copepods (*Mesocyclops, Thermocyclops*), but these are found in the nauplius or copepodite stage (adults are consumed). Cladocerans make up less than 50% of the biomass and are less than 2 mm in size. The only exception to this is found at the beginning of spring; rotifers are extremely abundant but too small to act as food for fish.

Fry ponds are likely to be different from this because the larvae are actively feeding on zooplankton. It is possible to direct production towards rotifers by applying Flibol E (Dipterex) at a rate of 1 mg l^{-1}: this concentration kills the copepods but does not affect rotifers or carp larvae.

The benthic fauna is permanent and largely composed of oligochaete worms (Tubificidae and Naididae); other worms (nematodes), molluscs or crustaceans are present in negligible quantities. The temporary benthos is made up mainly of insect larvae, especially Chironomids, often with an abundance of *Chaoborus*; this makes up 60–80% in number and 60–70% in total biomass (representing around 10–50 g m^{-2}). This biomass reaches its maximum in the autumn and drops sharply in spring because of the mass hatching of insects, and represents a loss to the aquatic ecosystem of over 200 kg ha^{-1} yr^{-1}.

2.6 Bacteria

The role of bacteria is poorly understood as methods for studying them have not been particularly well developed. However, it is known that their density goes from 7 to 19 million cells ml^{-1} in Polish fish ponds (Korinek *et al.* 1987). Biomass is between 2.6 and 5.6 mg ml^{-1} wet weight according to the same authors. These densities are far higher than those found in other ecosystems. While the number of faecal streptococcae is usually low (0–50 ml^{-1}), where effluents from a duck farm are being spread on the ponds they may reach 1000 ml^{-1}. Bacterial activity is often not appreciated because of problems with measuring instruments, but it is significant compared with that of phytoplankton. Costa-Pierce (1989) described the problems associated with estimating the production of microbial heterotrophs in ponds.

2.7 The need to manage water quality

There are many parameters which characterize the quality of the pond environment; these are interdependent and determine the survival of the cultured species. This is an identical situation to that in shellfish beds and suggests that benefits would be accrued from managing the quality of the water. It is essential that the optimum biomass which can be supported by a pond (or shellfish bed) should be determined. Water quality should be monitored. For further information on management in these systems the reader is referred to the work of Billard & Marie (1980) for pond fish culture and that of Héral (1990) for shellfish beds.

3. THE TRADITIONAL METHOD OF EXTENSIVE CULTURE OF FISH IN PONDS

The biological aspects of fish culture have been dealt with in the preceding four chapters. In this section, details of ongrowing in ponds will be added to the basic information. The object of pond aquaculture is to optimize the use of the trophic web described above for fish production. Details of different types of pond culture can be found in the works of Balvay (1980), Billard (1980), Billard & Marcel (1980), Marcel (1990), Michael (1987), Barnabé (1986 and 1990).

Conditions for ongrowing are relatively diverse, varying in relation to climate and to species. The principal types are summarized below.

In temperate regions the production cycle takes two to three years, but in the tropics the production of tilapia species and milk fish (*Chanos*) takes six to nine months (Jhingran 1989). The pond types described above are ongrowing ponds. They are fertilized, and in many cases supplementary feeding is given in the warm season. Fig. 6 shows the change in temperature, natural production and growth in weight in ponds in east Germany.

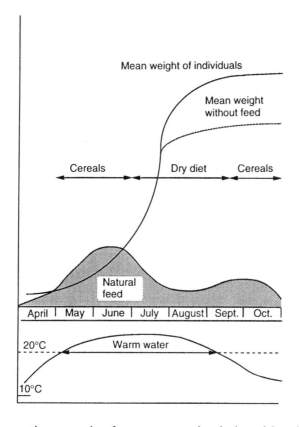

Fig. 6. Diagrammatic representation of temperature, natural production and the weight of fish in ponds with complementary feeding (adapted from Matena & Berka 1987).

During the first year, juvenile common carp, produced in fry ponds (Chapter 4) where they spend four weeks or less, are fished out and transferred to ongrowing ponds. The number of individual fish varies from 10 000 to 50 000 fish per hectare depending on the productivity of the ponds. These fish reach a weight of 25–80g at the end of their first year in culture (the mean weight varies according to latitude).

At the end of the second year in culture the carp reach a weight of 250–400 g in east Germany, but in warmer waters (Bulgaria) they reach 600–1200 g. The stocking rate of the ponds also affects growth rate as the available food is unchanged. There are formulae to help determine the stocking rate at the start of rearing (N). The formula given by Marcel (1989) is:

$$N = \frac{R}{P} \times \frac{100}{100 - x}$$

where R is the mean production of the pond in kilograms of fish (kg ha^{-1} yr^{-1}), P is the mean weight of individual fish at harvest (kg) and x is the loss through natural fish mortality in percentage of fish stocked.

Where production is 1000 kg ha^{-1} yr^{-1} (as in eastern Europe), the expected weight of individuals after one year in culture is 70 g (also the mean figure for eastern Europe) and the mortality is predicted at 20%. It is necessary to stock 17 857 fish per hectare to obtain 14 285 fish with a mean weight of 70 g. In the second year 2500 fish should be stocked in a pond capable of the same level of production and with the same mortality rate to obtain 500 g fish.

It is unusual to have a third rearing season except at high latitudes or altitudes where there is a long cold season. However, carp, the species most frequently cultured in Europe, reach 3 kg if grown for another year. In colder regions, pond fish culture becomes more complicated as it is necessary to construct wintering ponds with deeper parts where the fish can escape from frost, ice and snow.

4. POLYCULTURE

In polyculture, several species of fish are cultured together in the same pond, each exploiting a different ecological niche. In China four to five species are farmed together, feeding on the different types of living matter described above (phytoplankton, zooplankton, benthos, algae or other macrophytes). The following can be distinguished:

— Planktonophages which feed on either zooplankton (e.g. bighead carp) or phytoplankton (silver carp).
— Herbivores feeding on macrophytes (grass carp).
— Detritivores feeding on the bottom (common carp).

For this system to be successful, not only must suitable, complementary species be available, but the climate must permit optimal production throughout the year. There is little polyculture in Europe but it is common in India and China. Polyculture is carried out in salt water in Israel.

The ponds used in polyculture are relatively modest in size so as to allow emptying and removal of the fish. In China an area of 6000 m^2 is considered optimal. The preferred depth is between 2 and 2.5 m; shallower ponds have a lower production (4.5 t ha^{-1} yr^{-1} for ponds 1 m deep compared with 7.5 t ha^{-1} yr^{-1} for ponds 2.5 m deep). The percentage of the different species of fish, management of the fish and their harvesting are complex and variable (Marcel 1990; Jhingran 1990).

5. INTEGRATED AQUACULTURE IN PONDS

This term is usually applied to the integration of aquaculture with agricultural activities, although there are other forms of integration; the purification of waste waters in lagoons is a good example (Barnabé 1982).

Chickens, ducks, pigs and cattle are reared on the edge of ponds or (small animals) on grids above the ponds. This allows direct recycling of their excretory products in the aquatic ecosystem. For this to be effective, temperatures must be high; this is why polyculture is confined to warm countries. There must also be a 'harmony' between the quantity of animals farmed and the capacity of the various links in the trophic web of the pond, and the incorporation of these matters into the ecosystem.

The complexity of the interactions within this type of ecosystem is such that the fate of the manure introduced has never been followed precisely. An overall balance has, however, been worked out in the Philippines (Hopkins & Cruz 1982) during a large scale experiment.

The system combined pigs and fish, ducks and fish, and chickens and fish. The livestock were reared on the banks of ponds which were 400–1000 m^2 in area. The animal wastes were added daily to the ponds. The fish used were tilapia (*Oreochromis niloticus*) and carp (*Cyprinus carpio*), sometimes with predators (*Clarias batrachus* and *Channa striata*). Tilapia production of 15 kg ha^{-1} day^{-1} (or 5.5 t ha^{-1} yr^{-1}) and carp production of 4 kg ha^{-1} day^{-1} (1.5 t ha^{-1} yr^{-1}) were obtained with fertilization rates of 100 kg ha^{-1} day^{-1} of pig manure or poultry waste. The dissolved oxygen level reached 200% saturation during the afternoon but fell to 1 mg l^{-1} at dawn. The level of ammonia exceeded 2 mg l^{-1} when chicken droppings were used. Even in ponds which were treated in the same way there were great differences in the plankton populations. No parasites common to fish and man were found.

Castell (1989) reported details of a modern Chinese integrated aquaculture system. Eight species of fish were cultured in 70 hectares of ponds with a depth of 3 metres. This farm produced 800 tonnes of fish, 400 000 ducks (the livers of which were exported to France), 1060 pigs, 123 cows, and chickens were reared on another site. The mean production is 13.5 t ha^{-1} yr^{-1}. A hotel has been built on the site and all the products from the farm are served in it. Table wastes are fed to the carnivorous fish. This whole operation employs 837 people and can be described as an agricultural purification plant, recycling the animal wastes through the fish.

Sometimes ducks and fish are reared together in the same water body; production of fish can reach 9 t ha^{-1} yr^{-1} with over 30 t of duck. An added bonus is that the pond water can be used for the irrigation of agricultural land and, being rich in nutrients, increases production to over 10 t ha^{-1} (Schoonbee & Prinsloo 1988).

Tests have been carried out in South Africa with domestic waste water supplied to ponds with tilapia. Without any other form of food, production was between 3.9 and 8.1 t ha^{-1} for an experiment lasting between 210 and 226 days (Turner *et al.* 1986).

As well as being a means of recycling waste water, the production of fish in ponds offers a way of recycling solid wastes; Marcel (1989) showed that 6.8 kg of rice flour or 4 kg of soya cake would produce 1 kg of fish. The same quantity of fish can also be produced from 15 kg of land plants or 45 kg of aquatic plants.

6. SEA RANCHING AND RESTOCKING

This technique consists of controlled reproduction and managed larval rearing followed by release of juveniles into the natural environment where they grow, feeding on the production of the natural ecosystem. The fish are captured using the techniques of traditional fisheries or by taking advantage of particular behaviour. Salmonids which have grown to full size in the sea are recaptured when they return to their river of origin to spawn. The river is recognized by its smell.

Most restocking in rivers is to satisfy the requirements of the leisure fishery. However, the most spectacular examples are those of sea ranching, as carried out with Pacific salmon in Japan, Canada, the United States and Russia in the Pacific Ocean. Around 2 billion juveniles are released into rivers, migrate to sea where they grow, and then return to their river of release. The recapture rate varies between 1 and 5%; this type of sea ranching is therefore profitable in the main.

Other, less migratory, species have also been used for restocking. These include sturgeon and coregonids in Europe and bream, abalone and prawns in Japan. For the future, Japanese authors envisage the restocking of the world's oceans with tuna so as to transform oceanic pelagic production (phytoplankton, herbivorous zooplankton and small planktonic fish species) which is difficult or costly to exploit and convert to a desired product.

Fish culture for restocking is no small operation; 844 million juveniles were produced in the USA in 1980 to supplement natural populations.

7. OTHER ASPECTS OF EXTENSIVE FISH CULTURE

The culture of ornamental fish is an activity of considerable economic importance. In Japan, the middle classes often build houses around a pond containing valuable fish (Doumenge 1990).

The world market for live animals and aquarium products is around US$4 billion annually and involves 350 million people. There is an aquarium in 7% of all US households (fish-keeping ranks second to photography as a hobby) and figures of the same order are characteristic of several European countries (Germany): 80% of the market is for fish which have been reared in captivity.

One of the most active regions in the world for the production of aquarium fish is Florida (USA). The farms are made up of round-bottomed ponds excavated in the ground. The overall area is a few tens of hectares broken up into numerous small ponds of 200 m^2 area which are around 2 m deep. Some are covered with glass to reduce the mortalities which might occur in winter from storms and low temperatures. The management is similar to that of ponds used in other types of aquaculture; the cultured species (guppies, platies, mollies, etc.) are stocked in fertilized earth ponds. Reproduction is not controlled but takes place naturally in the ponds, even in nest-building species (Cichlida, Centropomida) but food is added, the growth of macrophytes regulated and predators controlled. As the ponds cannot be emptied, anaerobic sediments are removed by pumping (Winfree 1989).

Farmers tend to specialize in the rearing of a small number of species. Spawning and hatching take place in the earth ponds and the larvae are removed and placed in their own round-bottomed earth ponds when the yolk is resorbed. There are very few farms of this type for marine fish.

Because of this activity, 15 000–20 000 boxes, each containing 500–600 live fish, are air-freighted from Florida to the four corners of North America every year.

8. CONCLUSIONS

Throughout this chapter we have pointed out the problems encountered by man in directing the production of aquatic ecosystems towards the production of food fish, juveniles for sale and water purification.

Away from their role in fish production, ponds are important in the protection of animals and in the conservation of particular types of environment. They may also be a focus for tourists (either because of their visual attractiveness or for swimming, fishing or other watersports).

In spite of the high level of nutrients in these ponds, because they constitute a closed system, there are few problems with pollution. The pond can be a place of water purification as has been shown in many examples in this chapter. Lagoon systems which are used for water purification are nothing more than ponds in which the water to be purified circulates for a period of about one month. Although in Europe these ponds are not used for the production of fish for human consumption, their plankton production is mainly used for fish culture as the species produced are suitable for feeding to many different species of fish, including marine ones.

REFERENCES

Balvay G., 1980. Fonctionnement et contrôle du réseau trophique en étang. In: *La pisciculture en étang*, R. Billard (ed.). INRA Publ., Paris: 47–79.

Barnabé G., 1981. Collecte et utilisation de plancton. In: *Compte rendu des Activités Scientifiques du Centre de Recherches du Lagunage de Mèze*. Rapport 1980–1981: 112–132.

Barnabé G., 1982. Lagunage et aquaculture. In: *C.R. Colloque L'épuration par lagunage*. Communications Scientifiques, CERETE (Ed.), Montpellier: 52–59.

Barnabé G. (ed.), 1986. *Aquaculture*. Technique & Documentation (Lavoisier) (Publ.), Paris: 1123 pp.

Barnabé G. (ed.), 1990. *Aquaculture*. (2nd edn.), Technique & Documentation (Lavoisier) (Publ.), Paris: 1307 pp.

Billard R. (ed.), 1980. *La pisciculture en étang*. Actes du Colloque sur la Pisciculture en étang, Arbonne la Forêt, France, 11–13 Mars 1980. INRA Publ., Paris: 434 pp.

Billard R., Marcel J., 1980. Quelques techniques de production de poissons d'étang. *Pisci. Fr.*, **59**: 9–16 and 41–49.

Billard R., Marie D., 1980. Contrôle et qualité des eaux de l'étang. In: *La pisciculture en étang*. Actes du Colloque sur la Pisciculture en étang, Arbonne la Forêt, France, 11–13 Mars 1980, INRA Publ., Paris: 107–127.

Boyd C. E., Preacher J. W., Justice L., 1978. Hardness, alkalinity, pH and pond fertilization. *Proc. Ann. Conf. S.E. Assoc. Fish & Wildl. Agencies*, **32**: 605–611.

Boyd C. E., Lichtkoppler F., 1979. Water quality management in pond fish culture. Research and Development series, 22 (April 1979).

Castell J., 1989. An integrated fish farm. *World Aquaculture*, **20** (3): 20–23.

Costa-Pierce B., 1989. Stirring ponds as a possible means of increasing aquaculture production. *Aquabyte (ICLARM Ed.)*, **2** (3): 5–7.

Doumenge F., 1990. L'aquaculture au Japon. In: *Aquaculture*, G. Barnabé (ed.). Technique & Documentation (Lavoisier), Paris: 949–1068.

Héral M., 1990. L'ostréiculture française traditionnelle. In: *Aquaculture*, G. Barnabé (ed.). Technique & Documentation (Lavoisier), Paris: 347–397.

Hopkins K., Cruz E., 1982. The ICLARM-CLSU integrated animal-fish farming project. Final report. ICLARM techn. rep., 5: 96 pp.

Jhingran V. G., 1990. L'aquaculture en Inde. In: *Aquaculture*, G. Barnabé (ed.), Technique & Documentation (Lavoisier), Paris: 1117–1152.

Korinek V., Fott J., Fuska J., Lellack J., Prazakova M., 1987. Carp pond of central Europe. In: *Managed aquatic ecosystems. Ecosystems of the World*, G. Michel (ed.). Elsevier, **29**: 29–58.

Marcel J., 1989. La pisciculture en étang. In: *Aquaculture*, G. Barnabé (ed.). Technique & Documentation (Lavoisier). Paris: 615–652.

Marcel J., Doumenge F., 1990. L'aquaculture en Chine. In: *Aquaculture*, G. Barnabé (ed.). Technique & Documentation (Lavoisier), Paris: 1069–1086.

Martin J. F., 1987. La fertilisation des étangs. *Aqua-Revue*, **11**: 34–39.

Matena J., Berka R., 1987. Fresh-water fish pond management. In: *Managed aquatic ecosystems. Ecosystems of the World*, R. G. Michel (ed.). Elsevier: 3–27.

Michael R. G. (ed.), 1987. *Managed aquatic ecosystems. Ecosystems of the World*, **29**. Elsevier, Amsterdam: 166 pp.

Schoonbee H. J., Prinsloo J. F., 1988. The use of polyculture in integrated agriculture aquaculture production system aimed at rural community development. *J. Aquat. Prod.*, **2** (1): 99–123.

Turner J. W., Sibbald R. R., Hemens J., 1986. Chlorinated secondary domestic sewage effluent as a fertilizer for marine aquaculture. 1, Tilapia culture. *Aquaculture*, **53**: 133–143.

Valdeyron A., 1989. Fertilisations minérales et organiques. European Aquaculture Society, Special Publication No. 10: 253 pp.

Winfree R., 1989. Tropical fish. *World Aquaculture*, **20** (3): 24–41.

Wohlfarth G. W., Schroeder G. L., 1979. Use of manure in fish farming—a review. *Agricultural Wastes*: 279–299.

Wurtz-Arlet J., 1980. La fertilisation des étangs. In: *La pisciculture in étang*, R. Billard (ed.). INRA Publ., Paris: 99–106.

6

Ongrowing fish in intensive systems

1. GENERAL

It has become standard terminology in fish farming to speak of nursery stages for fish leaving the hatchery, weighing anything from a few hundred milligrams to several grams, and to reserve the term 'on growing' (or 'grow out') for fish of over 5–10 g (or even 50 g) in weight. Sometimes the nursery stage finishes at a particular stage of the fish's life cycle (transfer of salmonids from fresh water to sea water, transfer of marine fish from tanks to cages) rather than at a fixed weight; many variations are possible.

In nurseries, all juveniles become adapted to an inert diet in the form of granules (dry compound diet) or a moist or semi-moist diet. In practice, the weight of individual fish leaving the hatchery is never less than 300 mg (30 mm length) for marine fish and 500–1000 mg for salmonids, for example. Specialist articles can be found in the work on aquaculture edited by Barnabé (1990) and that edited by Laird & Needham (1988); these cover the procedures for the different farmed species.

In this type of rearing unit water provides the physical support for the fish, supplies oxygen and carries away wastes (CO_2, NH_4, urine and faeces). All the food is in the form of dry compound diet, the basis of which is fishmeal (Europe), or fish from industrial fisheries (Japan). The biological production of the environment plays no part in the rearing process.

The general objectives of this type of aquaculture have been defined in Chapters 1 and 2 of this part: some specific biological information can be added to this.

1. The market requires fish of an appropriate size (e.g. portion-sized trout). Producers may even have to display their fish in a special way, for example in Italy where they are bent back; this provides a gauge of their freshness. Sometimes the fish are sold alive (trout in Europe, shrimps in Japan, etc.), and sometimes as fillets (American catfish). Producing fillets leads to certain constraints: the fish should not be too fatty or stringy and be without intramuscular bones.

2. The market for aquatic products is often a temporary one (fish eaten on Friday, at certain feast days, shellfish at Christmas in certain countries). In theory, aquaculture is better able to respond to these demands than fishing but unfortunately the biology of the fish intervenes: smoltification and sexual maturation of salmonids and the breeding season for bivalve molluscs have consequences for the taste and flesh quality of the fish.

Trout reared in sea water must be sold before summer in places where water temperatures are likely to exceed 20°C, which would initiate mortalities. In spring, all farms put produce on the market at the same time, leading to a drop in prices or the need to freeze fish which are sold later but again at a lower price. All overproduction leads to a drop in price and, while the consumer gains in the short term, the viability of many farms is threatened. This has happened for salmon where the price fell from £6 to £3 per kilogram in 1989 because of the Norwegian overproduction which that year reached 117 000 tonnes.

2. RELATIONSHIP BETWEEN FISH AND THE ENVIRONMENT IN INTENSIVE REARING SYSTEMS

Intensive fish culture, whether of marine or fresh-water fish, is limited by two basic problems (once the availability of individuals for culture has been established): the maintenance of adequate environmental conditions, associated with the level of dissolved oxygen and water quality, and the provision of a high quality diet which is given in predetermined quantities at precise intervals. Here we shall not consider the problems associated with fish health in such rearing operations; these occur all too frequently in intensive rearing because of the density of the animals and are discussed in Part V of this book.

2.1 Temperature
We have already discussed the problems of balance between the temperature requirements of a species and the temperatures occurring naturally on typical farms (see Introduction, Part I, Chapter 2; this part, Chapter 2) and shown that solving the problems may be fundamental to the success of the operation. This is particularly true in intensive fish culture. There is no space here to list all the effects of temperature; and other than the examples described below, temperature has a role in maturation, the development of epizootics and feeding, etc., as can be seen throughout all the chapters in this part.

Table 1, adapted from Petit (1990) gives the optimum temperature values for several farmed species. The effect of temperature on the growth of trout has been shown in Fig. 1 of Chapter 2 in this part, as has the protein requirement (Table 2, Chapter 2, this part).

The influence of water temperature on growth is reflected in growth differences at different latitudes; this has been demonstrated in sea bass farms (Divanach, pers. comm. 1990). He has produced the growth curves shown in Fig. 1 from available information including his own results. A further effect of temperature concerns the levels of dissolved oxygen (see below).

2.2 Dissolved oxygen
The problems associated with nutrition have already been discussed in Chapter 2 of this part and will not be repeated here. However, they bear a relationship to the environment because of the fact that the process of digestion requires oxygen which must be extracted from the water. The quantity of oxygen dissolved in water varies in relation to temperature; dissolved oxygen levels drop when temperature and salinity increase (Chapter 1, Part I). Basal metabolism increases constantly with temperature, while maintenance

Table 1. Optimum and lethal temperatures for several farmed species (in part
from Petit 1990)

Species	Optimum for growth	Optimum for food conversion	Maximum lethal	Minimum lethal
Salmo trutta (Brown trout)	12–14	8–10	27	—
Oncorhynchus mykiss (Rainbow trout)	14–15	—	23	—
Salmo salar (Atlantic salmon)	15	—	28	—
Oncorhynchus nerka (Sockeye salmon)	15	12	—	—
Dicentrarchus labrax (Sea bass)	23–25	22	30–32	0
Sparus aurata (Gilthead sea bream)	25–26			5
Solea solea (Sole)	15	—	—	—
Setta maximus (Turbot)	19	16	—	—
Pleuronectes platessa (Plaice)	15	16–18	—	—
Anguilla anguilla (European eel)	22–26	—	33	—
Anguilla japonica (Japanese eel)	24–28	—	34	—
Cyprinus carpio (Common carp)	25–30	—	38	—
Mugil cephalus (Mullet)	28	—	—	4
Sarotherodon sp. (Tilapia)	28–30	—	38–42	5–11

requirements and those of growth only increase to an optimum temperature, above which
they drop rapidly (Fig. 2). In practice, the quantity of oxygen transported by the blood is
insufficient to support both basal metabolism and digestion. Table 3 in Chapter 2 (this
part) shows the influence of the oxygen level on protein metabolism.

The amount of oxygen in water is expressed in milligrams or millilitres; it is possible
to convert from one measurement to the other using the equation :

$$1.428 \text{ mg} = 1 \text{ ml}$$
$$0.7 \text{ ml} \quad = 1 \text{ mg}$$

For example, a trout weighing 250 g has a requirement for 50 mg O_2 h^{-1}. A litre of water
at 15°C contains 10 mg of O_2 which should allow the fish to survive for 12 min. In
practice the fish will die after 5–6 min because below 5.5 mg l^{-1} the fish is stressed and
below 3.3 mg l^{-1} it is unable to extract oxygen from the water. Thus, even though the
water contains 10 mg O_2 l^{-1}, the trout is only able to utilize 5 mg l^{-1}. The fish's oxygen
consumption is therefore distinct from the overall oxygen concentration in the water. The
concentration varies with temperature, and depends on whether aerators are being used or
liquid oxygen injected. Aeration or oxygenation increase the holding capacity of the fish
farm; they may only be used for a limited time in warm weather.

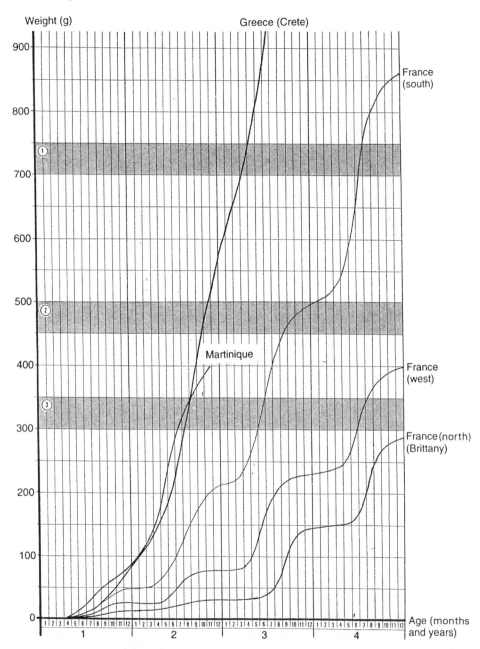

Fig. 1. Growth curves for farmed sea bass (Divanach 1990 pers. comm.). 1, 2 and 3 represent the best weight for market.

Digestion uses up a great deal of oxygen (220–300 g of oxygen are required for the digestion of 1 kg of dry diet by trout or sea bass). This leads to a marked drop in

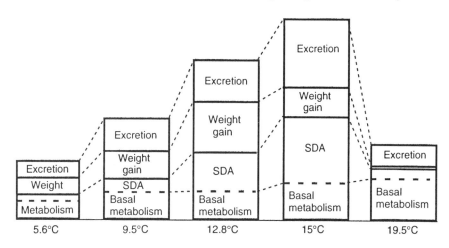

Fig. 2. Consumption and utilization of food by trout SDA = Specific Dynamc Action.

dissolved oxygen after meals on fish farms (Fig. 3). Every species has its own character-istic minimum level of dissolved oxygen; the knowledge of this level is fundamental to the success of rearing operations. In addition to increasing production capacity in aqua-culture, oxygenation can be justified for increasing feeding efficiency (conversion rate): Petit (pers. comm.) showed that the capital costs and the running costs of an aeration system could be recouped in a few months in trout farms where oxygen was the limiting factor because of savings through improved food conversion.

Petit (1990) made a detailed examination of all the interactions; these have been used in Fig. 4 which is also based on information from Liao (1971) and shows the relationship between temperature and oxygen for trout of varying size (we have seen that juveniles have relatively higher oxygen requirements than adults, Chapter 2, this part). In fish farms where water is reused, the young fish are always the first users of the water. The capacity of a fish farm using a given volume of water with a fixed amount of oxygen will vary depending on the size of the individual fish being farmed. There are other factors to be taken into account: increasing the speed of water through the tanks or ponds increases the oxygen consumption.

In practice, for salmonid culture, the water supply should have a mean dissolved oxygen level of 9 mg l^{-1}; the level should never fall below 7 mg l^{-1}. These oxygen requirements mean that it is essential to have a totally reliable supply of water; the level of reliability has been compared to that of a nuclear reactor (Priede & Secombes 1988). Because of this, gravity supply is preferable to the use of pumps which are susceptible to breakdown.

2.3 Ammonia and other limiting physical and chemical factors
There are other problems concerned with water quality besides the availability and the use of oxygen for the fish farmer. These include the level of ammonia, nitrites, carbon dioxide gas and dissolved organic matter which may vary between sites and water sources. Normally, when oxygenation is satisfactory, there are no problems with

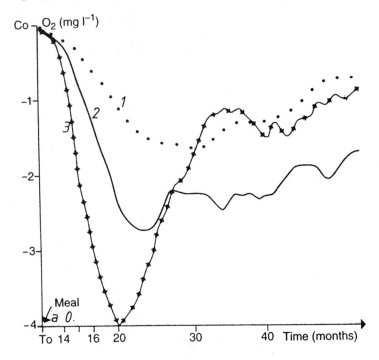

Fig. 3. Oxygen consumption: fluctuations in the level in a tank after successive meals 1, 2 and 3
(after Petit).

ammonia or other toxic compounds. However, when water is reused (passing through
several tanks or ponds in succession) or recycled through biological filters, there may be
problems; the complex interactions were analysed by Petit (1990).

One of the major problems limiting the expansion of fresh-water fish farming is that of
water supply to earth pond systems. Much of the problem comes from the fact that
ammonium hydroxide, which is highly toxic to fish, is the main form in which nitrogen is
excreted by fish. Thus autopollution occurs where water turnover is low or recycling is
inefficient.

Ammonium hydroxide (NH_4OH) dissociates in water forming the ammonium ion
(NH_4^+) and dissolved ammonia gas (NH_3) according to the following chemical equilibrium:

$$\underline{NH_3} + H_2O \leftrightarrow \underline{NH_4OH} \leftrightarrow NH_4^+ + OH^-$$

(Toxic forms are underlined)

Ammonium hydroxide is rare in the natural environment as it is only a transitory step in
the degradation of organic nitrogenous compounds, oxidized to nitrites and then nitrates
(see the nitrogen cycle, Alzieu 1990). These nitrates are utilized by plants and constitute
the starting point for living matter in aquatic ecosystems; production is limited by their
absence (Part I). However, in intensive fish culture ammoniacal toxicity may occur at

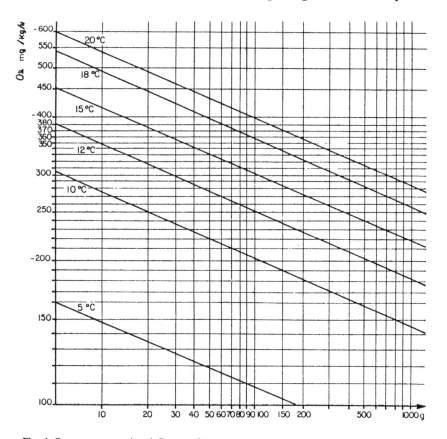

Fig. 4. Oxygen consumption: influence of temperature and weight of the individual, according
to Liao for trout.

what appear to be low concentrations (Table 2). Unionized ammonia, even at levels below those considered to be toxic, may have an adverse effect on food conversion (see Table 4, Chapter 2, this part).

The above equilibrium is pH dependent; Table 3 shows the dangerous effects of high pH. We have seen (Chapter 5, this part) that at pH 9 and above, because of the ionization of ammonia and the direct effects of pH, all animals die in sea water. This also happens in fresh water when the pH reaches or exceeds 10. This type of problem is unusual in fish aquaculture as recycling tends to have the effect of decreasing pH. Solbé (1988) produced a grid which helps calculate levels of toxic ammonia at different pH and temperature values (see Table 3).

Different species have differing susceptibilities to ammonia and these affect the technology of rearing (see Barnabé 1990 for details of the culture of a wide range of species; Laird & Needham 1988 for salmonids). Marine fish appear to be less sensitive to ammonia than fresh-water fish (for salmonids the critical level of non-ionized ammonia is 0.002 mg l^{-1}, other fresh-water species, 0.01 mg l^{-1}; and for marine species, 0.05 mg l^{-1}).

Table 2. Maximum levels of ammonia for a long duration exposure without affecting growth and conversion index

Species	Ammonia level in weight of nitrogen N–NH$_3$ (mg/l)	Authors (in Wickins 1980)
Oncorhynchus tshawytsha	0.06	Robinson-Wilson & Seim 1975
Oncorhynchus mykiss	0.4–0.14 after acclimatation	Schultre Wiehenbrauck 1976
Catfish	0.01–0.06	Robinette 1976
Sole	0.06	Alderson 1979
Turbot	0.11	Alderson 1979
Prawn	0.09	Wickens 1976
(*Macrobrachium rosenbergii*)	0.07–0.18	Aquacop 1977
Lobster (larva 4th stage) (*Homarus americanus*)	0.14	Deslestraty *et al.* 1977
Oyster (*Crassostrea virginia*)	0.06	Hartman *et al.* 1973

Table 3. Quantity of ammoniacal nitrogen giving a concentration of non-ionized ammoniacal nitrogen of 0.1 mg l^{-1}

pH	Fresh water			Sea water		
	10°C	15°C	20°C	10°C	15°C	20°C
6.5	170	116	80	217	147	102
7	54	37	25	69	47	32
7.5	17	12	8	22	15	10
8	5	4	3	7	4.8	3.3

There are other water quality factors which are important in aquaculture. These include suspended particles, low levels of metal ions and pesticides. Table 4 (from Billard 1989) shows the required composition of water for salmonid farming. This author also listed the levels of micropollutants which can be tolerated by farmed salmonids. Fish farms are extremely vulnerable to industrial and agricultural pollution.

2.4 Rearing density, flow rate and structures used in aquaculture

Intensive aquaculture is characterized by high biomasses per cubic metre (sometimes as high as 200 kg m^{-3} in silos or even 400 kg m^{-3} for eels). Such densities would never be

Table 4. Composition of water for salmonid rearing
(values mg l^{-1}) (Billard 1989)

Dissolved oxygen	> 70% saturation[*]
CO_2	0–10
Total alkalinity ($CaCO_3$)	10–400
pH	6.6–8
Calcium (Ca)	4–150
Manganese	0–0.01
Total iron	0–0.15
• ferrous ions	0
• ferric ions	0.5
Phosphorus	0.01–3.0
Nitrates	0–3.0
Nitrites	
• Ca > 50	< 0.2
• Ca < 50	< 0.1
Ammonia (NH_3)	< 0.020
Nitrogen (% saturation)	< 102–105
Suspended solids	< 25
BOD	< 10
COD	< 30

[*] A lower level of saturation may be compensated by an increased
flow of water. However, the level of dissolved oxygen should not be
below 5.5 mg l^{-1} for ongrowing and 7 mg l^{-1} for eggs.

found in the wild. The effect of density on food conversion is shown in Table 5 in
Chapter 2 of this part. This crowding favours the spread of infectious diseases; rearing
conditions are a compromise between biological and economic requirements; densities
are usually much lower than those above. Densities are low at the start of the rearing of
sea bass (1–3 kg m^{-3}) but may reach 30 kg m^{-3} by the end of the ongrowing stage
because of rapid growth. For most of the time densities are of the order of 10–15 kg m^{-3}.
In order to avoid frequent splitting of the stocks in tanks or cages, farmers calculate the
likely final biomass when carrying out the initial stocking. For many species, maximum
stocking densities are around 25 kg m^{-3}, although this may be raised as high as 40 kg m^{-3}
for sea bass in heated water or even to 50 for trout (at low temperatures or with oxygen
injection). However, Scottish salmon farmers are tending to reduce densities to a maximum of 10 kg m^{-3} and achieving better growth, food conversion and survival rates than
at the higher densities used previously (Laird pers. comm. 1992).

While the biomass of fish in a rearing unit is an important parameter, it should only be
interpreted in relation to the water turnover rate. The incoming water brings oxygen;
wastes are removed with outflowing water. There are two ways of considering the rate of
water flow: in cubic metres per hour ($m^3 h^{-1}$) or cubic metres per hour per tank (or other
rearing unit) ($m^3 h^{-1} m^{-3}$) which indicates the flow rate in relation to biomass.

Dosdat (1984) calculated flow rate in sea bass ongrowing units at a temperature of 20–25°C; at this stage oxygen consumption is 0.177 ± 0.035 mg O_2 g^{-1} h^{-1} and it is necessary to maintain a minimum concentration of 3 mg O_2 l^{-1} for this species. At a density of 30 kg m^{-3}, a relationship between biomass and flow rate of 1.67 kg m^{-3} per m^3 h^{-1} was established, although for this, supplementary oxygenation was required in summer (when the temperature reached 25°C), even when the relationship between density and flow rate is as low as 1 kg m^{-3} per m^3 h^{-1}. In practice the water turnover rate is between 0.5 and 4 times per hour (mean once per hour) for stocking densities between 1 and 15 kg m^{-3}.

The maximum biomass which can be supported in a fish farm where water is drawn from a stream or river should be calculated based on the lowest flow rate in the source. Other information to be taken into account includes mean temperature and oxygen levels. With a knowledge of the annual temperature and the growth rate of the farmed species it should be possible to calculate the potential annual production. It is always recommended to allow wide margins of error on the results from this calculation.

The main difference between the nursery and the ongrowing stages centres on the sensitivity and fragility of the larvae to biological and abiotic aspects of the environment. There is a well-defined gradient from larval to adult stages; fish become increasingly robust up to the time of sexual maturation, then once more they become fragile. When European bass are reared in the tropics, up to 90% of juveniles die before they are one year old at temperatures above 30°C but older individuals, stocked at high density in cages, survive without problem (Barnabé & LeCoz 1987).

The tanks used for intensive ongrowing are elongated rectangles known as raceways. Water enters at one extremity and leaves at the other through a grid and then an overflow system. Raceways can be anything between 3 and 30 m long and up to 80 m^3 in volume. They may be constructed from concrete or a plastic-type material and can be emptied completely. Some of them have a central wall giving them the appearance of a racetrack; others, used in the early stages of salmon rearing, are circular. Billard (1990), Brunel & Fuzeau (1990), Needham (1988) and Laird & Needham (1988) have described and illustrated this type of tank-based rearing system and the associated operating techniques. There are no fundamental differences between the rearing of fresh-water and salt-water fish in these systems (salmon, sea bass, eels).

Cages are seldom used for the nursery stages of fish culture because of the need to use small-mesh nets which become quickly fouled. However, they have been used successfully where the environmental and operating conditions are compatible with the survival of juveniles (frequent food distribution, little variation in environmental variables such as temperature and currents). It is likely that future developments will make the use of nursery cages more attractive and that they will replace land-based tanks.

True ongrowing is carried out either in tanks or cages depending on the characteristics of the site. Aquaculture is often limited by the availability of sites; there are few land-based sites where sufficient quantities of water are available. Aquaculture is therefore moving to natural or artificial waters such as reservoirs, estuaries, fjords, outer harbour areas, sheltered coastal areas and especially towards the open sea which covers the largest proportion of the planet's surface.

Hydrological factors as well as feeding, growth and factors affecting survival are similar, whether ongrowing takes place in tanks or cages. However, the comparison between small-scale trout rearing in cages of 600 m^3 volume in sheltered waters and large volume (3500 m^3) offshore cages where food is distributed automatically has shown that the fish reared in exposed offshore waters are in better conditions than those in the sheltered waters, as measured by several indices and a halved mortality rate (Sveäl 1988).

Wastes are eliminated from cages either by sedimentation (particulate matter) or diffusion in water (dissolved substances). The idea of flow rate is difficult to apply to cages but is no less important. Reports on the experimental rearing of sea bream at high densities (60 kg m^{-3}) in Israel suggest that swimming movements have the effect of creating current, causing the transfer of water from the outside of the cage to the inside and vice versa. However, these high-density experiments were carried out in very small cages, only one cubic metre in volume. Such an effect undoubtedly occurs in larger cages and at lower densities (10–20 kg m^3) but probably to a lesser extent. However, in the former Yugoslavia, mortalities have occurred in sea bass cages because of the crowding of cages and the lack of water turnover in the rearing site (inshore end of a marine bay). Submerged propellers which can induce water circulation and counteract these problems are commercially available and are used in salmonid culture in fjords where there is poor water exchange. Natural currents may sometimes be sufficient but must not be so strong as to deform the cage. Currents may be strong when the tide is ebbing or flowing but absent around high and low water.

The procedures for rearing salmonids in sea-water cages are described by Laird & Needham (1990) and are also applicable to other species of marine fish which have been reared in this way for over ten years (Lucet *et al.* 1984; Berg 1985). The technology of cages and their operation has been reviewed by Beveridge (1987) and many models suitable for use in the open sea are now being developed. There have been two projects for offshore rearing in the Mediterranean; in one it is proposed to rear 800 tonnes of sea bass in a converted ship anchored off the coast of Monaco and another, off the Costa del Sol, intends using a converted oil platform. A similar scheme in the Balearics in 1988 failed for financial reasons.

The relative merits of cages and tanks have been discussed elsewhere (Barnabé 1990); the general trend is for tanks (which cost more to buy, have higher manpower requirements and are totally dependent on a reliable supply of water) to be used for the early stages of rearing where large numbers of individuals can be kept in small volumes of water because individuals are small, and cages to be used for older fish. There are exceptions to this; in Galicia turbot are ongrown in tanks because of problems of theft from cages! However, the rapid expansion of salmon farming in northern Europe and sea bass farming in the Mediterranean has only been possible because of the use of cages. The warm waste water (produced, for example, from power stations) has been used for aquaculture; the advantages and disadvantages have been discussed by Brunel & Fuzeau (1990).

There are some methods used to keep fish other than land-based tanks and cages in the natural environment; these include areas separated off by nets and enclosed tanks fixed in natural water bodies, fed by pumped water (Barnabé 1979, 1990). There are other types of rearing system tailor made for sites which are not suited to tanks or cages.

3. GROWTH

3.1 Assessment of growth from the measurement of weight and length

Fish are weighed and measured as a way of assessing their growth rate and condition. Fish are sold by wet weight, usually of the whole animal but sometimes gutted. There are several standard measurements used to describe fish length. In fish farming total length, fork length (distance from the snout to the centre of the concave V in the caudal fin) and standard length are used (Fig. 5). Sometimes, however, the extremity of fins becomes worn and it is difficult to locate the end of the caudal peduncle and in some species the caudal fin does not have a pronounced fork.

Other measurements (Fig. 5) are used to calculate biometric indices which indicate the relative proportions of the body. The body shape of captive fish is different from that of wild ones; in sea bass, *Dicentrarchus labrax*, the index expressing the relationship

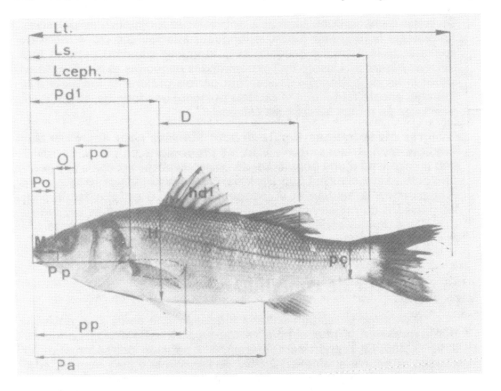

Fig. 5. Principal dimensions used to measure size and calculate the biometric indices in fish. Lt = total length, Ls = standard length, Lceph = head length, Pd[1] = pre-dorsal length, D = dorsal length, Po = pre-orbital length, O = orbital diameter, po = post-orbital length, M = superior maxillary length, Pp = pre-pectoral length, pp = 'post pectoral' length, Pp–pp = pectoral length, Pa = preanal length, H = body height, Pc = height of the caudal perduncle, hd[1] = maximal height of the first dorsal fin.

between the body depth, H and the length, L_t ($100H/L_t$) is lower for wild fish than for farmed fish. Individuals of the same species but from a different geographical area have different biometric indices.

3.2 Condition factor

This is an expression of the relationship between fish length and weight. The simplest equation is of the form $K = weight (g)/length (cm)^3 \times 100$. For an elongated thin fish K is below 1; for an average, well-fed fish 1.2; and for a female fish near to spawning, 1.4 or above. This index is only valid for fish which have the distinctive morphology of the adult stage. The use of the condition factor for comparative purposes is subject to several theoretical reservations which have been listed by Barnabé (1976).

3.3 Growth

Growth curves have a continually decreasing slope (except during the larval stage). The fish farmer needs to concentrate on the most steeply sloping part of the curve. Fig. 6 illustrates the growth of male and female sea bass (*Dicentrarchus labrax*) in the natural environment around Sète. The advent of sexual maturity is marked by a decrease in growth rate. This is a phenomenon well known to fish farmers and explains why there is a preference for farming female fish which have a better growth rate. All-female stocks of fish can be obtained by genetic manipulations, hormone treatment or environmental modification. Sea bass, *Dicentrarchus labrax,* and mullet do not mature in water of reduced salinity. Non-maturing fish concentrate all energy (and therefore weight gain) in somatic tissue; none is diverted to gonad; there is therefore an obvious advantage to farming in waters of low salinity.

Fig. 1 shows both growth differences between fish reared in different places and also seasonal differences. It can be seen that the winter growth check (linked to low water temperatures) causes the lower growth rates in northern farms. The slope of the curve during the active growth season is similar at all latitudes.

3.4 Coefficient of growth

We shall follow the example of Priede & Secombes (1988) and use the expression 'specific growth rate' to characterize the growth of fish in farms. This is defined as the daily gain in weight of fish expressed as a percentage of whole body weight (percentage daily growth rate). Fig. 7, adapted from Priede & Secombes (1988), illustrates the theoretical growth of a species with a constant specific growth rate. A specific growth rate of 2% per day only results in a daily gain of 0.2 g for a 10 g fish but results in a gain of 10 g for a 500 g fish. In compensation for this, the exponential part of the true growth curve is short and occurs during the juvenile stages.

Priede & Secombes give a simplified formula with which to calculate specific growth rate (G) over a period of several weeks, using logarithms. This assumes that there is little variation in G over the study period.

$$G = \ln \frac{W_t}{t} - \ln W_0 \times 100$$

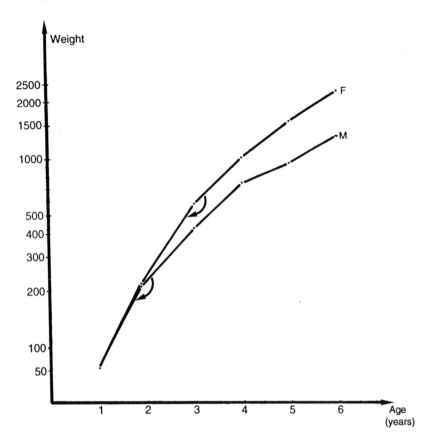

Fig. 6. Growth in body weight for males (M) and females (F) (semi-logarithmic coordinates). Arrows indicate the change in the growth of body weight with the advent of sexual maturity.

where W_t is the weight of the fish after t days; W_0 is the initial weight of the fish; ln is the natural logarithm; and t is number of days over which growth has taken place.

Other growth models have been suggested. Most of these are extremely complex and not suitable for use in aquaculture, as they require the measurement of a range of parameters such as temperature, food conversion rate or the quantity of oxygen consumed per kilogram of food ingested. These are not fixed values, but vary in relation to the behaviour of the fish which itself changes depending on the weather (sunny or cloudy, wind or rain) and with variations in the raw materials used in making up the diet (Fontaine 1990).

3.5 Heterogeneity of growth

Apart from the problems associated with the sex of the fish and sexual maturation, fish farmers, however competent, must cope with the fact that individuals in a stock of fish do not all grow at the same rate; this results in a range of sizes within that stock.

For this reason, grading is carried out periodically in fish farms, but, in view of the handling involved, perhaps not as frequently as necessary. There are always some

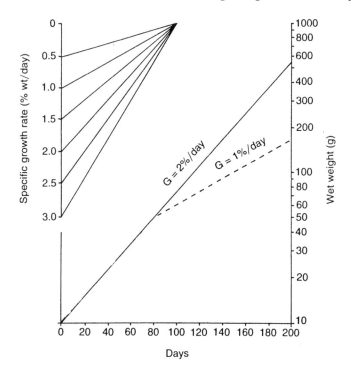

Fig. 7. Experimentally derived fish growth curves where the specific growth rate (*G*) is 2% (solid line) and 1% (dashed line), with semi logarithmic coordinates. Growth at a constant rate is shown as a straight line. Comparison with Fig. 6, also on semi-log coordinates, shows that real growth rates are not constant over long periods. At the top left of the figure are the slopes for several specific growth rates between 0.5% and 3% (adapted from Priede & Secombes 1988).

mortalities associated with grading in marine fish; some farmers grade when fish leave the hatchery and hardly at all during the ongrowing stage. When grading is carried out, fish are separated into three groups: the largest, fastest growing individuals, a median group and a 'tail' of small individuals. Fig. 8 demonstrates the spread in size of fish using data from a farm; however, there may be a much greater range of sizes. The problem is particularly acute in eel farming where, after a few months of culture, individuals weighing only a few grams may be found with others weighing over 150 g.

4. SURVIVAL

The survival rate of fresh-water and marine fish during the ongrowing stage is usually very high, compared with that in the wild.

Figures of 1–2% mortality are usual for standard farming of rainbow trout, unless there are accidental mortalities caused by pollution or pathogenic disease. In salmonids, smoltification (the process of morphological and physiological change associated with the passage from fresh to sea water) is often accompanied by significant mortalities. Normal mortalities have been estimated at 15% from weaning to smoltification and 20%

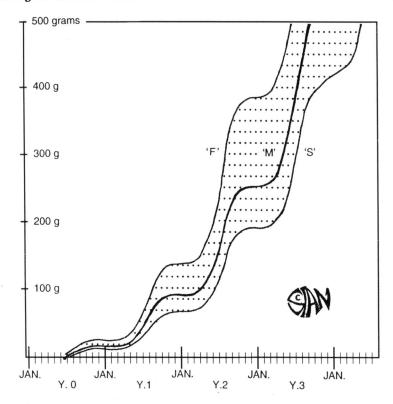

Fig. 8. Predicted growth curves for the growth of sea bass (*Dicentrarchus labrax*) on the Côte Basque (S.I.A.N., pers. comm.). F = fastest growers, M = mean growers, S = slowest growers.

for the rest of the rearing cycle. However, incorrect selection of smolts and poor handling can result in higher rates of mortality.

There is little precise information on mortality rates for marine fish: mortalities in the larval stage (in the hatchery) are high (Chapter 4, this part). The quality of the fry (swim bladder present, no dietary problems) is the determining factor. In 1983, some farmers achieved a survival rate of 70%. Conditions have been improved and survival rates in tanks and cages are now said to exceed 90%. However, accurate figures are unusual and some existing information is suspect. The poorest survival figures are often associated with less than ideal sites such as those with high summer temperatures in the tropics or Mediterranean or low winter temperatures in more northern latitudes.

Improvement in the standard of dry diets, particularly in their level of vitamins and trace elements, refinement of rearing conditions and procedures mean that the standards are now often higher than those in trout farming.

5. DIETARY PARAMETERS AND POLLUTION

In Chapter 2 of this part we have shown the role of dietary factors and their interaction with other environmental parameters. Nutritional requirements, frequency of meals, amount of food distributed daily and conversion efficiency vary between sites and species being farmed; precise details for different species can be found in works previously mentioned.

Tables 5 and 6 are based on feeding tables for rainbow trout and sea bass provided by two different feed manufacturers. The number of meals is not specified although the youngest stages should be fed many times (up to a dozen) each day and older fish less frequently (one to three times). Although automatic feeders are in common use, many smaller farmers and (increasingly with salmon) some larger farmers prefer to feed by hand as a better method of meeting the exact feed requirements of the stock. Hand feeding allows visual observation of the wellbeing of the stock.

Demand feeders, operated by the fish themselves, are not so widely used in intensive aquaculture as operators believe that this feeding method leads to major differences in growth rates and, therefore, a wide range of sizes. The advantages of demand feeders are that they are simple, can be large in volume and function without energy. At the other extreme there are automatic feeders controlled by computers and driven by compressed air which blow dry pelleted diets in the required quantity from large storage silos into the cage. There is, therefore, a range of types of feeders suitable for the needs of individual farmers.

In Chapter 2 of this part some information on pollution caused by aquaculture was given. Although there is no need to repeat this information here, it is interesting to compare the effluent from two types of aquaculture; one using the food resources of the site (mussels on a long line); the other fed on imported food (exogenous). Figs 9 and 10 (adapted from Folke & Kautsky 1989) show the fate of nitrogen and phosphorus and demonstrates that in both cases only 25% of the available nitrogen is retained while the rest is released into the environment, either in dissolved or particulate form (sedimentation under mussel ropes is three times greater than natural sedimentation; under salmon cages it can be 20 times higher). These figures clearly demonstrate that when nutrients are added to the system from outside (as with feeding fish in cages on pelleted diets) there is likely to be environmental pollution. However, the figures (for both examples) also demonstrate the importance of the biomass of cultured animals in the process of sedimentation.

6. IMPROVEMENTS IN THE TECHNIQUES OF PRODUCTION

6.1 Changing sex through hormonal manipulation

Many farms have problems with precocious maturation of male fish which retards their growth rate, which is already slower than that of females. For this reason there are advantages in farming all-female stocks of fish.

One technique consists of the direct reversal of sex by bathing the fish in a solution of female hormone. However, this procedure poses potential problems for the consumer and is hardly used. Hunter *et al.* (1986) used two treatments, each of two hours' duration, for newly hatched Pacific salmon; the rate of feminization varies between 66 and 96% and there are hardly any problems of residual hormones.

The most frequently used method involves the treatment of the parents rather than that of the offspring. Males are genetically different from females at the chromosome level.

Table 5. Feeding plan for rainbow trout (SARB diet)

SARB diet used	Super Alevin No. 0	Super Alevin No. 1	Super Alevin No. 2 or Alevin No. 2	Super Alevin No. 3 or Alevin No. 3	Alevin No. 4 1.7 mm	(Fingerling) 2.5 mm	(Fingerling) 3.2 mm	Trout 4.5 mm	Broodfish 7 mm
Number of fish per kg	5000	5000–2000	2000–400	400–200	200–66	66–33	33–11	11 → harvest 90 → harvest	Trout destined as broodfish from 300 g
Weight of a single trout in g	0.2	0.2–0.5	0.5–2.5	2.5–5	5–15	15–30	30–90	90 → harvest	300 g
Trout length (cm)	2.5	2.5–3.5	3.5–6	6–8	8–11	11–14	14–20	20	
Number of meals day^{-1}	10	8	7	6	4	3	2	2	2

Daily ration in kg per 100 kg live weight in relation to water temperature (6 days per week)

Temperature in °C

Water									
Low temperatures	0–4°	3.0	2.8	2.6	2.3	2.0	1.6	1.1	1.0	0.5
	6°	3.4	3.2	3.0	2.7	2.3	1.8	1.2	1.1	0.6
Optimal temperatures	8°	3.9	3.7	3.5	3.1	2.7	2.0	1.4	1.3	0.7
	10°	4.5	4.2	3.8	3.2	3.0	2.3	1.6	1.4	0.8
	12°	5.1	4.7	4.1	3.7	3.3	2.6	1.8	1.6	0.9
	14°	5.7	5.1	4.6	4.2	3.6	2.9	2.1	1.8	1.0
	16°	6.5	5.6	5.2	4.7	4.0	3.3	2.4	2.1	1.1
High temperatures	18°	5.0	4.5	4.0	3.7	3.3	2.6	2.1	1.8	1.0
	20°	3.0	2.8	2.6	2.3	2.0	1.6	1.1	1.0	0.5
Quantity of feed per 100 000 fish from a good source and in normal conditions		5 kg	25 kg	200 kg	300 kg	1 200 kg	2 000 kg	9 000 kg	17 000 kg	

Table 6. Ration table: Service Technique Aqualim. Daily rations in % live weight

	Temperature (°C)							
	13.0	15.0	17.0	19.0	21.0	23.0	25.0	27.0
Alevin 1: Bass < 0.5 g				Fed *ad lib*				
Alevin 2: Bass 0.5–1 g				Fed *ad lib*				
Alevin 3: Bass 1–3 g	0.4	1.5	2.6	3.6	4.2	4.7	5.1	5.0
Alevin 4: Bass 3–8 g	0.4	1.4	2.4	3.3	3.9	4.4	4.7	4.6
Bass granules 1.5 mm Bass 8–15 g	0.4	1.4	2.3	3.2	3.8	4.2	4.5	4.4
Bass granules 2 mm Bass 15–35 g	0.3	1.2	2.0	2.8	3.3	3.7	4.0	3.9
Bass granules 3.2 mm Bass 35–100 g	0.2	1.0	1.7	2.4	2.8	3.1	3.4	3.2
Bass granules 4.5 mm Bass 100–350 g	0.2	0.8	1.3	1.8	2.1	2.3	2.5	2.4
Bass granules 6 mm Bass 350–1000 g	0.1	0.3	0.6	0.8	1.0	1.2	1.4	1.3

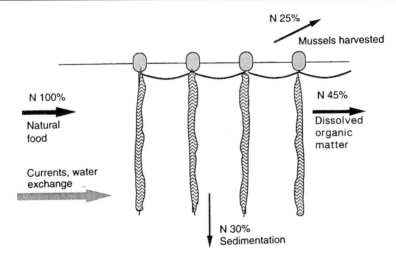

Fig. 9. Nitrogen transfer during mussel culture on long lines (after Folke & Kautsky 1989).

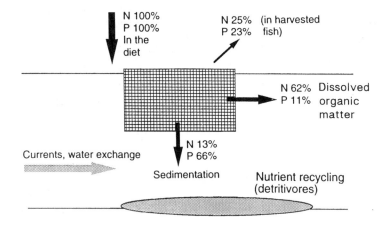

Fig. 10. Nitrogen and phosphorus transfer during a culture cycle in a salmonid cage (after Folke & Kautsky 1989).

Males are designated as having XY chromosomes and females as XX. Each sex produces a hormone which determines whether the germ cells develop into ovaries or testes and also influences the secondary sexual characteristics such as growth rate. It is possible to change a female (XX) into a male, capable of emitting sperm, but this sperm will only contain X chromosomes as the 'father' was genetically female. If a normal, genetically female ova (X) is fertilized by this sperm the resulting embryo is bound to be female (XX). The production of all-female eggs has become routine in trout farming.

Ingram (1988) described how the fish destined to be sex-reversed broodstock were fed on hormone enriched diet (3 mg kg^{-1}) from first feeding. The hormone is dissolved in ethyl alcohol and mixed with the food; the alcohol evaporates leaving the crumbs of diet coated in hormone. This treated food is fed exclusively for two months when the temperature is 10°C. The young fish are reared normally but away from the production fish in order to produce the eggs for all-female production fish.

6.2 Obtaining polyploid fish, gynogenesis

Triploid salmonids are sterile. They can be produced by heat shocking (26–32°C), cold shocking (1–1.5°C) newly fertilized eggs for several minutes or pressure shocking newly fertilized eggs. Pressure shocking is used routinely in salmon farms in Scotland because of the great benefits of producing large (4 kg+) sterile fish. Pressure shocking the ova results in the retention of the second polar body. The ova is therefore diploid (two sets of chromosomes) rather than haploid (one set of chromosomes) as is normal. Ideally, sperm from feminized male fish should be used to fertilize these pressure-shocked ova. The resulting all-female triploid salmon do not develop secondary sexual characteristics: although male triploids fail to complete sexual maturation, they show secondary sexual characteristics such as changes in body shape and colour.

Colchicine and cytochalasin B have also been used to induce triploidy in various species experimentally.

7. CONCLUSIONS

Intensive aquaculture is regulated by factors linked to sites. The availability of a supply of oxygen-rich water with no industrial or agricultural pollution is critical to the success of the operation. These characteristics are often variable and cannot be totally changed by the farmer, but may be altered by the use of equipment such as oxygenators, recycling systems and by heating. There is seldom enough information available on the biology of farmed species and research has to be carried out to find, for example, ideal rearing temperatures. The most important factor is that farmers should respond to the demands of the market. The food given to the fish is a further factor which determines the success of a rearing operation. It is difficult for the farmer to influence the quality of the food that is manufactured industrially. Only the label on the bag of food gives any form of guarantee of its contents.

From the ecological perspective, intensive fish culture is only a means of transforming one type of protein into another and is likely to involve pollution (see Chapter 2, this part). However, this type of aquaculture is becoming a major economic force in some parts of Europe, generating many jobs.

The 1980s saw the expansion of salmonid culture; the 1990s are going to be the years of sea bass and also sea bream in the south and east of the Mediterranean (Greece, Turkey and the former Yugoslavia) as well as the French Côte d'Azur and in Spain where turbot rearing is becoming important. Temperatures in the open sea are more favourable than those in coastal lagoons (minimum winter temperature 12–15°C in the sea), and the demands of a European market of 300 million people mean that there are good prospects for the more favoured species.

The 1992 predictions for the Mediterranean regions are for 27 000 tonnes of sea bass and sea bream, only a sixth of this figure will come from intensive production. However, in many countries, including France (350–400 tonnes in 1989), production is doubling every year. It is likely that production from aquaculture will soon exceed that of the capture fishery, as with salmon and yellowtails.

This explosion in production is not without its limits. The development and growth in size of offshore structures is a technological feat which shows that the colonization of the open sea with structures used in rearing is possible and has immense potential. However, the true limit is food supply; most farmed fish are fed on products derived from commercial fisheries (fish meal from species which are not consumed directly by man). Such species (capelin, sand eels) are already overfished in Northern Europe and catches are declining. Intensive fish farming is thus closely dependent on fishing.

For example, in order to produce the 80 000 tonnes of salmon farmed in Norway in 1988, 430 000 tonnes of fish from capture fisheries are used; this represents 15% of the primary production of the fishing grounds of the North Sea (Folke & Kautsky 1989). The overfishing of small species to feed farmed fish may also affect the stocks of their predators (cod, tuna); this demonstrates how these activities are all dependent on the same natural resources.

Despite this inter-relationship, all is not lost for aquaculture. There is a high market demand for fresh, luxury products leading to high prices. The products from extensive aquaculture can have 'value added'. It may also be possible to make increased use of products from the terrestrial ecosystem in fish feed (soya derivatives for example). One further technical problem of the development of fish culture is the increased pollution of the aquatic environment; the elimination of such pollution must be one of the objects of aquaculture development (Part I, Chapter 2).

At present, market demand is the major factor determining the expansion of fish production rather than the capacity of the ecosystem to supply food. In spite of the recent development of ecological conscience, there is little chance of a change in the economic perspective; extensive aquaculture also has its limits. Intensive culture will thus be constrained by events and remain independent of the production of natural ecosystems.

REFERENCES

Alzieu C., 1990. L'eau milieu de culture. In: *Aquaculture*, G. Barnabé (ed.). Tec. & Doc. Publ., Paris: 17–45.

Barnabé G., 1976. Contribution à la connaissance de la biologie du Loup *Dicentrarchus labrax* (L.) Poisson Serranidae. Thèse Doct. Etat, mention Sciences, Univ. Sc. Techn. Languedoc, Montpellier: 426 pp.

Barnabé G., 1979. Système d'aquaculture modulaire intégré en zone portuaire. *Aquaculture*, **16**: 369–374.

Barnabé G. (ed.), 1989. *Aquaculture*, Tec. & Doc. (Lavoisier) Publ., Paris, 2 vols: 1308 pp.

Barnabé G., 1990. Cages or tanks? Reality and perspectives for the Mediterranean sea. In: *Mediterranean aquaculture*. E. Horwood Publ., Chichester: 51–68.

Barnabé G., Lecoz C., 1987. Large-scale cage rearing of the European Sea-Bass *Dicentrarchus labrax* (L.) in tropical waters. *Aquaculture*, **66**: 209–221.

Berg L., 1985. Yougoslavie; développement de la culture du Bar. Rapport préparé pour le project 'Développement de la culture du Bar'. FAO, Rome, Réf.: FI TCP/YUG 2301 (Ma), Working Document 1, March 1985, 1 microfiche.

Beveridge M., 1987. *Cage aquaculture. Fishing.* News Books Ltd, Farnham, England: 351 pp.

Billard R., 1990. La salmoniculture en eau douce. In: *Aquaculture*, G. Barnabé (ed.). Tec. & Doc. Lavoisier Publ., Paris: 569–613.

Brunel G., Fuzeau P., 1990. L'aquaculture en eaux réchauffées. In: *Aquaculture*, G. Barnabé (ed.), Vol. 2. Tec. & Doc. Lavoisier, Paris: 889–910.

Dosdat A., 1984. Schéma de production d'une ferme de Loups. In: *L'aquaculture du Bar et des Sparidés*, G. Barnabé & R. Billard (eds). INRA Publ.: 361–372.

Folke C., Kautsky N., 1989. The role of ecosystems for a sustainable development of aquaculture. *Ambio*, **18** (4): 234–243.

Fontaine M., 1990. Préface. In: *Aquaculture*, G. Barnabé (ed.), Vol. 2. Tec. & Doc., Lavoisier, Paris: IX–XII.

Hunter G., Solar I., Baker I., Donaldson E., 1986. Feminization of Coho salmon and chinook salmon by immersion of alevins in a solution of oestradiol-17β. *Aquaculture*, **53**: 295–302.

Ingram M., 1988. Farming rainbow trout in fresh water tanks and ponds. In: *Salmon and trout farming*, L. Laird & T. Needham (eds). E. Horwood Publ., Chichester: 155–189.

Laird L., Needham T. (eds), 1988. *Salmon and trout farming*. E. Horwood Publ., Chichester: 271 pp.

Laird L., Needham T., 1990. Aquaculture des salmionidés en eau de mer. In: *Aquaculture*, G. Barnabé (ed.). Tec. & Doc. Publ., Paris: 653–674.

Lucet P., Balma G., Bonfils J., 1984. Elevage de Loups en cage à poissons dans l'étang de Thau (Hérault). In: *L'aquaculture du Bar et des Sparidés*, G. Barnabé & R. Billard (eds). INRA Publ., Paris: 380–394.

Needham T., 1988. Salmon smolt production. In: *Salmon and trout farming*, L. Laird & T. Needham (eds), E. Horwood Publ., Chichester: 87–116.

Petit J., 1990. L'approvisionnement en eau, le traitement et le recyclage en aquaculture. In: *Aquaculture*, G. Barnabé (ed.). Tec. & Doc. Publ., Paris: 47–182.

Priede I., Secombes C., 1988. The biology of fish production. In: *Salmon and trout farming*, L. Laird & T. Needham (eds). E. Horwood Publ., Chichester: 32–68.

Solbé J., 1988. Water quality. In: *Salmon and trout farming*, L. Laird & T. Needham (eds). E. Horwood Publ., Chichester: 69–86.

Svéal T., 1988. Inshore versus offshore farming. *Aquac. Engineer.*, **7**: 279–287.

Wickens 1980.

Winfree R., 1989. Tropical fish. *World Aquaculture*, **20** (3): 24–30.

Part V
Aquaculture diseases

F. Baudin Laurencin
M. Vigneulle

1

Diseases in aquaculture operations

Diseases can be defined as any modification of the normal anatomy or function of an organism. The particular disease is characterized by its causes, mechanisms and symptoms. During the progress of the disease, the host's defence mechanisms are overcome, at least temporarily, by the action of a pathogenic factor. This is likely to be associated with particular environmental conditions.

These principles apply to all living things, including aquatic animals. However, disease in aquatic animals has certain characteristics which depend both on the nature of the pathogenic factors in the aquatic environment and the biology of the species. Wild stocks are not exempt from disease, whether they be primitive invertebrates such as sponges (Lauckner 1980) or fish which are the object of commercial fisheries (Sindermann 1970). Natural mortality, as estimated by population dynamicists, is at least in part the consequence of disease. The natural environment is modified in culture but the farmer must do as much as possible to satisfy the biological requirements of the species for optimum production. Today there are many different types of aquaculture: new or traditional, intensive or extensive, fresh-water or marine, invertebrates or fish, etc. There are therefore many different aspects of aquatic diseases, some of which are described in this chapter.

1. PATHOGENIC FACTORS

Some manifestations of disease do not appear to be linked to any external pathogen. For example, certain types of tumours (neoplasms), which may be found in individual fish, may be absent from the rest of the farmed stock. However, in some mollusc stocks a significant proportion may be affected by haemocyte sarcomas (Farley 1978). Many of the types of tumours found in mammals have been described from fish, where similar tissues or organs are found (Balouet 1986).

However, in most cases, disease can be linked to environmental factors either through the absence of the conditions necessary for normal survival, or where physical, chemical or biological factors cause stress to the animal.

1.1 Environmental deficiencies

Lack of oxygen in the water or badly formulated diet are the two most frequently encountered deficiencies causing disease. A phytoplankton bloom may bring about a sudden
drop in the level of oxygen followed rapidly by mass mortality (Erard-Ledden & Ryckaert
1990). More often, hypoxia is less severe, chronic and occurs at night when there is no
photosynthesis. It is accompanied by a loss of appetite, slowing down of growth rate and
an increased susceptibility to infection. Many types of dietary deficiency have been
reported; lack of essential fatty acids (for example, in rotifers, Le Milinaire *et al.* 1982)
and trace elements (imbalance in the availability of the divalent ions zinc–calcium,
causes cataracts in trout; Ketola 1979). Vitamin C deficiencies cause malformation and
fractures of vertebrae, and, particularly in turbot, granulomatous hypertyrosinaemia characterized by the presence of nodules in many tissues (Messager 1986).

1.2 Adverse physical and chemical factors

In trout and salmon, the transition from fresh to sea water is a time of high physiological
stress. For most fish species, transport and changes in temperature are stressful events.

Any handling—for example, changing the nets on cages—may result in small wounds
on the surface of the fish, which may develop into cutaneous ulcers and then become
secondarily infected. Defective equipment may result in gas supersaturation which causes
gas bubble disease (embolisms, particularly in gill tissue).

Autopollution, the breakdown of metabolic wastes produced by the fish (into ammonia
and nitrites), may become excessive (Peters *et al.* 1984). Suspended solids become attached to the gills which become abraded and irritated; this reduces the efficiency with
which oxygen is extracted from the water and may lead to asphyxiation. Ammonia alters
the chemical equilibrium of the blood, reduces the affinity of haemoglobin for oxygen
and increases the permeability of the gills, which has the effect of reducing the ability of
the fish to osmoregulate. The nitrite ion causes the transformation of haemoglobin to
methylhaemoglobin, a pigment which does not have the capacity to transport oxygen,
thus decreasing the respiratory capacity of the fish.

Pollution may also come from external sources in the form of pesticides, hydrocarbons, disinfectants, etc. (Aldrin 1987). The pathogenic effects depend on the nature of the
chemical and its concentration in the water. Examples include inhibition of cholinesterase
by organophosphorous compounds, impairment of the transport of ions across the gills
and lysis of tissues by various metals, particularly zinc and copper, and anaemia induced
by hydrocarbons.

1.3 Bioaggressors: viruses, bacteria and parasites

Viruses

Viruses are obligate parasites which can only multiply within living cells. The metabolism of the host cell is altered to provide the environment for viral reproduction. The host
cell begins to degenerate and is finally destroyed over a short or long period of time,
liberating new virus particles which infect neighbouring cells. Fish and marine invertebrates are attacked by viruses belonging to most of the groups known at present; some of
these are responsible for major problems in aquaculture (Table 1).

Table 1. Several important pathogenic viruses

Virus	Common name for the disease	Susceptible farmed species
Rhabdovirus	Viral haemorrhagic septicaemia (VHS)	Salmonids, eels, sea fish
Birnavirus	Infectious pancreatic necrosis (IPN)	Salmonids, turbot
Herpesvirus		Carp, trout, turbot, catfish
Iridovirus	Lymphocystis	Plaice, sea bream, perch
Iridovirus (?)		Portuguese oyster
Picorna-like	Infectious hypodermic & haematopoietic necrosis	Shrimps (*P. stylirostris*, *P. monodon*)
Baculovirus	Penaeid baculovirus	Many species of penaeids

Bacteria

The bacteria which can be considered to be pathogens make up only a very small fraction of the whole aquatic bacterial flora. These bacteria are capable of adhering to healthy tissues such as the skin, gills and digestive organs and from here to spread and multiply inside the organs. They are able to resist the fish's defence mechanisms and can sometimes adapt to survive where there are deficiencies, for example, of iron (Crosa 1980). They produce endo- and exotoxins and various enzymes which often initiate the lesions and abnormalities observed in the host (Evelyn 1984; Baudin Laurencin 1987).

The most virulent bacteria are themselves the primary cause of some diseases (Table 2): these should be distinguished from other species which cause secondary bacterial infections to primary lesions caused by mechanical damage or parasitic infestation. Many types of aquatic bacteria are capable of multiplying on organisms without activating, or only slightly activating, any defensive response. Thus, invertebrate larvae or cultured fish constitute an ideal environment for the development of a heterogeneous bacterial flora. Bacteria are rarely the primary cause of disease in invertebrates: the fundamental role of a Vibrio bacterium in the 'brown ring' disease of clams is still disputed (Paillard *et al.* 1989); Gaffkaemia in lobsters mainly develops where wounds have already occurred (Sindermann 1988); the necroses which are found on the carapace of shrimps (bacterial shell or brown spot disease) result from the action of large numbers of chitinolytic bacteria (Lightner 1988a).

Other bio-aggressors belong to various groups of parasites (protozoa, helminths, crustaceans), and have a varied range of actions. It is usual, although not entirely justified, to consider alongside viral and parasitic diseases the vast range of diseases associated with parasites, even though their common denominator is that their agents, in common with all living organisms except viruses and bacteria, are made up of eucaryotic cells. Most parasites invade the host at a certain stage of their development. The term infestation is used; this contrasts with that of infection where multiplication (of microbes) takes place within the host.

Table 2. Some important pathogenic bacteria

Bacteria	Common name of the disease	Sensitive farmed species
Vibro anguillarum	Vibriosis	All species
Aeromonas salmonicida	Furunculosis	Salmonids
Yersinia ruckeri	Redmouth	Salmonids and other species
Renibacterium salmoninarum	Bacterial kidney disease	Salmonids
Flexibacter sp.	Myxobacteriosis	Many species
Vibrio sp.	Brown ring disease	Clams
Aerococcus viridans	Gaffkaemia	Lobsters

Table 3. Some important parasites

Parasite	Common name of the disease	Sensitive species
Fungus *Fusarium* sp.		Shrimps
Fungus *Saprolegnia* sp.		Fish (fresh water)
Fungus *Exophiala*		Fish (sea water)
Flagella *Costia*	Costiasis	All species
Ciliates *Ichthyophthirius*	White spot disease	Freshwater species
Ciliates *Cryptocaryon*		Marine species
Microsporidians	Milky or cotton disease	Shrimps
Myxosporidan *Myxobolus cerebralis*	Whirling disease	Rainbow trout
Haplosporidian *Marteilia refringens*		Oysters, mussels
Haplosporidian *Bonamia ostreae*	Bonamia	*Ostrea edulis*
Monogenean *Gyrodactylus* sp.		Salmonids
Monogenean *Diplectanum* sp.		Bass
Copepod *Caligus minutus*		Bass
Copepod *Lepeophtheirus salmonis*		Seawater salmonids
Copepod *Mytilicola orientalis*		Oysters, mussels
Isopod *Nerocilla*		Bass

It is often stated that parasites do not kill the host in 'normal' conditions, only in conditions which have become unfavourable due to disease and/or captivity (Baer 1971). Culture, particularly in intensive form, favours the development of certain parasites and the consequent development of disease. A small number of parasites (Table 3) have the potential to cause major disease problems in culture. However, most mycoses or protozooses (particularly those which occur during the larval stages of fish or inverte-brates) are considered to be secondary, occurring when environmental conditions are inadequate.

The disease-causing mechanism is sometimes linked to a despoilatory action—i.e. direct attack on the host tissue (haematophagous crustaceans, Kabata 1970 and some helminths). More frequently, the damage is through mechanical action; attachment to the skin, movements of some parasites (ciliates, flagellates, monogeneans) or simply their development (superficial mycoses). The damage caused impairs osmoregulation and pro-vides a route for secondary infection (Robertson et al. 1981). Some endoparasites cause encumberances or constrictions: if worms are present in large numbers they may impair the transit of food through the digestive system (acanthocephali) or blood circulation (ligula). The destruction of cartilages by *Myxobolus cerebralis* causes nervous constric-tions which are the origin of the characteristic symptoms of the disease (Halliday 1976). The secretion of irritant substances (*Argulus*, Kabata 1970), enzymes and anticoagulants may sometimes be a part of the action of the pathogen. Finally, even by their presence (as much as foreign bodies as by their antigenic nature), parasites as well as viruses and bacteria invoke reactions from the host which are themselves the basis of the observed symptoms.

2. HOST DEFENCE MECHANISMS

Aquatic animals have great powers of adapting to unfavourable physical and chemical conditions in the environment: a decrease in the availability of oxygen or an increase in autopollution is generally compensated for by a decrease in metabolic rate and a reduc-tion in food intake. Rainbow trout transferred from fresh water to sea water call on a complex system of hormonal and enzyme regulation which allows growth to continue as long as temperature and salinity are not too high (Aldrin et al. 1985). Oxidized food or food which is deficient in vitamins must be fed over a period of a few months before bass show clinical symptoms (Baudin Laurencin et al. 1990). Habituation to some pollutants has also been demonstrated (Dixon & Sprague 1981; Bradley et al. 1985).

However, host defences are usually considered in relation to bioagressors. These de-fences, the host's immune system, may be natural (innate) or acquired, cellular or hu-moral.

One of the starting points of modern immunology was the observation of phagocytosis in *Daphnia* (Metchnikoff 1884). Most of our knowledge of immune systems comes from the study of mammals but more recently there has been great progress with the study of invertebrates and lower vertebrates, particularly fish (Dorson 1984).

In invertebrates humoral defence systems (lysozyme, lytic enzymes) and cellular defences are relatively poorly understood but appear to be less specific than in vertebrates. Any damage to tissues is met by a localized inflammatory response (vasodilation, infiltration of haemocytes). The presence of foreign particles and, in particular, microorganisms, immediately brings about strong phagocytic activity which can be increased by opsonization. When the foreign bodies are too large haemocytes gather round them and begin the process of encapsulation (Götz 1986). Lectines are agglutinins which are calcium dependent and are inhibited by certain sugars and appear capable of a degree of antigenic recognition (Jeong *et al.* 1981; Van der Knaap *et al.* 1983). The number and significance of these lectines is very variable from one species to the next and their role in immune protection is poorly demonstrated. In arthropods the enzymatic complex from the activation of prophenoloxidase (responsible for the synthesis of melanin) appears to play an important role in the recognition of non-self. It has been shown that the enzymatic system present in granulocytes is the origin of opsonization, nodular formations, bactericide activity and haemocyte cooperation (Söderhäll & Smith 1986).

Fish are protected against bioaggressors by specific and non-specific defence mechanisms which are very similar to those found in higher vertebrates (Angelidis 1987). The outer layer of the skin, gills and digestive system constitutes the first barrier to the penetration by microbes. These surfaces are covered by a layer of mucus which contains most of the antibacterial substances found in the serum.

Lysozyme, haemolysins, C-reactive proteins and complement are, in teleosts, the main substances responsible for natural humoral immunity. The presence of interferon (a substance which causes cells successfully to inhibit the replication of viruses) has been demonstrated in trout (De Kinkelin & Dorson 1973). Complement may be activated directly or through antigen–antibody complexes (Nonaka *et al.* 1981); it plays a role in phagocytosis, the inflammatory response, lysis of cells and also participates in the acquired immunity. Fish antibodies are tetrameric immunoglobulins similar to the IgM of mammals (Marchalonis 1971). They have the characteristics of agglutination, precipitation and neutralization of viruses and bacteria. They are able to activate complement and, indirectly, bring about the lysis of cell membranes.

Macrophages, monocytes and neutrophils can have a phagocytic activity. These cells are able to ingest foreign particles before the specific immune response has begun. There are also pigmented macrophages, containing, among other things, melanin, lipofuchsins and haemosiderin which tend to aggregate and form nodules (Agius 1985). Antigenic microbes have been observed in these nodules (Balouet *et al.* 1986). Some lymphocytes have a cytotoxic activity towards foreign cells (Deschaux *et al.* 1983). The distinction between two types of lymphocytes (T cells and B cells) has been established recently, together with the existence of subpopulations, analogous to those of mammals (Cuchens & Clem 1977; Miller *et al.* 1987). The anterior kidney, thymus and spleen are the principal lymphoid organs, but lymphocytes are also present in mucus, particularly in the posterior intestine. Fish have a lymphatic circulatory system which is separate from the blood circulation but there are no lymph nodes or bone marrow as in mammals.

Immunocompetent cells are able to act together. After an antigen has penetrated the organism it is captured by macrophages and 'presented' to B or T lymphocytes. Substances similar to interleukins appear to be present (Grondel & Harmsen 1985).

3. CONDITIONS LEADING TO THE DEVELOPMENT OF DISEASE

Whether or not an outbreak of disease occurs is dependent on the aggressiveness of the pathogen, the defence mechanism of the host and the environmental conditions. These conditions are intrinsic, linked to the host, or extrinsic, depending on the environment.

3.1 Intrinsic factors

For each group or species of aquatic animals a set of optimal physical and chemical environmental conditions can be defined, outside of which there is a deterioration of the anatomy or functioning of the individual. In general, molluscs are able to accumulate, without any pathological signs, considerable quantities of chemical 'pollutants', particularly heavy metals. However, arthropods are very sensitive to certain compounds, for example organophosphates. As a consequence of this, these compounds can be used against crustacean parasites of farmed fish (Brandal & Egidius 1979; Egidius & Moster 1987).

Species which are related may have different biological limits: in Brittany the brown trout, *Salmo trutta*, can survive summer temperatures and salinities in the sea while the rainbow trout, *Oncorhynchus mykiss*, shows characteristic signs of disease in these conditions (Baudin Laurencin *et al.* 1983). Vitamin C deficiency induces granulomatous hypertyrosinaemia in turbot and sea bream (Tixerant *et al.* 1984), but not in sea bass and salmonids (Baudin Laurencin *et al.* 1990).

The concept of a disease which is confined to one or a few host species is mainly linked to infectious diseases, particularly parasitic infestations. For these latter the specificity is a characteristic of the parasite itself. Parasites may be associated with a single species of host, or even a certain site on the skin, fins or gills (mongenean trematodes, for example *gyrodactylus*; Baer 1971). In some cases the presence of a chemical attractant has been demonstrated in the host mucus, for example in *Solea solea* for larval *Entobdella solaea* (Kearn 1967). Other species of parasite have a much wider host range; the same parasite (for example *Trichodina*) may be found on a large number of different species living in the same environment.

Viruses and bacteria generally affect a wide range of species, but the clinical infection resulting may only be significant in a small number of species. Thus, while viral haemorrhagic septicaemia produces clinical symptoms in trout, other salmonids (for example the Atlantic salmon) can be carriers of the disease without showing any visible sign (De Kinkelin & Castric 1982). Similarly, coho salmon are much more sensitive to *Renibacterium salmoninarum* than brown or rainbow trout (Sanders & Fryer 1980). Even within the same species, some populations appear more sensitive than others to certain diseases (Lightner 1988).

In general, the incidence of diseases caused by infective agents decreases with age; the development of specific and non-specific immune defences limits infection or infestation. Farms rearing larval invertebrates or fish are the extreme example; the animals' defences are absent or extremely weak. The many bacteria present in the environment (but not themselves truly virulent) are able to develop rapidly on the living but immunologically deficient tissue. It appears that in marine fish specific immuno-competence develops soon after weaning (transition to feeding on manufactured diet). From this point on, infectious or parasitic diseases follow their true course. Juveniles are more sensitive than adults to bioaggressors. Most of the viral diseases found in shrimps tend to affect juveniles more than adults (Lightner 1988b, c). In trout, infectious pancreatic necrosis is classically limited to 1400 degree days (Dorson & Torchy 1981). Vibriosis in turbot is primarily confined to juveniles.

The physiological state of the animal is obviously an important part of the progress of disease. This state may relate to natural biological 'events' (smoltification, migration, reproduction) or unfavourable environmental conditions. At the end of the breeding season, trout kelts are particularly susceptible to fungal diseases and secondary bacterial infections. Stress is a general, non-specific response to aggression, whatever its origin. Initiated by the stimulation of the neuro-hypothalamo–hypophyseal axis, stress causes immunosuppression, particularly through the action of the secretion of corticosteroids. Stress brings about a reduction in the number of leucocytes, particularly lymphocytes and a drop in phagocytic activity (Angelidis *et al.* 1987). Flat oysters show a greater sensitivity to *Bonamia* after transport; this may be explained as a response to stress and a reduction in phagocytic activity.

3.2 Extrinsic factors

The sensitivity of animals to physical and chemical factors has been detailed at the beginning of this chapter. However, the environment may only act as a factor predisposing the animal to susceptibility to bioaggressors. Environmental conditions may alter the virulence of pathogens as well as the sensitivity of the host to these pathogens.

Parasites, viruses and bacteria all have optimum temperatures and salinities. The rhabdovirus responsible for viral haemorrhagic septicaemia, *Flexibacter psychrophylus* and *Renibacterium salmoninarum*, are all cold-water organisms. Depending on the strain, *Vibrio angillarum* may be adapted to mid-range temperatures and salinities (Bretonne 408 strain) or higher ones (Mediterranean 62 strain) (Breuil *et al.* 1988). It has already been shown that the specificity of parasites can be linked to environmental conditions. Suspended material may also support bacteria and favour their development; these bacteria cause gill disease by eroding the branchial epithelium and providing a means for the penetration of germs. There is competition within the biological environment, planktonic or bacterial; this regulates the development of pathogenic organisms.

A poor environment, whether because of physical and chemical factors, nutrition or husbandry (density too high, ineffective grading) can also be stressful with an adverse effect on the immune system. In addition to their directly toxic effects, many pollutants (e.g. copper, gas–oil) appear to induce symptoms of stress (Poirier *et al.* 1986).

4. SYMPTOMS OF DISEASE

The clinical signs of disease are determined partly by the tissue targeted by the pathogen and partly by its virulence and the capacity of the host to defend itself. These two last factors depend themselves on predisposing factors; every disease therefore has a range of symptoms and degrees of seriousness.

4.1 Systemic diseases

Pathogens may be transported in the circulatory system and reach most of the tissues in the body. These may be infectious agents; and the resulting disease is referred to as septicaemia. They may also be deleterious substances from the external environment (pesticides, heavy metals) or from inside the body resulting from a malfunction (hypertyrosinaemia caused by deficiency).

If the pathogen is extremely virulent it may quickly overcome the hosts' defences and result in an outbreak of the disease with a rapid and massive mortality and few clinical symptoms. Laboratory examination will reveal the destruction of a particular tissue. Accidental pollutions are often accompanied by sudden outbreaks of viral or bacterial septicaemias in aquatic organisms which have become highly susceptible.

Many diseases of fish are expressed in their acute form. The mortality rate increases exponentially but death is preceded by the appearance of symptoms such as loss of appetite, dark coloration of the skin, protrusion of the eyes (Fig. 1), congestive and haemorrhagic external (Fig. 2) or internal (Fig. 3) lesions. These symptoms are similar for many of the most frequently encountered diseases: rhabdoviruses, vibriosis, furunculosis, yersiniosis. Viral diseases of post-larval or juvenile penaeid shrimps also take the acute form with loss of appetite and a high mortality rate, alternating apathy and excitability and cuticular and muscular colour changes.

Other symptoms are found in subacute or chronic forms of diseases; however, these frequently fail to characterize the disease or identify its cause. Anaemia is the logical consequence of haemorrhages and explains the pallid appearance of gills and viscera. Some bacteria develop in the muscle causing focal necrosis 'furuncles' which are classically attributed to furunculosis, but which are also observed in other septicaemias (Fig. 4).

Renibacteriosis is found most frequently in the chronic form; necrosis and granulomatous lesions occur in several visceral organs, particularly the kidney (Fig. 5). Other pathogens favour the kidney, causing lesions which appear similar to form: these include mycoses associated with *Exophiala*, a fungal parasite of the liver and kidney. In turbot, the impairment of the metabolism of tyrosine due to vitamin C deficiency leads to hypertyrosinaemia and the formation of crystals, which then give rise to granulomatous nodules in the viscera, skin or muscles (Fig. 6). In all these chronic forms the lesions are mainly caused by the inflammatory response of the host rather than the direct action of pathogens.

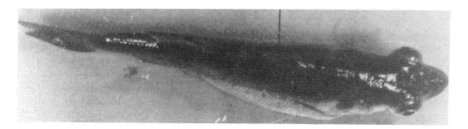

Fig. 1. Viral haemorrhagic septicaemia (experimental) in bass. Dark coloration, eyes protruding.

Fig. 2. Vibriosis in turbot. Congestive and haemorrhagic cutaneous lesions (arrow).

Fig. 3. Viral haemorrhagic septicaemia (experimental) in bass. Visceral and muscular haemorrhages (arrows).

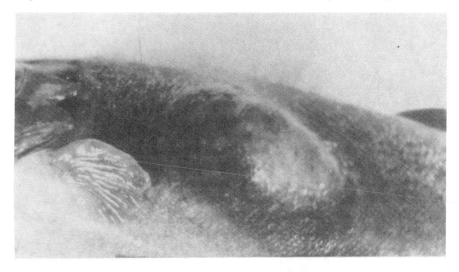

Fig. 4. Vibriosis in a rainbow trout. Inflammatory lesion localized in the musculature 'furuncle'.

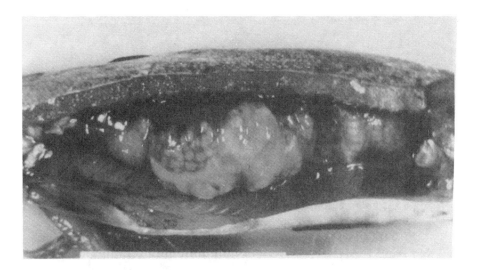

Fig. 5. Bacterial kidney disease in an Atlantic salmon. Nodular renal lesions.

Melanization of tissues generally occurs in chronic infections of crustaceans; it is an essential part of the defence mechanism associated with the activation of prophenoloxydase. This occurs during infections by chitinolytic bacteria and fungi (*Fusarium*, Fig. 7), as well as vitamin C deficiency and exposure to certain pollutants. The names of crustacean diseases, 'black spot', 'black gill syndrome' and 'black death disease', clearly demonstrate the association of the dark colour and disease.

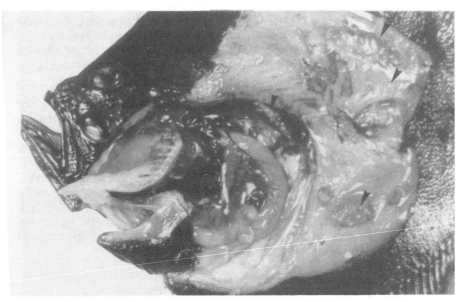

Fig. 6. Granulomatous hypertyrosinaemia. Granulomatous nodules in the viscera and muscles (arrow).

Fig. 7. Fusarium. Melanic lesions in *P. japonicus*.

4.2 Infections of certain tissues or organs

The outer coating of the body and the gills, while being the first line of defence of the organism, is also the most vulnerable to damage. The initial damage is usually followed by a localized inflammatory response, with cells infiltrating the wound in order to cleanse and repair the damage with the formation of replacement tissue.

However, in both fresh and salt water, the effects of osmotic pressure make healing difficult when wounds are large or multiple. The ulceration of the skin tends to enlarge or deepen, which opens the way to secondary infections (Fig. 8). In the gills (more than the skin covering the rest of the body) rapid production of large numbers of epithelial cells (hyperplasia) brings about the restoration of osmoregulation. However, if the irritation persists, the hyperplasia may become excessive, reducing the respiratory surface, leading ultimately to asphyxia and death. Some pathogens induce localized reactions: lymphocystis is an example of such a disease (Fig. 9).

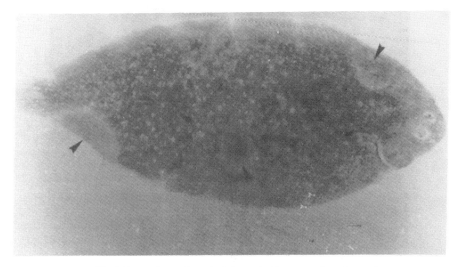

Fig. 8. External myxobacteriosis in a sole. Cutaneous ulcers (arrows).

In general, away from the skin, it is the true parasites (protozoa, metazoa or fungi) which affect particular tissues or organs. There are characteristic symptoms and lesions. *Myxobolus cerebralis* causes deformities in the cartilage and, indirectly 'whirling' and blackening of the tail in rainbow trout. The fungus *Exophiala* and myxosporidians induce lesions in the kidney and, indirectly changes the colour of the skin. A haemogregarine parasite in turbot is associated with a myeloid leucosis syndrome, with the appearance of readily recognizable nodular pseudo-tumours throughout the organism. Parasitic diseases of flat oysters of great importance in France. These take the form of either the degeneration of the digestive gland (*Marteilia refringens*, Fig. 10) or an ulceration throughout the gills (*Bonamia ostreae*, Fig. 11).

5. CONCLUSION

Knowledge of the mechanisms of disease allows the development and use of methods of diagnosis, therapeutic treatment and prophylaxis (prevention). Treatment and prevention eliminate the pathogenic factor, reinforce the host's defence mechanisms and change the environment to one which no longer encourages the development of disease.

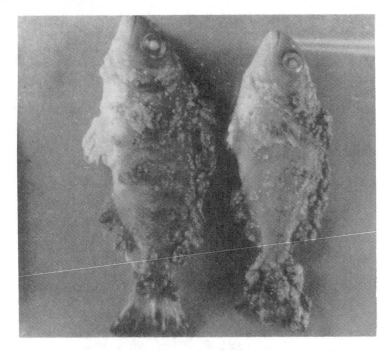

Fig. 9. Lymphosystis in a sea bream.

The visible signs of disease are, unfortunately, seldom instant indicators of the cause (Section 4) and only allow a preliminary diagnosis. The nature of the pathogen must be determined by specialized scientific methods: physical and chemical analysis, virology, bacteriology and parasitology. Fortunately the identification of the bacteria and parasites which are most likely susceptible to treatment is generally simple and rapid using microscopy and bacterial culture. The pathogenic agent can often be identified in under 24 hours and bacteria can be characterized for resistance and sensitivity to the permitted range of antibiotics.

Antibacterial and antiparasitic treatments always have their own toxic properties, for the fish farmer (during their use), for the treated animals, for the environment and for the consumer (through the residues of therapeutants remaining in the flesh of the animal).

Incorrect use of antibiotics leads to the development of resistance which may be transmitted to bacteria which are pathogenic to man. All treatments using drugs and chemicals should therefore not only be based on accurate diagnosis but also follow the rules laid down for administration, bearing in mind the problems cited above.

The prevention of a disease is always preferable to the treatment of an established one. However, prevention is limited by our insufficient knowledge of the causes of some diseases and by the economic pressures governing production.

Reinforcement of the host defences against most infective agents by vaccination is possible in theory, at least in fish. However, the 'classical' methods of vaccine preparation (inactivation of bacteria or use of a less virulent strain) may end up with products of low immunogenicity and their preparation in sufficient quantities to deal with protozoa

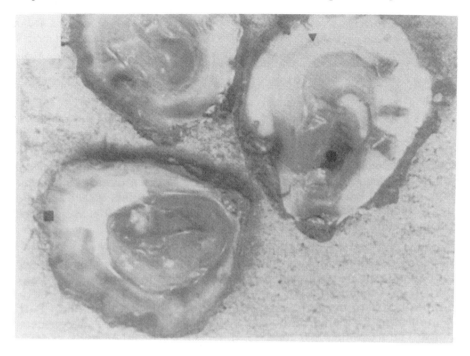

Fig. 10. Abers disease in the flat oyster. The edges of the shells are dark ■; the mantle is transluscent ▲; and the digestive gland thin, and visible through the mantle ●. (Photo & legend H. Grizel.)

Fig. 11. Haemocyte disease in the flat oyster. Gill ulcers. (Photo H. Grizel.)

and viruses may entail excessive costs. More precise identification of antigens and the use of new biotechnological methods should lead to better vaccines for aquaculture in the future, particularly those based on recombinant techniques. At present, only vaccines against certain bacterial fish diseases (vibriosis, enteric redmouth and, most recently, furunculosis) are on the market. For crustaceans, vaccines against shrimp vibriosis and gaffkaemia in lobsters have been successful experimentally and in trials.

Maintenance of good conditions and healthy stock is as important in aquaculture as in any other type of farming. Good hygiene consists of reducing the occurrence and development of pathogens and factors which diminish the host's defence mechanisms. Methods used must be adapted to the type of production. For the future, selection, crossing and genetic manipulations may result in the production of strains of fish resistant to certain diseases.

Disease is an important and, unfortunately, almost obligatory aspect of aquaculture. In principle, the mechanisms of disease and the causes are similar in aquaculture to those on land. Because of this, fish farmers can benefit greatly from the knowledge acquired for agriculture. Recent progress in basic biology and its applications (new technologies) must be used to help solve the problems of disease in aquaculture.

REFERENCES

Agius C., 1985. The melano-macrophage centres of fish: a review. In: *Fish immunology. Symposium proceedings*, Plymouth, July 11–13, 1983, Manning and Tatner (eds), pp. 85-105.

Aldrin J. F., 1987. Pollution chimique et sensibilisation des poissons aux agents pathogènes. *Ichtyophysiologica Acta*, **11**: 164–186.

Aldrin J. F., Barbou L. G., Baudin Laurencin F., Guillaume J., Messager J. L., Mével M., Stéphan G., Tixerant G., 1985. Quelques conséquences physiopathologiques de l'élévation de la température et de la salinité chez la truite arc-en-ciel, *Salmo gairdneri* Richardson. *Ichtyophysiologica Acta*, **9**: 175 pp.

Angelidis P., 1987. Eléments du système immunitaire du bar (*Dicentrarchus labrax*) et de la truite (*Salmo gairdneri*). *Thèse de Doctorat de l'Université de Bretagne Occidentale*, No. 35: 171 pp.

Angelidis P., Baudin Laurencin F., Youinou P., 1987. Stress in rainbow trout (*Salmo gairdneri*): effects upon phagocytes chemiluminescence, circulating leucocytes and susceptibility to *Aeromonas salmonicida*. *J. Fish. Biol.*, **31** (Suppl. A): 113–122.

Baer J. G., 1971. Les parasites animaux. *L'univers des Connaissances*. Hachette, Paris: 256 pp.

Balouet G., 1986. Tumors in marine organisms: nosology, incidence, importance in marine aquaculture. In: *Pathology in marine Aquaculture*, C. P. Vivarès, J. R. Bonami & E. Jaspers (eds). European Aquaculture Society, Special publication No. 9, Bredene, Belgium: 299–309.

Balouet G., Baudin Laurencin F., 1986. Granulomatous nodules in fish: an experimental assessment in rainbow trout, *Salmo gairdneri* Richardson and turbot, *Scophthalmus maximus* (L.). *J. Fish Dis.*, **9**: 417–419.

Baudin Laurencin F., 1987. Pathogénie de la vibriose: un exemple des relations entre organisme pathogène, hôte et environnement. *Océanis*, **13** (1): 115–123.

Baudin Laurencin F., Aldrin J. F., Messager J. L., Tixerant G., 1983. Summer pathology in marine cultured rainbow trout. Communication EAFP, Plymouth 1983. In: *Fish and shellfish pathology*. A. E. Ellis (ed.). Academic Press, London: 211–221.

Baudin Laurencin F., Messager J. L., Stéphan G., 1990. Two examples of nutritional pathology related to vitamin E and C deficiencies. In: *Advances in tropical aquaculture*, Tahiti, February, March 1989. IFREMER, Actes du Colloque, **9**: 171–181.

Bradley R. W., Duquesney C., Sprague J. B., 1985. Acclimatation of rainbow trout, *Salmo gairdneri* Richardson, to zinc: kinetics and mechanism of enhanced tolerance induction. *J. Fish Biol.*, **27**: 367–379.

Brandal P. O., Egidius E., 1979. Treatment of salmon lice (*Lepeophtheirus salmonis* Kroyer, 1838) with Neguvon. Description of method and equipment. *Aquaculture*, **18** (2): 183–188.

Breuil G., Baudin Laurencin F., 1988. La vibriose du loup. Fiche technique. *Equinoxe*, **21**: 10–13.

Crosa J. H., 1980. Some aspects of trout gill structure in relation to Egtved virus infection and defence mechanisms. In: *Fish diseases*. Third COPRAQ-Session (W. Ahne, ed.). Springer-Verlag, Berlin, Heidelberg, New York: 18–22.

Cuchens M. A., Clem L. W., 1977. Phylogeny of lymphocyte heterogeneity. II. Differential effects of temperature on fish T-like and B-like cells. *Cell. Immunol.*, **34**: 219–225.

De Kinkelin P., Castric J., 1982. An experimental study of the susceptibility of Atlantic salmon fry, *Salmo salar* L. to viral haemorrhagic septicaemia. *J. Fish Diseases*, **5**: 57–65.

De Kinkelin P., Dorson M., 1973. Interferon production in rainbow trout (*Salmo gairdneri*) experimentally infected with Egtved virus. *J. Gen. Virol.*, **19**: 125–127.

Deschaux P., Perroy-Cordier L., Peres G., 1983. Mise en évidence des cellules 'naturelles tueuses' (NK) chez le loup d'élevage (*Dicentrarchus labrax*). Influence de la température d'élevage. *Comp. Immunol. Micro-biol., Infect. Dis.*, **6**: 95–99.

Dixon D. G., Sprague J. B., 1981. Acclimatation to copper by rainbow trout (*Salmo gairdneri*)—a modifying factor in toxicity. *Can J. Fish. Aquat. Sci.*, **38**: 880–888.

Dorson M., 1984. Immunologie appliquée des poissons. In: *Symposium sur la vaccination des poissons*. P. De Kinkelin & C. Michel (eds). Off. Int. des Epiz., Paris: 41–82.

Dorson M., Torchy C., 1981. The influence of fish age and water temperature on mortalities of rainbow trout, *Salmo gairdneri* Richardson, caused by a European strain of infectious pancreatic necrosis virus. *J. Fish Dis.*, **4**: 213–221.

Egidius E., Moster B., 1987. Effects of Neguvon and Nuvan treatment on crabs (*Cancer pagurus*, *C. maenas*), lobster (*Homarus gammarus*) and blue mussel (*Mytilus edulis*). *Aquaculture*, **60**: 165–168.

Erard-Ledden E., Ryckaert M., 1990. Trout mortality associated to *Distephanus speculum*. In: *Toxic marine phytoplankton*, E. Graneli *et al.* (eds). Elsevier Sc. Publ., 137.

Evelyn T. P. T., 1984. Immunization against pathogenic vibrios. In: *Symposium on Fish Vaccination*, O.I.E., Paris: 121–150.

Farley C. A., 1978. Neoplasms in estuarine mollusks and approaches to ascertain causes. *Ann. NY Acad. Sci.*, **298**: 225–232.

Götz P., 1986. Encapsulation in arthropods. In: *Immunity in invertebrates. Cells, molecules, and defense reactions*, Michel Brehélin (ed.). Springer-Verlag, Berlin, Heidelberg, New York, Tokyo: 153–171.

Grondel J. L., Harmsen E. G. M., 1985. Do fish have interleukins? In: *Fish Immunol.*, M. J. Manning & M. Tatner (eds). Academic Press, London: 261–262.

Halliday M. M., 1976. The biology of *Myxosoma cerebralis*: the causative organism of whirling disease of salmonids. *J. Fish Biol.*, **9**: 339–357.

Jeong K. H., Sussman S., Rosen S. D., Lie K. J., Heyneman D., 1981. Distribution and variation of haemagglutinating activity in the haemolymph of *Biomphalaria glabrata*. *J. Invertebr. Pathol.*, **38**: 256–263.

Kabata Z., 1970. In: *Diseases of fishes*, Vol. 1. *Crustacea as enemies of fishes*, S. F. Snieszko & H. R. Axelrod (eds). THH Publications, Jersey City, N.J., 171 pp.

Kearn G. C., 1967. Host-finding and host specificity in the monogenean *Entobdella solae*. *Parasitol.*, **57**: 585–590.

Ketola H. G., 1979. Influence of dietary zinc on cataracts in rainbow trout (*Salmo gairdneri*). *J. Nutr.*, **109** (6): 965–969.

Lauckner G., 1980. Diseases of Porifera. In: *Diseases of marine animals*, Vol. 1, Otto Kinne (ed.). John Wiley and Sons, Chichester: 139–165.

Le Milinaire C., Gatesoupe F. J., Stephan G., 1982. Composition en acides gras du rotifère *Brachionus plicatilis* nourri avec différents aliments composés: influence sur la croissance et la teneur en acides gras essentiels de la larve de turbot (*Scophthalmus maximus*). Publication CNEXO (Actes et Colloques), No. 14: 275–290.

Lightner D. V., 1988a. Bacterial shell (Brown spot) disease of penaeid shrimp. In: *Disease diagnosis and control in North American marine aquaculture*, C. J. Sindermann & D. V. Lightner (eds). Elsevier, *Developments in Aquaculture and Fisheries Science*, **17**: 48–51.

Lightner D. V., 1988a. BP (*Baculovirus penaei*) virus disease of penaeid shrimp. In: *Disease diagnosis and control in North American marine aquaculture*, C. J. Sindermann & D. V. Lightner (eds). Elsevier, *Developments in Aquaculture and Fisheries Science*, **17**: 16–21.

Lightner D. V., 1988b. Baculoviral midgut gland necrosis (BMN) disease of *Penaeus japonicus*. In: *Disease diagnosis and control in North American marine aquaculture*, C. J. Sindermann & D. V. Lightner (eds). Elsevier, *Developments in Aquaculture and Fisheries Science*, **17**: 26–29.

Marchalonis J. J., 1971. Isolation and partial characterization of immunoglobulins of goldfish (*Carassius auratus*) and carp (*Cyprinus carpio*). *Immunology*, **20**: 161–173.

Messager J. L., 1986. Influence de l'acide ascorbique sur l'hypertyrosinémie granulomateuse du turbot d'élevage, *Scophthalmus maximus* L. In: *Pathology in marine aquaculture*, C. P. Vivares, J. R. Bonami & E. Jaspers (eds). European Aquaculture Society, Special publication No. 9, Bredene, Belgium: 381–390.

Metchnikoff E., 1884. Ueber eine Sprosskrankheit der Daphnien. *Arch. Pathol. Anat.*, **96**: 177–195.

Miller N. W., Bly. J. E., Van Ginkel F., Ellsaesser C. F., Clem L. W., 1987. Phylogeny of lymphocyte heterogeneity: identification and separation of functionally distinct subpopulations of channel catfish lymphocytes with monoclonal antibodies. *Dev. Comp. Immunol.*, **11** (4): 739–747.

Mitchum D. L., Sherman L. E., Baxter G. T., 1979. Bacterial kidney disease in feral populations of brook trout (*Salvelinus fontinalis*), brown trout (*Salmo trutta*), and rainbow trout (*Salmo gairdneri*). *J. Fish Res. Board Can.*, **36**: 1370–1376.

Nonaka M., Yamaguchi N., Natsuume-Sakai S., Takahashi M., 1981. The complement system of rainbow trout (*Salmo gairdneri*). I. Identification of the serum lytic system homologous to mammalian complement. *J. Immunol.*, **126** (4): 1489–1494.

Paillard C., Percelay L., Le Pennec M., Le Picard D., 1989. Origine pathogène de l'anneau brun' chez *Tapes philippinarum* (Mollusque, bivalve). *C.R. Acad. Sci. Paris*, **309** (3): 235–241.

Peters G., Hoffmann R., Klinger H., 1984. Environment-induced gill disease of cultured rainbow trout *Salmo gairdneri*. *Aquaculture*, **38**: 105–126.

Poirier A., Baudin Laurencin F., Bodennec G., Quentel C., 1986. Intoxication expérimentale de la truite arc-en-ciel, *Salmo gairdneri* Richardson, par du gas–oil moteur: modifications hématologiques, histologie. *Aquaculture*, **55**: 115–137.

Robertson D. A., Roberts R. J., Bullock A. M., 1981. Pathogenic and autoradiographic studies of the epidermis of salmonids infested with *Ichtyobodo necator* (Henneguy 1883). *J. Fish Dis.*, **4**: 113–125.

Sanders J. E., Fryer J. L., 1980. *Renibacterium salmoninarum* gen. nov., sp. nov., the causative agent of bacterial kidney disease in salmonid fishes. *Int. J. Syst. Bacteriol.*, **30** (2): 496–502.

Sindermann C. J., 1970. *Principal diseases of marine fish and shellfish*. Academic Press, London: 369 pp.

Sindermann C. J., 1988. Gaffkaemia of lobsters. In: *Disease diagnosis and control in North American marine aquaculture*, C. J. Sindermann & D. V. Lightner (eds). Elsevier, *Developments in Aquaculture and Fisheries Science*, **17**: 232–235.

Söderhäll K., Smith V. J., 1986. The prophenoloxidase activating system: the biochemistry of its activation and role in arthropod cellular immunity, with special reference to crustaceans. In: *Immunity in invertebrates. Cells, molecules, and defense reactions*, M. Brehélin (ed.). Springer-Verlag, Berlin, Heidelberg, New York, Tokyo: 208–223.

Tixerant G., Aldrin J. F., Baudin Laurencin F., Messager J. L., 1984. Syndrome granulomateaux et perturbations du métabolisme de la tyrosine chez le turbot (*Scophthalmus maximus*). *Bull. Acad. Vét. France*, **57** (1): 75–85.

Van Der Knaap W. P. W., Boots A. M. H., Van Asselt L. A., Sminia T., 1983. Specificity and memory in increased defence reactions against bacteria in the pond snail *Lymnaea stagnalis*. *Dev. Comp. Immunol.*, **7**: 435–443.

Conclusions

This has been a very rapid review of the fundamental biological and ecological aspects of aquaculture. It is now worth examining how these scientific and technical matters relate to the overall economic context in which aquaculture is placed.

1. AQUACULTURE, A NEW SCIENTIFIC SPECIALIZATION

Aquaculture has now emerged as a specialized subject in Asian and European universities (Stirling, Aberdeen, Wageningen, Ghent, Cork).

Major directed research programmes always provide the background information for the development of new types of aquaculture; this has happened in Japan for shrimps (work carried out by Hudinga), and in Europe for sea bass and sea bream, as well as for species of molluscs in many parts of the world. The transition from the laboratory to the pilot scheme and then to full-scale production is seldom straightforward, as has been demonstrated for sea bass and sea bream by Professor Barnabé; since 1970 there have been 18 theses and 200 publications from his laboratory on the subject.

Aquaculture combines a wide range of biological and ecological research, the significance of which depends on the species (e.g. nutrition, genetics).

It is also true that aquaculture can supply useful information to scientists. Aquaculture provides the access to large numbers of juvenile marine fish, molluscs and crustaceans, for example, and to salmon in their marine phase; these are normally unobtainable. They provide useful material for physiologists, geneticists and biochemists. Experiments in large volume containers (tanks, cages) can simulate natural ecosystems; results from these can be combined with those from the wild and smaller-scale laboratory trials.

2. PRODUCTION OF LIVING MATTER AND THE ECONOMIC IMPORTANCE OF AQUACULTURE

Details of this have been reviewed elsewhere (Laubier & Barnabé 1990); only the major details are mentioned here.

Some aquaculture systems, where animals are farmed, produce, in terms of yield, around ten times the production achieved from the best plant production systems in agriculture (primary production) over a similar time and area. In the rias of northwest Spain, mussels reared in suspended culture give a production of 200–250 tonnes wet weight per hectare per year: this can be compared with 2 tonnes of chickens or 0.3 tonnes of cattle produced from 1 hectare of maize or prairie respectively. In the best-developed systems where aquaculture is integrated with agriculture, production reaches 9 tonnes per hectare per year of fish and over 30 tonnes of ducks on the same water body. In addition to this, the nutrient-rich waters of the ponds can be used to irrigate the land and increase its annual production to over 10 tonnes per hectare (Schoonbee & Prinsloo 1988). Water-treatment lagoon systems produce 50–70 tonnes of microalgae (dry weight) per hectare per year. This huge production is merely a byproduct of water recycling, using only radiant solar energy.

Another characteristic of aquaculture is that it uses volume rather than surface area as the medium for culture: this means that the potential for production is vast.

The economic impact of aquaculture production is also distinctly different from that of agriculture and fisheries in that production is growing. Larkin (1989) predicted that the production from aquaculture will exceed that from fisheries by 2020. In France, more people are employed in aquaculture than in fisheries; numbers of fishermen have dropped by 30% in 15 years.

A further benefit of aquaculture is the possibility of integration with other rural activities (fishing, farming).

3. THE ROLE OF AQUACULTURE IN THE PROTECTION OF THE ENVIRONMENT

The aquatic environment is probably more threatened than any other by human activities; it is well known and accepted that all wastes finally end up in the sea. The situation in the Mediterranean and the Baltic, enclosed seas, is already critical.

3.1 Aquaculture production processes and the environment
Aquaculture is an economically based activity relying on ecological and biological processes in that it is dependent on their management and manipulation. It is often the case that economic and ecological objectives are conflicting; the many different forms of pollution and global warming are there to remind us.

3.1.1 Intensive fish culture
We have shown (Part IV, Chapter 6) that in industrialized countries the development of intensive aquaculture (comparable to land-based farming), where water is merely the supporting medium, is undergoing a very rapid development. At present, market demand is the factor determining the expansion of production; no account is taken of the ability of the aquatic ecosystems to supply the feed required.

Intensive fish culture generates pollution; other types of aquaculture can have the reverse effect, acting as water purification systems.

Uncontrolled developments of intensive aquaculture can entail an increase in pollution in enclosed water; eutrophication by intensive salmon rearing units may be the starting point for the development of toxic algal blooms. This type of rearing system and its size should be taken into account in the management of a given body of water, together with the runoff (fertilizers) and other, more extensive activities. Usually, the necessary regulations do not exist because there are not enough facts to go on.

3.1.2 Other types of aquaculture

Aquaculture offers one of the only ways where economics and ecology can be integrated; some forms of aquaculture can extract polluted material and recycle it to production which can be utilized by man; the production of microalgae (Part I, Chapter 3), fish culture in ponds (Part IV, Chapter 5) are fundamentally purification systems. These are simple processes, requiring little management (Grobbelaar 1979).

The rearing of pigs and poultry in China provides organic manure which is added to ponds where different species of carp are cultured. Another example of combined culture is that of rice with shrimps or shallow-water fish, with the addition of rice bran to fertilize ponds (Indonesian tambaks). Asian countries have thus acquired an excellent system for the management of pond ecosystems to dispose of terrestrial wastes (Marcel 1990, Marcel & Doumenge 1990; Jhingran 1990). Similar systems are found in the temperate countries of eastern Europe. These are genuine examples of ecosystems which are managed for the benefit of man and, although simple and practically based, result in the production of millions of tonnes of animal protein. Some of the water bodies managed in this way are of a significant size; examples include the large American Lakes described by Barraclough & Robinson (1972).

Closer to home, purification of domestic water in lagoon systems can be integrated with fish culture as was demonstrated at Meze in the early 1980s (Barnabé 1981, 1984). The use of manure from terrestrial production units for the production of microalgae is also a technique of proven value (De Pauw et al. 1984). In the long term, the development of such systems will be an important weapon in the fight against nitrate pollution.

3.2 Sensitivity of organisms and the aquatic environment

Whilst the sensitivity of organisms to infinitesimal concentrations of various substances can change the subtle equilibria which regulate the biological processes of production, the fact that these substances can be concentrated along the food chain leads to changes (sometimes toxic) in the natural composition of aquatic organisms which are used for human food. Aquaculture is therefore usually confined to non-polluted waters and the prohibition of the use of organo-tin based anti-fouling paints around the coasts of many countries (including France and the UK) has demonstrated the interest in protecting the quality of the environment.

This capacity to concentrate has applications in water purification, huge quantities of mussels are used to filter water in the Mersey estuary (Liverpool) and are said to have made it cleaner than for many years. These filter feeding molluscs are not used for human consumption.

Aquaculture is therefore a means of protection of the environment. However, its potential is not fully utilized because of the lack of fundamental research. Aquaculture operations can suffer because of their close dependence on the environment; toxic waters caused by plankton blooms can, for example, be a major hazard for aquaculture and fisheries.

3.3 Implications on a global scale

Away from the usual restricted developments, aquaculture may be integrated into major management projects or coastal protection; this has happened in Holland as part of the scheme for the protection of low-lying land which has been integrated with mollusc culture (Persoone & Sorgeloos 1982). The more frequent method of protection of coasts through the use of piles or rocks can also be combined with aquaculture (Barnabé 1979): fish reared in cages have benefited from the shelter provided by these structures around the French Mediterranean coasts.

Even offshore industries can be integrated with aquaculture as has been shown in southern Spain, off Monaco, and in Japan.

It may not be long before aquatic ecosystems are manipulated on a grand scale. Martin *et al.* (1990) suggested a method of stopping atmospheric warming (caused by the increased levels of carbon dioxide in the atmosphere) by enriching Antarctic waters with iron (1 million tonnes, or several tanker loads) to stimulate the growth of phytoplankton which is limited in the absence of iron. This would increase the fixation of carbon from the atmosphere. Calculations show that this ocean (3.2 million km^3) with a nitrate concentration of 25 nmol kg^{-1} water could extract 6.4 billion tonnes of carbon, in the form of CO_2 or around the same quantity each year as added to the atmosphere by the burning of fossil fuel or deforestation. This fixed carbon would find its way to deeper water with dead plankton and faeces. This is not science fiction; a few weeks after the publication of this work an experiment was carried out on a reduced scale (Concar 1990).

According to Charlson *et al.* (1987), the essential source of the condensation nuclei for clouds over oceans appears to be the dimethylsulphide produced by planktonic algae which is oxidized to sulphates in the atmosphere. The reflection of solar energy is sensitive to the density of these clouds and therefore biological regulation of climate could be possible through the effects of temperature and solar radiation on phytoplankton and therefore the production of dimethylsulphide. Thus, to fight the global warming caused by the increase in carbon dioxide in the atmosphere it is necessary to increase the nuclei which give rise to clouds by altering the densities of phytoplankton.

Will the fertilization of the upper layers of the ocean allow man to manage the climate? If the ocean plays the role of the sink for carbon and the generator of dimethylsulphide, it is plainly at the centre of the problems facing the planet.

These are large-scale effects; there are many others. Suspended culture systems and the use of artificial reefs are already well-known proposals, but according to Assat (1985) the mixing of surface and deep waters in the Mediterranean would produce, for each kilowatt of energy used, 20 m^3 precipitation on the neighbouring land.

4. MARINE EXPLOITATION AND AQUACULTURE

4.1 Relationship between fishing and aquaculture

For some farmed species, culture is directly dependent on natural populations. This is where farming is based on the removal of juveniles from the wild (mollusc spat, wild milkfish, yellowtail or eel juveniles). Hatcheries also have a requirement for plankton and other food organisms harvested from the wild; the extreme examples is that of the Artemia gathered from around salt lakes (Part I, Chapter 3).

Aquaculture activities such as marine salmon ranching interfere directly with marine and estuarine fisheries. The production of Canadian and Alaskan hatcheries contributes around 30% of the salmon fishery in the region and in Japan, sea ranching leads to the harvest of 75 000 chum salmon.

The interactions between aquaculture and fisheries also include the use of fishmeal from industrial fisheries to produce the diet fed to fish in intensive culture systems. However, there remains an essential difference between aquaculture and fisheries: the concept of ownership. Fishing is based on a common resource, aquaculture on a private one. In Japan, aquaculture operators are proprietors of their own piece of the sea.

The possibility of joint exploitation of the aquatic environment by aquaculture and fisheries is well illustrated by the example of salmon ranching (juveniles produced in freshwater, released to migrate to sea and recaptured on their return migration to freshwater to spawn). In order to safeguard the interests of the hatcheries and increase the rate of recapture Needham (1990) proposed that the fishery in Canada could be better managed through privatization.

Aside from the local aspect, this author proposed taking aquaculture away from sheltered coasts and the overdependence on costly factors (cages, fishmeal), integrating fisheries into the global management of the ocean as part of a system of fisheries based on aquaculture. This form of 'privatization' would entail a significant reduction in the cost of fishing and would provide a way of managing major aquatic ecosystems. This link between aquaculture and fisheries need not be limited to the sea ranching of salmon: other species are likely to become important when the difficulties of producing the juvenile stages have been overcome. Aquaculture will make possible the stocking of huge areas of the ocean; the rights of fishing over these areas will then be allocated. This has already happened for scallops in Japan. The sea bass, a coastal fish which shows little in the way of migratory behaviour, may be a species for management in this way. However, there are many problems connected with such a plan, not all of them connected with the technology of fish production.

In spite of the potential described above, the production capacity of the aquatic environment is not without its limits; problems of over-exploitation of fisheries are already known. Intensive aquaculture relies on such fisheries (Part IV, Chapter 6). To feed one tonne of salmon produced in a cage takes three tonnes of fish captured in the industrial fishery: this represents the production of a square kilometre of the Baltic Sea or the North Sea (40–50 000 times the area of sea filled by the cage) (Faulke & Kaustsky 1989).

Intensive fish culture or fishing is not the only form of exploitation with the capacity to modify ecosystems: the culture of mussels (often held up as an environmentally friendly method of production) is in competition with zooplankton for the available phytoplankton. This must have an effect on the other parts of the food web and also increases the sedimentation rate by a factor of three in relation to natural sedimentation.

5. THE NEED FOR A COHERENT RESEARCH EFFORT

There are many consequences of the various human activities on aquatic ecosystems and altering one factor is bound to have an effect on many others. Aquaculture sets in play biological and ecological processes within the natural ecosystem and allows their manipulation to the benefit of man in ways including:

- biological production for economic purposes
- biological and chemical depuration (cleansing)
- biological recycling
- protection against further human-derived pollution

Fisheries and aquaculture should work together in a global approach to the ecosystem of the ocean, taking into account the potential of the producing systems, new technologies and their possibilities and the recycling of wastes.

In order for this to be successful, more information on the complexities of aquatic ecosystems are required; too much is still based on trial and error or on accidental observations. There is a requirement for basic research on physical and biological oceanography as well as on ecology, aquaculture on the dynamics of fish populations. However, most important of all is the promotion of coherent, worldwide interdisciplinary research, bringing together all aspects of aquatic science. The survival of a healthy environment in the freshwaters and oceans of the world is essential for our own futures.

REFERENCES

Assaf G., 1985. Artificial sea mixing. *J. Earth Sc.*, **34** (2–3): 110–112.

Barnabé G., 1979. Système d'aquaculture modulaire intégré en zone portuaire. *Aquaculture*, **16**: 369–374.

Barnabé G., 1981. Collecte et utilisation de plancton. In: *Compte rendu des Activités Scientifiques du Centre de Recherches du Lagunage de Mèze*. Rapport 1980-1981: 112–132.

Barnabé G., 1984. Utilisation de plancton collecté pour l'élevage de masse de poissons marins. In: *Actes du Colloque L'Aquaculture du Bar et des Sparidés*, G. Barnabé & R. Billard (eds). Publ. INRA, Paris: 185–207.

Barraclough W. E., Robinson D., 1972. The fertilization of Great Central Lake. II—Effect on juvenile sockeye salmon. *Fish. Bull.*, **70**: 37–48.

Charlson R. J., Lovelock J. E., Andrae M. O., Waren S. G., 1987. Oceanic phytoplankton, atmospheric sulphur, cloud albedo and climate. *Nature*, **326** (6114): 655–661.

Concar D., 1990. Filing plans for algae. *Nature*, **346**: 691.

De Pauw N., Morales J., Persoone G., 1984. Mass culture of microalgae in aquaculture systems: Progress and constraints. *Hydrobiologia*, **116/117**: 121–134.

Folke C., Kautsky N., 1989. The role of ecosystems for a sustainable development of aquaculture. *Ambio*, **18**(4): 234–243.

Grobbelaar J. U., 1979. Observations on the mass culture of algae as a potential source of food. *South Africa Journal of Science*, **75**: 133–136.

Hanson, J. E., 1983. Ecology, Aquaculture and Space Colonies (pers. comm. 1983).

Jhingran V. G., 1989. L'aquaculture en Inde. In: *Aquaculture*, G. Barnabé (ed.). Tec. & Doc. (Lavoisier), Paris: 1117–1152.

Koike I., Shigemitsu H., Kazuki T., Kazuhiro K., 1990. Role of sub-micrometre particles in the ocean. *Nature*, **345**: 242–244.

Larkin P., 1989. *Mariculture and fisheries: prospects and partnerships*. Symposium on the ecology and management aspects of extensive mariculture. ICES, Nantes, EMEM, No. 58.

Laubier L., Barnabé G., 1990. Conclusions. In: *Aquaculture*, G. Barnabé (ed.). Tec. & Doc. (Lavoisier), Paris: 1279–1293.

Marcel J., 1989. La pisciculture en étang. In: *Aquaculture*, G. Barnabé (ed.). Tec. & Doc. (Lavoisier), Paris: 615–652.

Marcel J., Doumenge F., 1989. L'aquaculture en Chine. In: *Aquaculture*, G. Barnabé (ed.). Tec. & Doc. (Lavoisier), Paris: 1069–1092.

Martin J. H., Gordon R. M., Fitzwater S. E., 1990. Iron in Antarctic waters. *Nature*, **345**: 156–158.

Needham T., 1990. Canadian aquaculture—Let's farm the oceans. *World Aquaculture*, **21**(2): 76–80.

Persoone G., Sorgeloos P., 1982. Perspectives in maricultural technologies. *Phil. Trans. R. Soc. Lond., A*, **307**: 363–375.

Pshenichnyj B. P., Vershinskij N. V., 1985. Possibilities of increasing marine biological productivity by artificial upwelling. *Aquaculture*, **46**: 77–80.

Rotschild B. J., 1986. *Dynamics of marine fish populations*. Cambridge: 277 pp.

Schoonbee H. J., Prinsloo J. F., 1988. The use of polyculture in integrated agriculture aquaculture production system aimed at rural community development. *J. Aquat. Prod.*, **2**(1): 99–123.

Simard F., 1986. Un nouveau plan de développement pour les pêches au Japon: le Marinovation. *La Pêche Maritime*, April 1986: 258–270.

Troadec J. P., 1986. Aménagement des pêche et des cultures marines en zones littorales: perspectives et axes de recherche. *La Pêche Maritime*, April 1986: 271–275.

Von Walhert G., Von Walhert H., 1989. Extensive rural mariculture in the tropics; experiences and issues. In: *Aquaculture, a biotechnology in progress*. European Aquaculture Society, Bredene, Belgium: 97–102.

Yamame T., 1986. Le programme national d'implementation de récifs artificiels au Japon: efficacité des récifs au plan économique. *C.R. Coll. Fr. Japon. Océanogr., Marseille*, **6**: 7–31.

Index